4/9 95

S0-AZK-425

507.24
Ani

192060

Animals in research.

The Lorette Wilmot Library
Nazareth College of Rochester, N. Y.

Animals in Research

Animals in Research

New Perspectives in Animal Experimentation

Edited by
David Sperlinger
Bexley Hospital, Bexley, Kent

WITHDRAWN

JOHN WILEY & SONS
Chichester·New York·Brisbane·Toronto

LORETTE WILMOT LIBRARY
NAZARETH COLLEGE

192060

Copyright © 1981 by John Wiley & Sons Ltd.

All rights reserved.

No part of this book may be reproduced by any means, nor
transmitted, nor translated into a machine language without the
written permission of the publisher.

British Library Cataloguing in Publication Data:

Animals in research.
  1 Laboratory animals
  I. Sperlinger, David
  507′.24   QL55   80-49974

ISBN 0 471 27843 2

Photoset by Thomson Press (India) Limited, New Delhi and
printed in the United States of America.

Acknowledgements

The editor and the publisher wish to thank the following for their kind permission to reproduce previously published material:

British Psychological Society;
Hans Huber Publishers, Bern;
N. Tinbergen

List of Contributors

DON BANNISTER — *Medical Research Council External Scientific Staff, High Royds Hospital, Ilkley, West Yorkshire, England*

MARIAN DAWKINS — *Departmental Demonstrator in Animal Behaviour, Animal Behaviour Research Group, Department of Zoology, University of Oxford, England*

CORA DIAMOND — *Associate Professor, Department of Philosophy, University of Virginia, Charlottesville, Virginia, USA*

ROBERT DREWETT — *Lecturer, Department of Psychology, University of Durham, England*

RICHARD ESLING — *Education Officer, Life Sciences Sector, Technician Education Council, London, England*

MICHAEL FESTING — *Head of the Genetics Department, Medical Research Council Laboratory Animals Centre, Carshalton, Surrey, England*

MICHAEL FOX — *Director, Institute for the Study of Animal Problems, Washington, DC, USA*

LOUIS GOLDMAN — *Medical Journalist, West Byfleet, Surrey, England*

HAROLD HEWITT — *Honorary Consultant Pathologist, Department of Morbid Anatomy, King's College Hospital Medical School, London, England*

WALIA KANI

Graduate Student, Department of Psychology, University of Durham, England

DAVID MACDONALD

Ernest Cook Fellow, Animal Behaviour Research Group, Department of Zoology, University of Oxford, England

HEATHER MCGIFFIN

Research Associate, Institute for the Study of Animal Problems, Washington, DC, USA

MARY MIDGLEY

Lecturer, Department of Philosophy, University of Newcastle-upon-Tyne, England

MARGARET MORRISON

Legislative Associate, The Humane Society of the United States, Washington, DC, USA

DAVID PATERSON

General Secretary, British Union for the Abolition of Vivisection, London, England

JENNY REMFRY

Assistant Director, Universities Federation for Animal Welfare, Potters Bar, Hertfordshire, England

ANDREW ROWAN

Associate Director, Institute for the Study of Animal Problems, Washington, DC, USA

RICHARD RYDER

Clinical Psychologist, The Warneford Hospital, Oxford, England

DAVID SPERLINGER

Clinical Psychologist, Bexley Hospital, Bexley, Kent, England

Contents

Animals in Research
Edited by David Sperlinger
© 1981 John Wiley & Sons Ltd.

Introduction

DAVID SPERLINGER

The use of live animals as the subjects of research in science and medicine has a long history. The history of those who have protested against this practice is almost as long. The arguments of the two sides have changed very little since the original experiments were conducted. In many respects this is not surprising. Scientific theories and practices may have developed enormously, but the ethical and social questions raised by experiments on animals remain unchanged. On the other hand, discussions about animal experimentation often appear unproductive and ritualistic; the two sides expounding their views *at* each other, with little sense of any attempt to engage in serious debate or to examine what areas of agreement they might share.

In terms of the numbers of animals involved it may seem strange that such strong passions are aroused by experiments on animals. Thus in Britain, for example, over *fifty* times as many animals are slaughtered for food as are used for experimental purposes. However, debates over the fate of farm animals rarely produce the strength of feeling found in discussions about experimental animals. Part of the reason for this seems to lie in the particularly difficult and distinctive moral issues raised by animal experimentation. Thus, from the anti-vivisectionist point of view, the over-riding concern is that pain is deliberately inflicted on many experimental animals; while many vivisectionists are equally convinced of the importance of their work for human welfare.

The two sides in the debate can be sketched in the following way:

The anti-vivisectionist position, includes arguments such as:

(1) No matter how important the aim of the experiment may be, one cannot be justified in inflicting pain on animals in order to obtain it. (A weaker version of this argument would stress that many animal experiments are for unimportant purposes, which cannot justify the suffering inflicted on the animals. This version would allow that there *might* be some purposes for which the infliction of pain was justified.)

1

(2) Many experiments are of little or no relevance to humans and their welfare; little of benefit to humans has come from animal experiments, that could not have been arrived at by other means.

(3) Much experimental use of animals could be eliminated by the use of various alternative techniques.

(4) Humans have no right to use animals for practices such as scientific research.

The vivisectionist position, includes arguments such as:

(1) Most animal experimentation is directed towards increasing human knowledge and/or human welfare. Major advances in human welfare would not have been achieved without such experiments.

(2) Few animal experiments involve serious pain to animals and experimenters strive to minimize any pain that is inflicted on animals.

(3) The use of animals in research could not be reduced by the use of alternative techniques and scientists, in any case, use such techniques whenever they are available.

(4) Humans have the right to use animals in order to achieve important human ends.

The positions characterized above obviously represent the extremes in the debate. It is disturbing, however, how frequently the issues are presented in such a polarized way, with the arguments (on both sides) including not only statements about ethical positions but also assertions about the actual and potential benefits to be derived from animal experiments. Such discussions often seem to turn into morality plays, where Reason is pitted against Emotion—with the scientists taking the former role and those concerned with animal welfare the latter. (See, for example, from the scientific side, Shuster (1978) or the papers given to a conference of the American Public Health Association (1967) which were designed to document 'the benefits to men and animals that result through research using animals'. On the animal welfare side, Ruesch's (1979) book is explicitly designed as a text to show the reader how he can, and why he should, put a stop to all animal experiments.)

However, there have been signs recently of a willingness by both sides to look for areas of possible agreement. Thus in Britain, for example, several anti-vivisection organizations have established research funds to encourage scientists to develop alternative techniques which do not involve (or require fewer) animals (see, as one example, the report of a symposium on this topic organized by the National Anti-Vivisection Society (1976)). Similarly the Royal Society for the Prevention of Cruelty to Animals organised a symposium, in 1978, attended mainly by scientists, which looked at the reduction and prevention of suffering in animal experiments. On the other side, the Research Defence Society, which has vigorously championed the right of scientists to experiment on animals, commissioned a book examining alternatives to animal experiments (Smyth, 1978).

This is not to say that agreement is about to be reached. But it does suggest that there is the possibility of a more reasoned and fruitful debate on the issues involved—and there *are* major issues raised by the use of live animals in research. For example, assuming that it is accepted that we should not cause suffering to animals needlessly, what purposes can be regarded as justifying what kinds of experiments? Given that we have to balance the interests of humans and animals, what interests of humans should be counted in the balance—efforts to understand and cure cancer, the production of new drugs to treat athelete's foot, increasing our knowledge about how the brain works, developing a new cosmetic?

This book aims at contributing to the more reasoned debate which is developing. The contributors were asked to have as basic premises: (i) that some current animal experimentation in medicine and science produces important benefits to humans and that some such experimentation will be necessary for the foreseeable future and (ii) that amongst the current uses of animals in experiments there will be cases where there is over-use or mis-use of animals or where the experiments appear to be for relatively trivial ends. This book does not aim, therefore, to catalogue (or to deny) the many ways that animals are often brutally treated in experiments without any apparent justification (in terms of significant human interests); this has already been done in other places (see, for example, Pratt, 1976; Ruesch, 1979; Ryder, 1975; Singer, 1976). Nor does it aim to offer easy justification for all current scientific practices with animals. It aims instead to tackle the important issues involved in a serious and critical way. It attempts to evaluate, in the context of particular scientific disciplines, areas where the use of animals in experiments is producing significant benefits and is contributing to the tackling of major issues, but also to raise questions about the areas where there are no such benefits or where the experiments result in a large degree of suffering to the animals involved. The focus of the book is research in medicine and the main scientific disciplines, since many of the more difficult problems about the use of animals concern work in these areas. For scientific and medical research *may* involve important human interests, although the labels 'scientific' or 'medical' in themselves give little indication of the value of any particular piece of research.

The book is divided into three parts. Part I aims to place the scientific activity of animal experimentation in the wider social, political, and legal context in which it occurs. Part II examines some of the most important areas of science and medicine in which animals are used. Part III looks at some general issues which are common to many fields of animal experimentation.

The use of animals in experiments is determined not only by scientific theories and practices, but also by what is socially and legally permitted in any given society. Part I explores how different countries have attempted to control animal experimentation, how this legislation has been implemented in practice, and recent proposals for the reform of such legislation. It also considers some recent changes in attitudes to animals generally. It begins with Ryder examining the legal and political framework in which animal experiments in Great Britain take place. The *Cruelty to Animals Act 1876* is the earliest example of attempts to legislate on animal experimentation, and it remains in operation over 100 years later. There are, however, as Ryder discusses, two proposals before the Houses of Parliament, at the time of writing, to reform this Act. (These recent attempts at legislative reform appear to reflect an increased public concern about this issue. A concern which was taken up in the manifestos of the two major political parties in the 1979 British general election, which referred to the need to reform the law relating to animal experimentation. It is also shown by the popularity of two recent novels (Adams, 1977; Kotzwinkle, 1976) which have experimental animals as their 'heroes'.) Esling, in his chapter, describes the enormous range of legislation relating to animal experimentation that is currently in operation in the rest of Europe. He draws attention to the trend in more recent legislation to take account of the increased public concern for the welfare of experimental animals. This is seen, for example, in the stress laid on the experimenter's obligation to make use of alternatives to animals where ever possible. Morrison, in her chapter on legislation in the United States, points out that in spite of the huge government investment in animal experiments, the present federal laws are limited in scope and ineffectively enforced. All three of these writers point to areas where the current legislation needs to be reformed if the welfare of animals in scientific laboratories is to be improved. But scientists, as well as being influenced by the legal framework in which their experiments take place, also have to take account of attitudes to animals generally. For such attitudes will play a part in forming the views of the experimenters themselves about their animal subjects, as well as being a significant factor in determining the general public's attitude to the use of animals in research. Sperlinger, in his chapter, looks at some of the major strands in current attitudes to animals, in particular he focuses on views about the nature of the relationship between humans and the other animals.

Part II looks at some of the main fields of science and medicine where animals play a significant part in research. The areas examined are: medicine (by Goldman), biological sciences (by Remfry), cancer research (by Hewitt), behavioural research (by Drewett and Kani), and ethology (by Macdonald and Dawkins). The space available to the contributors did not allow them to provide highly detailed reviews of the experimental literature. In addition, they were asked to try to make these chapters accessible to the non-specialist, to those who may be concerned about the issues involved but who may lack the technical knowledge required to evaluate work in a particular area. Each of these chapters picks out the major issues which are raised by experiments in their particular field. They give examples of particular pieces of research which illuminate the problems to be faced. They examine cases where the experiments appear to be of value and also look at experiments which appear to be questionable (on the grounds of, for example, the suffering caused to the animals, or because of the pointless or trivial nature of the experiments themselves). These contributions bring out very clearly the complexity of the issues involved and the absurdity of attempts to praise or condemn 'animal experimentation' as a whole. It is only by examining particular experiments, *in their scientific context*, that it is possible to arrive at any balanced assessment of how much value to humans (or other animals, see particularly the chapters by Remfry and by Macdonald and Dawkins for discussions of this topic) is to be derived from any particular experiment or how much the animals' suffering has been for quite worthless ends. This suffering may, of course, not only involve painful or unpleasant experimental procedures but may also arise from the inevitable restraints of laboratory life. This issue is referred to in the chapters by Drewett and Kani and by Macdonald and Dawkins. The latter authors, as well as exploring issues relating to ethological experiments with animals, also point out that ethology can be used as a tool for *understanding* animals. It seems essential, if we are to reduce the suffering of animals both in the animal houses prior to experimentation and in the experiments themselves, that we have a thorough understanding of the animals which are being used. We cannot assume that other animals will react to conditions as would a human being. Ethology provides us with a methodology, which is potentially much less intrusive and damaging to the animals concerned than many of the procedures examined in the other chapters of Part II, to enable us to begin to gain some knowledge about our animal subjects.

The contributors in this section of the book also look at the extent to which animals could be replaced by alternatives in their particular fields. This issue is also raised in the last two chapters of Part II, on the use of animals in schools in Great Britain (by Paterson) and in the United States (by Fox and McGiffin). These two chapters also emphasize how crucial experiences with animals in schools may be in forming 'scientific' attitudes to the use of animals. If using live animals as the subjects of research involves scientists learning to inhibit to some degree their natural reactions to animals (see Sperlinger's chapter for discussion of this point), experiences at school seem likely to play a significant role in this process. More humane attitudes to animals in schools may be a vital first step in any attempts to introduce more humane methods in animal experiments and to reduce the numbers of animals used.

Part III turns to look at some more general issues raised by the whole area of animal experimentation. It begins with Rowan examining what has become known as the question of 'alternatives'; that is, the various ways in which animals may be dispensed with in experiments or, where they are used, smaller numbers of animals can be involved or any suffering that is inflicted can be reduced. In his chapter Rowan focuses particularly on toxicity testing, but the issues to which he draws attention apply to most fields of animal experimentation. He emphasizes that alternatives will not provide a quick and simple solution to the problems involved in using animals in research, although they do have an important role to play. He, like the authors in Part II, stresses the necessity of looking at alternatives in the context of answering *particular* research problems; they do not provide a general answer, which can be simply duplicated from one area of research to another. Festing, too, looks at the ways in which fewer animals might be used in research. He describes several ways in which this aim could be achieved *and* scientific validity enhanced—e.g. by improvements in the quality of animals used, in the design of experiments, or in the analysis of experimental data. Rowan and Festing, in their respective chapters, concentrate on some of the ways in which current uses of animals may be invalid, wasteful, inefficient, etc., and they point to some practical changes which could be made to improve current practices.

The remaining chapters in Part III take a rather broader perspective, looking at some theoretical and philosophical issues raised by animal experimentation. Bannister's chapter focuses on the use of animals in psychological research, but, unlike Drewett and Kani, he does not attempt a detailed evaluation of work in this area. Rather he examines the theoretical assumptions underlying the use of animals in psychology and argues that the use of animals is linked to a particular (and, in his view, limited) definition of psychology. Bannister confines his argument to psychological work, but the issues he raises concerning the links between methodology and theory obviously have a much wider application. Thus, for example, it is clear that the use of animals in some medical research is tied to a particular approach (emphasizing cure rather than prevention) to disease and its treatment. If a model of medicine were generally adopted which emphasized the ways in which disease might be prevented, rather than searching for ever newer forms of treatment to *cure* disease, this alone could have an enormous impact both upon the numbers of animals experimented upon *and* upon human health. Midgley, in her chapter, examines one of the most central defences that is offered as a justification for animal experiments—namely, that the experiment will advance human knowledge. She points out some of the difficulties involved in such a defence and explores what would be needed to make such a defence a meaningful one. In the final chapter, Diamond looks in some depth at two characteristic and opposed views of animal experimentation. She discusses how these two views differ not only in their attitudes to the experimental animals but even over the fundamental question of whether or not there is a moral issue to be faced in this area. She also draws attention to some of the underlying assumptions of these two opposed views, noting their similarities as well as their differences.

The contributors to this book do not share a unified position towards animal experimentation. Indeed, they encompass a very diverse range of views about the validity and morality of much research on animals. Nonetheless, certain common concerns and themes do emerge from many of the chapters. Firstly, there is the desire to tackle the issues raised by the use of live animals in experiments in a fresh and undogmatic way, trying to get away from the strait-jacket of the old arguments on this subject. Secondly, there is an emphasis on evaluating the ethical and scientific problems, which are undoubtedly raised by animal experiments, in relation to *particular* experimental questions. Evaluations of wide areas or of 'animal experimentation' as a whole have often, in the past, led to the vociferous proclamation of unsupported arguments and sweeping generalizations (by both pro- and anti-vivisectionists). Focussing on narrower areas of animal experimentation is likely, from the evidence of these chapters, to produce a more detailed and reasoned debate. Thirdly, there is considerable agreement that much could be done to reduce both the number of animals involved in experiments and the amount of suffering that is caused to those animals which are experimented upon. And fourthly, and perhaps most importantly, a consistent theme throughout the book is that such a reduction in the numbers of, and the suffering to, experimental animals will not only be of benefit to the animals but will also lead to improvements in the meaningfulness and validity of the experiments themselves.

This book does not pretend to offer any easy solutions to the dilemmas raised by the use of live animals as experimental subjects in science and medicine. It will, however, have served its purpose if it contributes to changing the terms of the debate on this issue and brings forward recognition of the fact that 'animal welfare' and 'science' are not necessarily irreconcilably opposed. Many fewer animals could unquestionably be used in experiments, with a resulting benefit not only to the animals but to science (and thus to humanity) as well. Science itself is ill served by badly conceived, poorly designed, and trivial experiments and such experiments must cause particular concern when they involve taking the lives of, or causing suffering to, animals. However, even if the arguments put forward in this book were to be acted upon (with a subsequent reduction in the use of animals in research), the controversy over the experiments that remained would, rightly, still continue. But at least the issues to be faced would be clearer for all to see; the debate would be a real one.

REFERENCES

Adams, R. (1977). *The plague dogs*, Allen Lane, London.

American Public Health Association (1967). 'Vivisection—vivistudy: the facts and the benefits to animal and human health', *American Journal of Public Health*, **57**, 1598–1626.

Kotzwinkle, W. (1976). *Doctor Rat*, Aidan Ellis, Henley-on-Thames.

National Anti-Vivisection Society (1976). *The moral, scientific and economic aspects of research techniques not involving the use of living animals*, National Anti-Vivisection Society, London.

Pratt, D. (1976). *Painful experiments on animals*, Argus Archives, New York.

Ruesch, H. (1979). *Slaughter of the innocent*, Futura Publications, London.

Ryder, R. (1975). *Victims of science*, Davis-Poynter, London.

Shuster, S. (1978). 'The anti-vivisectionists—a critique', *New Scientist*, **77**, 80–82.

Singer, P. (1976). *Animal liberation*, Jonathan Cape, London.

Smyth, D. H. (1978). *Alternatives to animal experiments*, Scolar Press, London.

Part I

Animal Experimentation:
The Legal and Social Context

Animals in Research
Edited by David Sperlinger
© 1981 John Wiley & Sons Ltd.

Chapter 1

British Legislation and Proposals for Reform

RICHARD D. RYDER

INTRODUCTION

The United Kingdom can claim no monopoly of reforming zeal in the field of animal experimentation and, indeed, as in so many other areas where the British once led the way, it can be argued that the last half of the twentieth century sees Britain being overtaken by more progressive nations, some of which have introduced legislative concepts theoretically in advance of some British welfare statutes of the 1980s.

Nevertheless, there are two reasons why the British example is worthy of special treatment. The first is that its history is the longest and the best documented (see, for example, French, 1975; Ryder, 1975; Vyvyan, 1969, 1971). The second reason is that the decade of the 1970s saw an eruption of public concern over 'vivisection' and an organized reform movement unprecedented for at least a century associated with a campaign for animals' rights (see, for example, Paterson and Ryder, 1979).

By the end of the decade the British reform lobby based its position on four fundamental requirements:

(1) that pain and distress be prohibited in all animal experimentation; *and*
(2) that humane 'alternative' techniques be developed and used wherever feasible; *and*
(3) that no animal experimentation be allowed unless it be for the alleviation of suffering or for the saving or prolonging of life; *and*
(4) that government control over animal experimentation be effective, open, accountable, and amenable to public and welfare influences.

The first comprehensive legislation to protect animals in Britain was passed in 1822 and this was amended in 1835 and again in 1849. When, in 1874, the Royal Society for the Prevention of Cruelty to Animals (RSPCA) wished to prosecute for cruelty to experimental animals it was under the 1849 Act and was undertaken against a Frenchman, Eugene Magnan, and three English medical men who had arranged for Magnan to demonstrate publicly the effects upon dogs of injections of absinthe. Magnan fled the country and the prosecution failed because the Norwich magistrates found that the three Englishmen had not themselves been involved in the alleged cruelty; but a defence claim for costs was rejected.

Public concern about vivisection had been increased following newspaper reports in 1863 that atrocious procedures were being carried out upon horses in the Veterinary School at Alfort near Paris. Students were being taught to practice repeated surgical operations upon old horses without the use of anaesthetics.

In response to the ensuing campaigns of Frances Power Cobbe and others, the General Committee of the British Association for the Advancement of Science in 1870 formulated its own guidelines 'to reduce to its minimum the suffering entailed by legitimate physiological inquiries'. These guidelines were as follows:

'i. No experiment which can be performed under the influence of an anaesthetic ought to be done without it.

ii. No painful experiment is justifiable for the mere purpose of illustrating a law or fact already demonstrated; in other words, experimentation without the employment of anaesthetics is not a fitting exhibition for teaching purposes.

iii. Whenever, for the investigation of new truth, it is necessary to make a painful experiment, every effort should be made to ensure success, in order that the suffering inflicted may not be wasted. For this reason, no painful experiment ought to be performed by an unskilled person with insufficient instruments and assistance, or in places not suitable to the purpose, that is to say, anywhere except in physiological and pathological laboratories, under proper regulations.

iv. In the scientific preparation for veterinary practice, operations ought not to be performed upon living animals for the mere purpose of obtaining greater operative dexterity.'

But after the Magnan affair, public opinion was not in a mood to leave it to the scientists to keep their house in order, and in 1875 Frances Cobbe approached the RSPCA suggesting that the Society should press for new legislation having three aims:

(i) to prohibit painful experiments except by registered persons in authorized laboratories; and

(ii) to prohibit painful experiments in illustration of lectures; and

(iii) to render liable to prosecution the publishers of journals printing accounts of cruel experiments.

The RSPCA failed to rise to the occasion, and so Cobbe and her allies themselves approached Parliamentarians. A Bill was drafted and presented to the House of Lords on 4 May 1875 by Lord Henniker. This Bill proposed that all animal experiments without anaesthesia should be conducted only in registered and inspected laboratories and by licensed persons.

Only eight days later, on 12 May 1875, a rival Bill, aimed at protecting the scientists' position, was introduced in the Commons by Lyon Playfair. This Bill was remarkably similar to the Henniker Bill and allowed experiments without the use of anaesthetics only if all the following conditions were complied with:

(i) that the experiment is made for the purpose of new scientific discovery, and for no other purpose;

(ii) that insensibility cannot be produced without necessarily frustrating the object of the experiment, and that the animal should not be subjected to any pain which is not necessary for the purpose of the experiment;

(iii) that the experiment be brought to an end as soon as practicable;

(iv) that if the nature of the experiment be such as to seriously injure the animal, so as to cause to it after-suffering, the animal shall be killed immediately on the termination of the experiment;

(v) that a register of all experiments made without the use of anaesthetics shall be duly kept, and be returned in such form and at such times as one of Her Majesty's Principal Secretaries of State may direct.

Both Bills were rather poorly worded and raised a number of issues of definition. As neither Bill would have made administerable law and as both were introduced late in the session the Government announced a Royal Commission which duly reported on 8 January 1876, and recommended that experiments should be performed under licence and on registered and inspected premises. Licences should be bound by conditions 'to ensure that suffering should never be inflicted in any case in which it could be avoided, and should be reduced to a minimum where it could not be altogether avoided'.

As part of its evidence to the Commission the RSPCA had presented its own draft proposals suggesting that licences should be available only to physicians, surgeons, anatomists, physiologists, and those with medical qualification. No vivisections were to be allowed for teaching purposes or for attaining manual skill. No experiment 'of a nature to produce pain or disease in any animal, as well as the cutting or wounding of any living animal was to be allowed' without anaesthesia which produces 'complete insensibility to pain'.

After approaches by Cobbe and Lord Shaftesbury, the Disraeli Government invited their suggestions for legislation. These were incorporated in a Government Bill introduced by Lord Carnarvon on 15 May 1876. This Bill completely prohibited experimentation upon dogs and cats. Experiments without anaesthesia were to be permitted only under special certificate, and experimentation generally was allowable only with a view to the advancement of 'knowledge which will be useful in saving or prolonging human life, or alleviating human suffering'.

Again, the scientific lobby moved into opposition. The General Medical Council demanded:

(i) that cats and dogs be made available for research;

(ii) that inspectors under the Bill be 'scientifically competent';

(iii) that experiments be allowed for the 'advancement of science', and for the alleviation of *animal* suffering as well as for the alleviation of *human* suffering.

The British Medical Association took a similar line, adding that research should not be confined entirely to registered laboratories.

As a result of this pressure, the Government introduced an amended Bill on 9 August 1876 which made the following concessions:

(i) private places could be used for purposes of research on animals;

(ii) the special certificates to experiment upon dogs, cats, and equines (horses, mules, and asses) would be required only if such experiments were to be performed without anaesthesia;

(iii) prosecutions against licence-holders could only be undertaken with the permission of the Secretary of State;

(iv) the Act would only apply to warm-blooded animals.

In debate the Bill was further amended so as to apply to all vertebrate rather than merely warm-blooded animals, thereby covering for example, the much-used laboratory frog.

The Bill was then rushed through all its remaining stages, receiving the Royal Assent on 15 August, becoming 39 and 40 Victoria, cap 77.

THE CRUELTY TO ANIMALS ACT 1876

The Act is in two main parts: the first part makes several humane *restrictions*, and the second part counteracts most of these restrictions by providing the experimenter with special *certificates*. (In practice, further restrictions can be imposed through Conditions which the Secretary of State is empowered to attach to licences.)

Properly equipped with licence and certificate, an experimenter may inflict severe pain without fear of prosecution.

Restrictions

The restrictions in the first part of the Act, at a glance, appear to be humane. They are contained in Section 3 of the Act, and are:

(1) 'The experiment must be performed with a view to the advancement by new discovery of physiological knowledge or of knowledge which will be useful for saving or prolonging life or alleviating suffering'; and

(2) 'The experiment must be performed by a person holding such licence from one of Her Majesty's Principal Secretaries of State'.

(3) 'The animal must during the whole of the experiment be under the

influence of some anaesthetic of sufficient power to prevent the animal feeling pain'; and

(4) 'The animal must, if the pain is likely to continue after the effect of the anaesthetic has ceased, or if any serious injury has been inflicted on the animal, be killed before it recovers from the influence of the anaesthetic which has been administered'; and

(5) The experiment shall not be performed as an illustration of lectures in medical schools, hospitals, colleges, or elsewhere; and

(6) The experiment shall not be performed for the purpose of attaining manual skill.

Other restrictions appear in different sections of the Act, and are as follows:

(7) Any exhibition to the general public, whether admitted on payment of money or gratuitously, of experiments on living animals calculated to give pain shall be illegal (Section 6).

(8) An experiment calculated to give pain shall not be performed without anaesthetic on a dog or cat (Section 5).

(9) An experiment calculated to give pain shall not be performed on any horse or mule (Section 5).

(10) Curare shall not for the purposes of this Act be deemed to be an anaesthetic (Section 4).

Certificates Which Allow Exemption from Restrictions

The second part of the Act counteracts the first part by providing for certificates which allow exemption from restrictions 1*, 3, 4, 5, 8, and 9 above.

The exemption from restriction 3 is given under *Certificate A*, and allows an experiment to be performed without anaesthesia. This certificate should state that 'insensibility cannot be produced without necessarily frustrating the object of such experiments' (Section 3, Provision 2). Since 1910 approximately 90 per cent of all experiments have been performed without anaesthetic under Certificate A.

The exemption from restriction 4 is given under *Certificate B*. This is required 'if experiments are to be performed under anaesthesia but the animal is not killed before the anaesthesia has passed off' (Form of Application, Home Office).

The exemption from restriction 5 is given by *Certificate C*, which allows experiments 'to illustrate lectures' (Form of Application, Home Office).

* In theory, a Certificate D can be issued to permit the testing of a particular *former* discovery, stating that such testing is absolutely necessary for the effectual advancement of such knowledge as is defined in Section 3(1). This still means that all experiments must be performed with a view to the advancement 'of physiological knowledge or of knowledge which will be useful for saving or prolonging life or alleviating suffering' whether or not the experimenter holds a Certificate D. In practice, the Home Office regards Certificate D as being redundant, and none have been submitted for many years.

The exemption from restriction 8 is given under *Certificate E*, which allows experiments on unanaesthetized dogs and cats, and *Certificate EE*, which allows experimental cats and dogs to recover from anaesthetic.

The exemption from restriction 9 is given under *Certificate F* which allows experiments upon horses, asses, and mules.

It is laid down (Section 5) that certificates E, EE, and F should certify that 'the object of the experiment will necessarily be frustrated' unless it is performed on the particular sort of animal specified.

(Exemption from restriction 10 is not allowed, but curare and similar drugs which can paralyse an animal can be used with permission from the Secretary of State.)

Section 11 requires a certificate to be signed by one or other of the persons who must be signatory to an application for a licence; allows it to be given for such a period or for such series of experiments as the signatories think expedient; requires the applicant to send a copy to the Secretary of State and renders the certificate ineffective for a week thereafter; and empowers the Secretary of State to suspend or disallow any certificate (except that given by a judge in connection with an experiment essential for the purposes of criminal justice).

It can be seen that six out of ten of the Act's humane safeguards can be dispensed with by these certificates. The most important restrictions, those dealing with anaesthesia, are nearly always dispensed with, so that the vast majority of experiments, nine out of every ten, do in fact take place without the use of anaesthetics.

It is important to realize that only about 3 per cent of all experiments take place under licence alone. All the rest are performed under Certificate as well as licence.

Licences

Section 8 empowers 'one of Her Majesty's Principal Secretaries of State' (in practice this appears invariably to have been the Home Secretary) to grant a licence for such period and on such conditions, not inconsistent with the Act, as he thinks fit, and to revoke a licence at his discretion. Section 12 empowers a judge to grant a licence to perform experiments essential for the purposes of criminal justice. (It seems that this only once occurred, in 1885). Section 11 requires an application for a licence to be signed by one or other of the presidents of 13 specified learned societies (mostly medical) and also by a professor of physiology, medicine, anatomy, surgery, medical jurisprudence, or *materia medica*, unless the applicant is himself such a professor.

Registration of Premises for Experiments

Section 7 requires that the Secretary of State shall approve and register any place where experiments are performed for class instruction; and empowers him to require the registration of any place where other experiments allowed under the Act are performed.

Reports

Section 9 empowers the Secretary of State to 'direct any person performing experiments under this Act from time to time to make such reports to him of the result of such experiments, in such form and with such details as he may require'.

Inspection

Section 10 requires the Secretary of State to cause all registered places to be visited from time to time by inspectors for the purpose of securing a compliance with the provisions of the Act, and empowers him to appoint inspectors.

Penalties

Section 2 imposes a penalty on conviction for unlawful experiment of £ 50 for a first offence and £100 or three months' imprisonment for a subsequent offence. Anyone may prosecute an unlicensed experimenter, but a licensee can only be prosecuted with the written consent of the Secretary of State (Section 21). Moreover, Section 15 contains an unusual provision in that it gives a person charged with an offence against the Act (and for which a penalty of more than £5 can be imposed) the right to elect to go for trial on indictment.

A magistrate, on sworn information that there is reasonable ground to believe that experiments in contravention of the Act are being performed by an unlicensed person in any unregistered place, may issue a warrant authorizing the police to enter and search such place, and to take the names and addresses of the persons found therein. The penalty for obstructing the police under these circumstances must not exceed £ 5 (Section 13). (No conviction under this Act of *any licencee* has been recorded.) (These penalties were increased under the *Criminal Law Act 1978*.)

Extent

The Act applies to Great Britain and Northern Ireland.

Application to The Crown

The Act does not bind the Crown. Unless the contrary is expressly stated or necessarily implied, the provisions of all statutes bind neither the Sovereign nor any body or persons acting as servants or agents of the Crown. (This means that Government bodies such as the Ministry of Defence and the Department of Health and Social Security are not bound by the Act, and so may experiment upon animals without licence and without their experiments being counted in the Home Office returns.*)

Scope

The Act covers 'the performance on any living animal of an experiment calculated to give pain' (Section 3). It does 'not apply in invertebrate animals' (Section 22).

(The Act does not define key words such as 'experiment', 'living', and 'pain'. The interpretation of these words, in practice, is left to the Home Office, and can be variable. With mounting criticism of the enormous numbers of animals being used in research, there may be, therefore, increasing pressure brought to bear upon the official licensing authorities and the compilers of statistics to exclude thousands of procedures using animals, on the grounds that such procedures are not defined as experiments or that any suffering they cause is not actually pain.)

Home Office Conditions Attached to Licences

Section 8 of the *Cruelty to Animals Act 1876* states that there may be attached to any licence 'any conditions which the Secretary of State may think expedient for the purpose of better carrying into effect the objects of this Act, but not inconsistent with the provisions thereof'.

It should be noted that it is not necessary that any conditions be attached to licences. It is left to the discretion of the Secretary of State, who can add or subtract conditions as he sees fit.

The main licensing authority has been the Home Office, and over the years a more or less standard body of conditions have been evolved:

Condition 1 authorizes the place or places at which experiments must be performed. These are usually, but not always, registered places. (Section 7 of the Act expressly provides power for inserting this condition.)

Condition 2. 'No experiment under any Certificate held by the Licensee may be performed until he/she has been notified that the Certificate has not been disallowed by the Secretary of State'.

Condition 3. 'Unless otherwise provided below, the following conditions are to be observed in all experiments under any Certificate A (whether or not accompanied by a Certificate E or F) or under any Certificate B (whether or not accompanied by a Certificate EE or F):

(a) If an animal at any time during any of the said experiments is found to be suffering pain which is either severe or is likely to endure, and if the main result of the experiment has been attained the animal shall forthwith be painlessly killed.

(b) If an animal at any time during any of the said experiments is found to be

---

* The Home Office states that, by an 'administrative arrangement', all government departments and research councils comply with all the provisions of the Act.

suffering severe pain which is likely to endure, such animal shall forthwith be painlessly killed.

(c) If an animal appears to an Inspector to be suffering considerable pain, and if such Inspector directs such animal to be destroyed, it shall forthwith be painlessly killed.'

(Conditions 3(a), 3(b), and 3(c) apply to all 'recovery' experiments, but are not at all humane, although at first reading they may appear to be. In fact they provide no real safeguard against suffering whatsoever, and actually condone the infliction of 'severe pain'.)

Condition 4. 'Unless otherwise provided below, the following condition is to be observed in all experiments under any Certificate A (whether or not accompanied by a Certificate E or F):

'No operative procedure more severe than simple inoculation or superficial venesection may be adopted in any of the said experiments.'

(Condition 4 is deceptive. Most experiments do not involve major surgery anyway. Much more often they involve the injection or force-feeding of animals with poisons, drugs, cosmetics, infective substances which cause diseases, and so on. So a 'simple inoculation' can lead to prolonged suffering and death. After injection, animals under Certificates A and B remain subject to Condition 3.)

Condition 5. 'Unless otherwise provided below, the following conditions are to be observed in all experiments under any Certificate B (whether or not accompanied by a Certificate EE or F):

(a) All operative procedures in connection with the said experiments shall be carried out under anaesthetics of sufficient power to prevent the animal from feeling pain.

(b) The animals upon which experiments are performed shall be treated with strict antiseptic precautions, and if these fail and pain results, the animals shall forthwith be painlessly killed.'

(Conditions 5(a) and 5(b) only cover those animals allowed to recover from anaesthetic under Certificate B. This affects only approximately 10 per cent of all experiments. In other words, these conditions do not cover those animals anaesthetized under licence alone, nor those animals operated on without anaesthetic under Certificate A.)

Condition 6. 'Unless otherwise provided below, the following condition is to be observed in all experiments under any Certificate C (whether or not accompanied by a Certificate F):

'On the completion of any such experiment the animal shall be killed forthwith by, or in the presence of, the licensee.'

Condition 7. 'No experiment in which curare or other substances having similar curare-form effect upon the neuromuscular system is used shall be performed without the special permission of the Secretary of State; and forty-eight hours' notice of the performance of every experiment or series of similar experiments so permitted shall be given to the Inspector of the District. This condition shall not apply to experiments on a decerebrated animal in which the cerebral hemispheres and basal ganglia have been destroyed.'

Condition 8. 'The Licensee must keep a written record of all his/her experiments, which shall be open to examination by an Inspector at any time: and he/she shall send to the Secretary of State within fourteen days at latest of the close of each year a report on the number and nature of all experiments performed during the year, and from time to time such other reports as may be required.'

Condition 9. 'In the event of descriptions of any experiment performed by the Licensee and requiring a Licence under the Act appearing in any Medical, Scientific or other Journal or Magazine or in a report of any lecture delivered by the Licensee printed for publication or private circulation, the Licensee shall transmit to the Secretary of State, as soon as practicable after its appearance, the said Journal or Magazine, or the fullest description of such printed publications or reports of Lectures, accompanied by a letter drawing attention to the experiments performed by him/her and stating when and where the experiments were performed.'

Condition 9a. 'The Licensee shall not permit any cinematograph film to be made which shows any animal, or a part of it, undergoing an experiment performed by him under this Licence except with the prior consent in writing of the Secretary of State and unless the person or body in whom the copyright of the film when made will be vested has before the film is made agreed as part of the consideration for permission to make the film to observe such conditions respecting the use and exhibition of the film as the Secretary of State may have specified to the Licensee in granting his consent as aforesaid.'

(Condition 9a is an illustration of the way in which the Home Office has actively encouraged secrecy about the widespread use of animals in research. The Home Office also vets some television science programmes and has encouraged the locking of the doors to animal laboratories.)

Other conditions are sometimes added, but not so regularly as Conditions 1–9a. When students and technicians are licensed, they are sometimes subject to the condition that their work is supervised.

Condition 10. 'The eyes shall not be used for administration of substances, implantation or withdrawal of body fluids.'

(This condition is not always attached and there is evidence that the eyes have been used for growing tumours and testing cosmetics (see Ryder, 1975, pp. 48–49).)

Some *Notes on Delegation* are also attached to licences by the Home Office emphasizing that non-licensed persons cannot take part in experiments except 'to administer anaesthetics to an animal' (Note 2) and to 'carry out mechanical duties' (Note 3).

Home Office Notes for Guidance

In addition to the Conditions attached to licences, the Home Office has issued 'Notes for guidance in completing forms of application under the *Cruelty to Animals Act 1876*'. These suggest that various procedures are acceptable without anaesthetic and these include administration of substances, withdrawal of body fluids, variations of diet, exposure to infections, exposure to radiation, electric shock, and gases, needle biopsies, and implantations (Notes, p. 3.) Some ambiguously worded sentences could lead licensees to believe (Note 4, p. 6) that provisions are made for those who wish to paralyse an animal without giving it an anaesthetic of sufficient power to prevent pain. Provisions are also made (Note 9(V), p. 8) for a licensee who makes a request 'for amendment to or cancellation of any of the Secretary of State's special conditions on pages 2–4 of the licence'.

PROSECUTIONS

It seems that there have only been two prosecutions ever brought under the 1876 Act and only one of these was successful. This is hardly surprising because the general public and members of animal welfare societies have no right of access to laboratories and so are unlikely to witness any breaches of the law; nor are the experimenters' published accounts of their actions likely to reveal any of their own misdemeanours that might have occurred. It is thus highly unlikely that any outsider should get to know of a breach of the Act within the statutory limit of 6 months. Even if this were to happen, the Secretary of State's permission is still required before a prosecution can be undertaken.

The Home Office Inspectors annually report a number of irregularities that they have detected but until recently no licences or certificates were ever suspended and certainly there have never been any Home Office prosecutions. For many years the Home Office concluded its report on the irregularities with the identical paragraph:

> 'It was concluded that in none of these cases had there been any deliberate intention of evading the requirements of the Act and that it would suffice to send the person concerned a letter of admonition from the Secretary of State stressing the need for strict observance of all the requirements of the Act at all times.'

(On a purely anecdotal level, however, senior scientists are quite prepared to admit that both accidental and deliberate infringements of the law are far from being uncommon.)

In evidence before the second Royal Commission (1906–12) Mr Scott for the RSPCA told the story of the first known prosecution under the 1876 Act. The Act had come into operation on 15 August 1876:

> 'Three days later Dr. Abrath, of Sunderland, issued a large placard headed "The Balham Mystery", announcing his intention to deliver a lecture at Sunderland on "Antimony", when he would perform experiments on animals to show the effects of

poisons, and to demonstrate his theory that Mr. Bravo was not killed by that drug. The Branch of the Royal Society for the Prevention of Cruelty to Animals at Sunderland immediately reported the matter to the Secretary in London, and prompt measures were taken to prevent the learned gentleman from performing his experiments. According to the sixth section of the statute named above, an offence had already been committed by the announcement of 'a public exhibition of experiments on animals, and Dr. Abrath, having spoken contemptuously of the new Statute at his lecture, and provoked the derision of a portion of his audience against it, instead of apologizing for his projected defiance of the Act, was summoned to appear yesterday before the Sunderland Borough Bench,' at the instance of the RSPCA, when he was fined for publishing the illegal placard alluded to. (Royal Commission, 1876, 19568)

The second prosecution was brought in 1881 by the Victoria Street Society against David Ferrier alleging that he had 'mangled' the brains of monkeys while inappropriately licensed to do so. The prosecution failed when witnesses gave evidence that G. F. Yeo, and not Ferrier, had been the actual experimenter. Nevertheless the case caused widespread alarm in scientific circles (French, 1975, p. 202).

There have, however, been four prosecutions of licence-holders (two of them successful) under other cruelty legislation. The first appears to have been *Dee v. Yorke* (*The Times Law Reports*, **XXX**, 29 May 1914). The respondent, Dr W. Yorke, was a licensee under the 1876 Act, but was prosecuted unsuccessfully under the *Protection of Animals Act 1911* by RSPCA Inspector H. L. Dee. It was proved that Dr Yorke administered a drug to a donkey which brought on gradual paralysis. The animal was then turned out into a field used exclusively by the Runcorn Research Laboratories and the Liverpool School of Tropical Medicine. On 30 June 1913, 10 days after the drug was given, the donkey was reported to be lying in a ditch, unable to rise. On 3 July, a police-sergeant examined the animal and found 'two open wounds on the jaw and several wounds on the body, from which yellow matter had been oozing'. It was contended on behalf of the RSPCA that the suffering undergone by the donkey while exposed to the unshaded glare of the sun and to the attacks of flies was unnecessary, but the magistrates found to the contrary

'and as the respondent was a great medical authority engaged in research work for the benefit of mankind, and moreover as he was licensed to conduct an experiment on the ass by the Home Secretary, they were of opinion that he could not be treated or regarded as having knowingly or intentionally caused unnecessary suffering to the animal, and accordingly dismissed the information.' (*Ibid.*, p. 553).

This case differs from the prosecution of Professor E. G. T. Liddell of Oxford in 1946 in that Yorke's donkey was claimed to be under experiment until it died on 5 July 1913, whereas Liddell's cats were stock animals not under experimentation at the time of the alleged offence. Liddell was convicted under the 1911 Act.

The other two relevant cases are recent. The first involved a World Health Organization scientist who lost his licence under the 1876 Act following a successful prosecution by the RSPCA under the *Protection of Animals Act 1911*. The animals in question were kept upon a farm belonging to the accused, and their possible use in research was not relevant to the case.

The second case occurred in Scotland.

On 20 March 1978 at Cupar Sheriff Court, a lecturer at St Andrews University was charged under the *Protection of Animals (Scotland) Act 1912* that on numerous occasions together with a student psychologist: 'He cruelly ill treated, tortured or terrified 178 canaries, 160 laboratory mice, 17 goldfish and 2 rats'.

The series of experiments involved offering the animals as live bait to several cats and observing their pre-catching behaviour. These experiments were unlicensed under the 1876 Act and the Home Office had never been approached to discuss them.

On 23 March, Sheriff John C. McInnes found the case proved and imposed the then maximum fine of £ 50, adding—'this sum is ridiculously low for the cruelty involved'. (See unpublished report by Clive Hollands of the Scottish Society for the Prevention of Vivisection.)

Under the *Criminal Law Act 1978*, the maximum fine under the 1911 and 1912 Acts was raised to £ 500.

THE SECOND ROYAL COMMISSION

In 1906 the Government set up another Royal Commission which reported in 1912 recommending:

(i) an increase in the Inspectorate;

(ii) greater supervision of the use of curare;

(iii) pithing to entail the total destruction of the brain and to be performed under adequate anaesthetic;

(iv) 'No animal should be allowed to live in severe pain which is likely to endure'. Inspectors should be empowered to order the destruction of animals which are suffering.

The Government accepted these recommendations which required no legislative amendments.

THE LITTLEWOOD REPORT, 1963–65

Another half-century was to pass before the next official inquiry into the subject. On 23 May 1963 a Departmental Committee was appointed under the chairmanship of Sir Sydney Littlewood. It reported to the Home Secretary on 19 February 1965 and was first debated on 11 June 1971. It consists of 15 general findings and 83 recommendations. Fourteen years later the Government could claim that only about 20 of these recommendations had been implemented administratively, either wholly or in part (Hansard, Lords, 25 October 1979, column 255.)

The report is divided into eight parts dealing with the background to the enquiry, the existing legal provisions and their administrative practice, pain in animals, the review of evidence, wastage of animals, the scope and organization of control, the supply of animals, and a summary of conclusions and recommendations for reform.

The Committee recognized that pain was a central issue and it seems to me to be worthwhile quoting their relevant deliberations:

Pain in Animals

'Paragraph 174. While the expression of pain in an animal is not inhibited as it may be in man by familiarity or voluntary effort, it is very much obscured by the animal's relatively limited powers of communication.'

'Paragraph 178. In a situation which is, and must remain, so full of doubt, it may perhaps be that the problem is best approached pragmatically . . . this being so, the only basis on which to proceed would appear to be that of practical human experience. On this basis, it would be assumed that any procedure that would cause pain or distress in man would do so also in the animal, and it would follow that the animal should be protected from these possible ill-effects by measures analogous to those which would be applied to a human being who had been subjected to a similar procedure. For example, in experiments involving surgery with subsequent recovery, this principle would require not only adequate anaesthesia during the operation but also proper post-operative care—a point that has been brought to our attention by several witnesses; furthermore it would require that an animal subjected to a procedure that is initially so trivial as to require no anaesthetic but which, as a later result, may produce pain, should be carefully observed so that it may receive adequate analgesia or be painlessly destroyed if and when such pain ensues.'

In Appendix VII of the Littlewood Report a section of the *Report of the Committee on Cruelty to Wild Animals* (1951) is reproduced in which that Committee argues that:

(1) Some mammals are known to have, and all may be presumed to have, the nervous apparatus which in human beings is known to mediate the sensation of pain.

(2) Animals squeal, struggle, and give other behavioural evidence which is generally regarded as the accompaniment of painful feelings.

(3) Pain is 'a sensation of clear-cut biological usefulness'.

Using these three types of evidence the Committee states: 'We are satisfied, therefore, that animals suffer pain in the same way as human beings.'

VICTIMS OF SCIENCE, 1975

The publication of *Victims of science* (Ryder, 1975) shortly preceded another eruption of public interest in the subject (see Hampson, 1979).

In *Victims of science* a proposal was made that reforms should include '(a) the prohibition of all cruel non-medical experiments and (b) governmental backing for the development of alternatives to animals in research' (p. 249).

In the campaign that followed the book's publication, the author sent out several thousand leaflets calling for reform in more detail. These were:

'(1) *The existing Home Office Advisory committee should be revitalised*
This committee should become an active standing committee representing a wide range of scientific, veterinary and animal welfare interests. It should be a professionally staffed body with real powers. It should consider the justification for experiments and publish its own reports. It should have access to public opinion.
(2) *Strengthening the Home Office Inspectorate*
There should be 25 Inspectors of whom the majority should be veterinarily qualified. (At present there are only 14 inspectors to control about 19 000 licence-holders.)
(3) *Channelling funds into the further development of the humane alternatives to research animals*
A central office (perhaps a Medical Research Council sub-committee) should collate and disseminate information; this body should communicate regularly with the new Advisory Committee (1). Development Units should be established in existing laboratories. (Humane alternatives mostly are in their infancy, but in some cases are already cheaper and safer than using animals.)
(4) *Improving Home Office information to the public*
MPs and the tax-payer (who sponsors much of the research being done) receive very little information. According to the Littlewood Report (p. 164) "there has been an appearance of secrecy about the practice of animal experimentation". The tax-payer has some right to know just how many of the over 5 000 000 licensed experiments on animals each year are for strictly medical purposes and how many are for other purposes such as the testing of cosmetics, toiletries, weed-killers, oven-cleaners and fire-extinguishers. Animals are being poisoned to death with such substances, and are being used in the testing of weapons, and in behavioural and "stress" experiments, which are not always medical.
(5) *New and up to date legislation*
This should encourage humane practices such as anaesthesia, analgesia (pain-killing) and euthanasia. (The present law (the *Cruelty to Animals Act 1876*) allows the infliction of "severe pain" and the wholesale use of animals in tests which are not truly medical. The Danish, Swedish and German laws are, in some respects, more humane and up to date; Britain needs to catch up.)

Any increased funds required for the implementation of these reforms could be derived from licence fees and registration fees.'

Within the next two years, the Home Office made cautious administrative improvements in several areas, and the Secretary of State gave undertakings generally in line with 1, 2, and 4 above.

THE 'HOUGHTON–PLATT MEMORANDUM', 1976

In May 1976 a group of parliamentarians, scientists, and welfarists, under the leading names of Lord Houghton of Sowerby and Lord Platt (a former President of the Royal College of Physicians), submitted a paper to the Home Secretary calling for ten principal reforms:

(i) The expansion of the annual Home Office Returns so as to give more information, e.g. on the purposes for which animals are being used.

(ii) A tightening up of the system of licences and certificates. Sponsors should have to confirm that alternatives to animal experiment have been considered; that the question of pain, misery, and suffering has been taken into account; that the conditions of housing and experimentation are satisfactory and appropriate; and that the proposed experiment is not an unnecessary repetition and will either add importantly to physiological knowledge or to knowledge which will protect or improve health.

(iii) 'A determined effort should be made to restrict the use of animals in procedures which are not strictly for human medical therapeutic purposes such as the testing of cosmetics and toiletries, the testing of weapons and behavioural research'.

(iv) 'There should be strong, constant emphasis upon the need to restrict procedures likely to cause suffering of any sort, including stress'.

(In 1976 the Home Office undertook to review the wording of the 'Pain Condition' attached to licences.)

(v) The Inspectorate should be enlarged; not less than half should have veterinary qualifications and some should be trained in behavioural science.

(The Inspectorate was enlarged to 14 in 1976 and to 15 in 1979.)

(vi) Government funds should be made available for the development and use of humane alternative techniques. There should be a central unit to collate and disseminate information in this field, and to co-ordinate research programmes established in existing laboratories. And there should be progressive restrictions on the use of animals in favour of non-sentient alternatives.

(In 1977 the Prime Minister, Mr Callaghan, announced that the Home Office should integrate matters in this field and stated that it was his Government's policy 'to move to alternatives to animal experiments as quickly as possible' (Hansard, 8 December 1977, column 1644.) Furthermore, the Home Secretary, Mr Merlyn Rees sent out an exhortation to all licencees urging them to use alternatives wherever feasible. The Conservative Government in 1979 pledged to continue this exhortation.)

(vii) 'All breeding and supply establishments should be registered and inspected'.

(viii) 'Greater importance should be attached to the need for training in, and the application of, the techniques of anaesthesia, analgesia and euthanasia'.

(ix) Protective legislation should be expanded to apply to or cover (a) any research procedure which 'interferes with the animals' ordinary state of health or well-being', and (b) certain invertebrate species, and (c) stock animals, and (d) the Crown.

(x) An Advisory Committee be established with enlarged composition and functions. It should have the power of investigate on its own initiative, to advise the Secretary of State and to publish its reports.

The Memorandum pointed out that the Committee must serve just two public interests, on one hand *science* on the other *humanity*:

> 'The balance to be struck between two interests—both of them public interests; the public interest in suppressing or controlling cruelty to and the abuse of animals; and the public interest in medical research, safety of drugs and safeguards against the use or marketing of harmful substances. No *private* or commercial interest in the use of living animals for these purposes can be admitted. It is *all* within the legitimate scope of the *public* interest.
> 'Therefore, those concerned in these operations must accept this, and accept with it that their desire for unhampered activity and investigation and freedom from "interference" has to be reconciled with public opinion.'

THE GENERAL ELECTION COORDINATING COMMITTEE FOR ANIMAL PROTECTION, 1978–79 (GECCAP)

Under the direction of its Chairman, Lord Houghton, and its Secretary, Clive Hollands, this joint committee representing a wide range of animal welfare bodies and supported by the RSPCA, succeeded in persuading the three major political parties in Britain to find a place for animal welfare in their official policies. GECCAP's own document 'Putting animals into politics' reiterated five of the reforms called for in the Houghton – Platt Memorandum (nos (iii), (iv), (vi), (viii), and (ix) above). The 1979 election manifestos of the Conservative, Liberal, and Labour parties, for the first time in history, all contained clear commitments for reform in the field of animal welfare generally.

Relevant to laboratory animals the manifestos stated:

Conservative Party: 'We shall update . . . the legislation on experiments on live animals'.
Labour Party: 'Under Labour's new Council of Animal Welfare we will have stronger control . . . on experiments on living animals'.

A background policy paper was published by the Labour Party in July 1978 (Labour Party, 1978). On the subject of laboratory animals it included the following proposals:

(i) The legal requirements on animal experimentation should be updated in view of the scientific advances of the last hundred years and should include regulations governing procedures, which although not strictly experimental within the terms of the 1876 Act, impinge on the health and wellbeing of animals (transplantation of ova, breeding of abnormal animals, breeding of animals for production of sera and vaccine, etc.).

(ii) The Littlewood recommendations should be updated and speedily implemented.

(iii) The Government should take measures to encourage a reduction in the use of animals for experimentation by assisting in the development and promotion of alternative methods as outlined above.

(iv) Specifically, the Government should encourage and assist research to find an alternative to the LD_{50} test as well as looking at the whole area of toxicology.

(v) Experiments should be restricted to proven medical research where it can be demonstrated that no alternative method is available.

(vi) The legislation should be more strictly enforced, through increased inspections and more severe penalties for offenders.

The Conservative Research Department published a booklet by Charles Bellairs in the following year (Bellairs, 1979) which, although not an official policy document, mentioned some further suggestions for reform including:

(i) 'that those licensed to carry out experiments on animals under the 1876 Act should have adequate training in anaesthesia and the use of analgesics';

(ii) 'laboratories should be allowed to obtain animals only from dealers who are licensed, having been checked on welfare grounds, their dealings being open to inspection by the RSPCA';

(iii) 'It is to be hoped that the review of the 1876 Act will result in restricting the purposes for which painful experiments on animals can be carried out to the saving of human life and the prevention of serious human suffering. It is questionable, for example, whether painful experiments should be performed on animals for the purposes of testing cosmetics, herbicides or substances modifying plant growth';

(iv) 'It is hoped that the newly constituted Advisory Committee on Animal Experiments will initiate a drive with the aim of preventing the repetition of experiments. The animal welfare organisations believe that if repetition could be cut out there would be a worthwhile reduction in the number of painful experiments carried out'.

A statement of Liberal views in February 1979 endorsed GECCAP's recommendations on laboratory animals.

Between 1966 and 1976 almost a dozen relevant Private Members Bills were presented to Parliament; none of them reached the Statute Book.

Notable among them were Richard Body's, introduced on 30 November 1970, Douglas (later Lord) Houghton's of 29 November 1972, and Philip Whitehead's of 9 April 1975. These sought (in the words of the Houghton Bill) to make it law 'that no experiment on a living animal shall be performed under the authority of that licence if the purpose of the experiment could be achieved by alternative means'. A similar Bill was introduced by Ivor Stanbrook on 17 December 1975.

On 1 November 1973, Lord Willis introduced a Bill that failed at its Second Reading on 10 December in the House of Lords. Its aims were wider and more general.

On 11 February 1975, Baroness Phillips presented a Bill to ban the testing of non-medical cosmetics on animals. This was blocked by the Government having had its Second Reading on 27 June.

The year 1975 had been widely felt to mark the beginning of progress by the reformers. In this year the Government made five minor concessions to public pressure:

(1) Four lay persons were appointed to the Home Office Advisory Committee which had been asked to investigate the case of the ICI 'smoking beagles'.

(2) The question of 'stress' in experiments would be examined.

(3) The question of the amount of government information provided about experiments would immediately be reviewed.

(4) The effectiveness of the contact between Inspectors and experimenters would be considered (Hansard, 14 May 1975).

(5) The Secretary of State assured Parliament that in future no more smoking experiments upon dogs would be allowed.

Further improvements followed with the publication of the new format for Home Office annual statistics starting in 1978. In the same year, following representation from the Committee for the Reform of Animal Experimentation (CRAE), the Home Secretary (Mr Merlyn Rees) asked his Advisory Committee to investigate the LD_{50} test, and in response to a request from the author (as RSPCA Council Chairman), Mr Rees sent out an exhortation to all licensed experimenters urging them to use alternative methods wherever available.

During this period the Home Office Inspectorate was also increased. In 1885 there had been only one Inspector; in 1930 there were two; in 1950 there were four; in 1960 there were five; in 1970 there were 12 and in 1979 they were increased to 15.

Until 1979 CRAE and the RSPCA were content to see reforms taking the form of changes in the administration of the 1876 Act which, as we have seen, gives very considerable powers to the Secretary of State. A sympathetic Home Secretary could, theoretically, refuse to licence any research which was not approximately 'medical', in accordance with a more exact interpretation of Section 3(1) of the Act. He could also, by means of Conditions imposed at his discretion, very much improve the restrictions on pain and, by entirely reconstituting the Advisory Committee, he could allow access to public opinion and to the legitimate interests of the welfare movement.

THE HALSBURY BILL

Alarmed at the administrative moves towards the welfare position, and by the sympathy shown in the media and even in scientific circles, the Research Defence Society decided to attempt to arrest the slide by introducing their own legislation.

The Society's President, Lord Halsbury, after consultation with scientific bodies, published his *Laboratory Animals Protection Bill* on 16 July 1979.

This Bill was a well drafted and major contribution to the field, and it contained various attractive features from the animal protection point of view. Basically, however, it failed on the four cardinal points of reform: (i) increased public accountability, (ii) proper control over pain, (iii) satisfactory constraints upon experimenters to use alternative methods wherever feasible, and (iv) restriction of experiments to worthwhile medical purposes.

The Bill covers procedures 'which can reasonably be expected, in the absence of anaesthesia, to cause distress, pain, or ill-health' (Clause 1). Those that carry out such procedures must be licensed and must work on licensed premises (Clause 2). Such procedures (defined in Clause 28 as 'any stressful or potentially stressful act or series of acts done to a living animal for the purpose of testing a hypothesis, collecting information, testing or producing a substance, registering an effect of a substance on an animal or for educational purposes') must be for one or more of the following objects:

(i) the advancement of biological knowledge;

(ii) the provision of a substance or knowledge intended for saving, prolonging, or improving life, or avoiding, preventing, or alleviating the suffering of man or any animals (whether or not animals of any of the species mentioned in Schedule 1 to this Act);

(iii) the imparting or demonstrating of such knowledge as aforesaid to colleagues or students of the biological sciences;

(iv) the making of visual recordings for teaching purposes;

(v) the teaching or learning of surgical or other techniques;

(vi) the fulfilment of any statutory requirement or duty or any other governmental requirement under other legislation, including any foreign legislation; or

(vii) any other object approved by the Secretary of State after consulting the Advisory Committee

(Clause 2(c)).

It can be seen that in comparison with the 1876 Act such provisions very considerably widen the range of purposes for which potentially painful procedures can be allowed. Almost any experiment can be said to be 'for the advancement of biological knowledge' (i); most consumer substances can be claimed to be for the 'improving' of life (ii)—even cosmetics, toiletries, and other inessential luxury products; subsection (v) would seem to allow procedures for 'learning' manual dexterity; and subsection (vi) would allow any testing required under foreign law, so, for example, if a foreign country required that a luxury product be force-fed to 100 chimpanzees and to 200 dogs before being imported, then the British exporter or producer of the luxury product would be allowed to carry out these tests in Britain.

The Bill provides rather vague controls over pain. Clause 2(d) states:

'during any procedure liable to cause pain or severe distress the animal must be afforded the same degree of anaesthesia, sedation or pain relief as would be afforded in accordance with contemporary standards of veterinary practice unless otherwise authorised by the Secretary of State';

'if an animal, or any offspring born of an animal subjected while pregnant to a procedure, shows any sign of, or can reasonably be supposed to be, suffering pain or distress it shall be killed as soon as the object of the procedure is achieved and if a procedure results in an animal showing signs of severe pain or distress that is likely to endure, the animal shall be killed forthwith or other effective steps shall be taken to alleviate the pain or distress' (Clause 2(h)).

The Bill extends cover to the fetuses of mammals (Schedule I) and allows the Secretary of State, after consultation with the Advisory Committee, to add or delete species to be covered (Clause 3).

The Secretary of State is empowered to vary, suspend or revoke licences. He may 'determine the standards of premises' and may 'make conditions in any such licence having the effect of limiting the procedures that may be carried out' (Clause 5) or any other conditions or restrictions as he sees fit (Clause 6(7)). Licences (except in the case of procedures where there is no recovery from anaesthesia) shall specify the species and the procedures permitted (Clause 6(4)).

Applications for licences must carry the signatures of two sponsors acceptable to the Secretary of State (Clause 6(2)); the first having knowledge of the applicant and the laboratory, and being the holder of a senior appointment at that laboratory; and the second being an 'independent professor of a biological science'. The first sponsor must certify that the applicant 'has been instructed in the handling of animals in the proposed procedure' and the second must certify that the procedure 'is justified in the circumstances after having considered the possibility of replacing the procedure with an alternative method using non-sentient material or using a species of animal having a less developed nervous system', (Clause 6(2(d)) and 6(2(e))).

The Secretary of State is empowered to 'make regulations to control the procurement, breeding, transport, accommodation, management, veterinary supervision and care of animals for laboratory use' (Clause 7), and shall publish a 'Guide to good laboratory animal practice' (Clause 8). He shall appoint Inspectors with powers of entry to licensed premises (Clause 11) and shall cause to be published an annual report showing the purposes of the procedures, the species used, and any alleged infringements (Clause 13).

The Bill proposed the establishment of an Advisory Committee to report annually and with general duties to review the use and care of animals under the Act, and to advise the Secretary off State of trends, on 'novel or controversial procedures', on the information to be made available to the public, and on 'procedures considered by the inspectors appointed under this Act to be especially painful or distressing to the animals used' (Clause 14(b)). Justices of the Peace are given powers of entry on suspicion of contravention (Clause 18) and prosecutions under the Act are to proceed before a Magistrates Court (Clause 19).

Unfortunately, the Bill suggested that the Advisory Committee (Schedule 3) be composed perhaps entirely (other than its legally qualified chairman) of scientists and industrialists.

This Bill received its Second Reading in the House of Lords on 25 October 1979 and was committed to a Select Committee (see footnote, page 38).

In his speech for the Government on this reading, Lord Belstead pointed out that legislation in this field 'must maintain a proper balance between the legitimate requirements of science and industry and those of animal welfare'. He implied some criticism of Clause 2(c)(ii) which would legitimise procedures for 'the improvement of life of man or animal' on the grounds that this 'could allow all manner of abuses' and might require 'further refinement and restriction' (Hansard, 25 October 1979, column 258). He went on to suggest that in some respects the Halsbury Bill might 'compare unfavourably in concern for the welfare of experimental animals with the provisions of the 1876 Act which they are intended to replace. That is particularly so as regards the purposes for which animals may be used' (Hansard, 25 October 1979, column 258). He also suggested that the Bill's proposal to allow the use of live animals for the purpose of gaining manual dexterity 'is surely to many people morally repugnant' and needed careful consideration. He went on

'our informed non-scientists would also be likely to find difficulty with Clause 2(h) which deals with the infliction of pain . . . I am not convinced that the drafting of Clause 2(h) would adequately deal with the problem. It would, for example, allow prolonged pain if the experimenter could satisfy himself that the degree of pain was not too severe; and it is left to the judgement of the experimenter as to what is severe pain or distress.' (Hansard, 25 October 1979, column 259)

THE FRY BILL

The other Private Member's Bill of 1979 was also a major document and was introduced by Peter Fry on 27 June and passed its Second Reading on 16 November in the House of Commons.

This Bill, entitled the *Protection of Animals (Scientific Purposes) Bill* was a lengthy document, running to 39 Clauses and three Schedules.

Although the RSPCA had presented Fry with its own suggestions for legislation, his Bill showed little resemblance to these. The RSPCA had suggested a Bill prohibiting painful experiments entirely and restricting the remainder to therapeutic purposes only; experimenters would be required to use alternative methods wherever feasible and would be under the control of the Secretary of State advised by an Advisory Committee composed equally of scientists and welfarists.

The Fry Bill, like the Halsbury Bill, contained certain improvements on the welfare side but fell short on fundamentals. On balance, its bad points probably outweighed its good ones, leaving the animals in some respects worse off than under the 1876 Act.

The Bill allowed procedures to be licensed for a wide range of purposes including the attainment of manual skill (Clause 2). Experimenters under the Act would require a licence, graded as to severity (Clause 6), showing species and procedures permitted, on their application being signed by two sponsors (Clause 8). The first sponsor was to be an officer concerned with the work being proposed and the second an officer of a listed learned society (Schedule I). The first sponsor was to certify as to the suitability of the applicant and the premises, and that he 'has been instructed in the handling of animals for the purpose of the operative procedures to which the application relates and that

'in his opinion, the purpose of those procedures, or a sufficiently similar result, cannot be achieved by an alternative procedure using non-sentient material or forms of life other than animals and, in the case of an application relating to an operative procedure which, if carried out on an animal, is likely to cause the animal to suffer pain or distress, the purpose of that procedure, or a sufficiently similar result, cannot be achieved by an alternative operative procedure which is not likely to cause such suffering.' (Clause 8(2c))

The second sponsor was required to certify that he had no reason to dissent from the opinions certified by the first sponsor (Clause 8(4)).

Licences could be varied or revoked by the Secretary of State (Clause 10) who was empowered to register premises used for the breeding and supplying of animals as well as for the performance of procedures under the Act (Clauses 11–15). However, the decisions of the Secretary of State to licence or to certify registration, or to revoke the same, could be quashed by the High Court (Clause 22).

The Secretary of State could issue a Code of Practice providing practical guidance in the care and handling of animals (Clause 17) and the Act would cover the Crown (Clause 31).

The Fry Bill seemed to offer both improvements and disimprovements as far as welfare was concerned. In comparison with the 1876 Act the advantages included:

(1) The *registration and inspection of breeding units* in addition to laboratories (Clause 11).

(2) The Bill applied to a *wider range of procedures* and would, for instance, cover the production of biological substances (such as vaccines) not previously covered (Clause 1). (But, at the same time, it would withdraw these from the protection of the 1911 Act.)

(3) Animals to be used under the Act would have to be obtained from registered premises (Clause 2(5). (Unfortunately, the Bill then allowed the Secretary of State to waive this restriction; this had worrying implications especially as regards cats and dogs, which may be stolen for research.)

(4) Licences would have to be paid for (Clause 8(8)).

(5) The Act would apply to the Crown (Clause 31); but again, unfortunately this could be waived by the Secretary of State.

(6) Clauses 4 and 5 allowed the making and lawful showing of recordings of procedures carried out for other purposes (i.e. 'ordinary' research procedures).

(7) The Secretary of State must annually report to Parliament on species and purposes (Clause 21) and other matters.

(8) The Bill made prosecutions easier (Clauses 1(2) and 26).

(9) Inspectors were given increased powers (Clause 19).

The drawbacks of the Bill, from the welfare viewpoint, included:

(1) *Pain*. The Bill allowed under Grade IV licence the infliction of 'pain or distress' and 'without the administration of an anaesthetic or analgesic agent or for the use of any other means for the prevention or relief of pain or distress' (Clause 6(2)(d)); admittedly such pain or distress must not be allowed 'unnecessarily' (Clause 7(1)(c))—but who would decide what is necessary? Also, any such pain shall be 'the least pain practicable in the circumstances' (Clause 7(1)(h))—but again, on whose judgement? In practice, the licencee would decide.

The only safeguard against pain is that a Grade IV licence is issued solely on the advice of the Advisory Committee and if the Secretary of State is satisfied that the purpose of the research cannot be achieved by an alternative procedure and if the procedure 'is necessary for the advancement of biological science concerning the mechanisms of pain or the relief of pain in human beings or animals' (Clause 9(a) and (b)). This wording, especially the last ten words, appeared ambiguous. Would a Grade IV licence be allowable, for example, for the toxicity testing of cosmetics or other products which might cause pain if accidentally ingested by the consumer?

By attaching Conditions to licences under the 1876 Act, the Secretary of State already controlled pain and could, at any time, tighten the wording of these Conditions to outlaw pain completely. Fry's Bill, however, could have had the effect of enshrining in statute the basic acceptability of pain in research.

In the Fry Bill the watchdog which could make things less painful for the animals might be the Advisory Committee. But the Bill did not say very much about this Advisory Committee and if it was to continue (as in the past) to be predominantly composed of scientists, then Grade IV licences might be all too easy to obtain.

Less than satisfactory was the wording restricting pain under Grades I, II, and III licences (Clause 7(1)(f)). This allowed pain or distress provided it was not actually 'severe' or 'continuing'. This seemed to make a nonsense of the distinction between Grade IV and the other grades (which were supposed to be innocuous experiments). And who decides what is 'severe' or 'continuing'?

This was no advance on the protection offered by the 1876 Act, which could be increased at any time by the improved wording of the Pain Condition.

Clause 6(2)(e) allowed the Secretary of State to licence an ungraded procedure which might, presumably, *allow painful experiments without any safeguard.*

(2) *Purposes of research.* At present, under the 1876 Act (Section 3(i)), experiments can only be licensed if they are 'performed with a view to the advancement by new discovery of physiological knowledge or of knowledge which will be useful for saving or prolonging life or alleviating suffering'.

Under the Fry Bill, procedures could be licensed for a range of less high-minded purposes. They must be 'calculated to lead to the saving or prolonging of life or the preventing or alleviating of suffering, in human beings or animals' and they must be for 'the advancement of biological science' (Clause 2(1)(a)); but biology is defined to include the dental, veterinary, and behavioural sciences (Clause 34(1)). This would probably mean procedures carried out for the agricultural industry which are often euphemistically described as 'veterinary', as well as behavioural experiments by zoologists and psychologists—the latter being some of the most cruel that we know of. The 1876 Act questionably excludes psychological research on animals (e.g. experimental blinding or deafening, deprivation of food and water, separation of infants from mothers, interference with the brain, severe electric shock administered as 'punishment', driving animals mad with drugs or inescapable stress such as intense noise, rearing them for months in total isolation, etc.) and this has had a restraining effect upon work in this country. The Home Office has acted responsibly in this respect and has not felt able to licence the more extreme types of psychological experiments. But the Fry Bill might oblige them to do so, despite the findings of opinion surveys which repeatedly have shown that the British electorate does not want the continuation of 'non-medical' experiments on animals.

The Bill allowed procedures for the 'identification' of certain substances which may have positive effects (Clause 2(a)(i)).

Professor D. H. Smyth (late Chairman of the Research Defence Society) admits that only one substance in about 3000 tested in laboratories turns out to be useful. Presumably the Fry Bill would not prevent any of this wasteful 'identification' process being carried out on animals, including testing the 2999 useless substances.

Far more worrying, the Bill proposed to legitimize (following the example of Lord Halsbury perhaps) any procedure legally prescribed by any *foreign* legislation (Clause 2(2a)(ii)). So, to take an extreme example, if a foreign government made a regulation that a type of paperclip sold in that country must have been force-fed to 500 cats in order to determine how large a dose will cause half of the animals to die within 14 days of dosing, then this Bill would presumably allow a UK paperclip exporter to do this research.

Just as worrying was the proposal to allow (with the Secretary of State's consent, which may be given 'generally'—Clause 2(6)) procedures for the 'purpose of attaining manual skill' (Clause 2(2)(c)). This was an issue fought tooth and nail by the Victorian reformers following the revelation of revolting practices at French veterinary colleges where old horses, quite unanaesthetized, were subjected to up to 60 practice surgical operations each before dying. The 1876 Act specifically prohibits this. So the Bill would quite simply be a step backwards in this respect.

(3) *Public accountability.* The need would seem to be for more public accountability.

Allegedly animal experimentation is carried out for the public good and is often done at the public expense. The public therefore have some right to know what is going on and to have a say in what is to be tolerated. The Secretary of State is made accountable in the Fry Bill—but he has *no more discretion than he already has under the 1876 Act*, which gives him considerable powers should he care to use them. (Victorian Home Secretaries used these powers fairly freely. See French, 1975, pp. 184–189.)

There is a need for a powerful Advisory Committee, but the Fry Bill hardly refers to it and makes no suggestions as to its composition.

(4) *Alternatives.* The Bill required the first sponsor for a licence application to certify that in his opinion the result 'cannot be achieved by an alternative procedure' (Clause 8(2)(c)). But we know from Lord Platt's evidence that licences have often been signed 'blind'. Without sanctions against sloppiness, will sponsors bother to find out about alternatives? If they are signing for a licence that will last 5 years, how can they realistically foresee all the procedures and alternatives possible under that licence? The Advisory Committee is asked for its advice about alternatives (Clause 9(b)) and this is to be welcomed, but it is only for Grade IV licences. But although the Bill asked the sponsors to give an undertaking as regards alternatives it contained no satisfactory requirement on the licensee himself to use alternative methods.

(5) *Miscellaneous.* There were various other possible disadvantages, for instance:

(i) There is no detailed requirement that licensees should show competent knowledge of humane techniques such as analgesia, tranquilization, animal-care, euthanasia, and (where appropriate) anaesthesia. Such a requirement would be widely supported in scientific circles and was suggested to Fry by the RSPCA.

(ii) The final sentence of Clause 1(1) was ambiguous. It could be read to mean that veterinarians could carry out experiments of any sort without coming under the Act.

(iii) The complexities and apparent contradictions in the Bill would make its administration most difficult.

(iv) The Bill made less restrictions on the re-use of animals than are in Halsbury or the 1876 Act (Clause 7(1e).)

CRAE'S PROPOSALS FOR CHANGE, NOVEMBER 1979

In response to the unsatisfactory Halsbury and Fry Bills in preparation during

1979, a section of the animal welfare movement put its own proposals into a document prepared by the Committee for Reform of Animal Experimentation (CRAE).

This paper was submitted to the Home Secretary and to Members of Parliament prior to the Second Reading of the Fry Bill in November 1979.

It is entitled 'Proposals for change in the legislation governing the use of live animals in research, experiments and other laboratory purposes' and was issued by CRAE's Secretary, Clive Hollands.

The document makes specific proposals for new legislation which are cross-referenced with the Council of Europe draft convention (see Esling's chapter in this volume).

On pain, CRAE states:

'CRAE considers that the physiological disturbances associated with the infliction of pain or distress may actually invalidate the results obtained in an experiment. Therefore the principal reform which we wish to see in new legislation is that infliction of pain or distress shall not be permitted. However there may be occasions when, in the course of an experiment, an animal may be found to be suffering pain or distress, and we therefore recommend that the following pain clause be inserted in all licences:

Pain clause
All animals being used for any purpose under the Act which has the potential for causing pain or distress must be kept under regular supervision. If, despite all reasonable precautions both before, during and after experimentation, an animal is found to be suffering pain or distress, it must be forthwith humanely killed, and the procedure shall not be re-employed.'

CRAE suggests that legislation should apply to the Crown and should cover mammals, birds, reptiles, amphibia, fish, and certain cephalopods (e.g. octopus), including fetal and larval stages.

'Licences should be issued only for research procedures or for experiments performed with a view to the advancement by new discovery of knowledge in the medical, dental or veterinary sciences, which is likely to be useful for the saving or prolonging of human or animal life or the alleviation of suffering and this only where no alternative non-animal method can be used.'

CRAE recommends that the Advisory Committee become a statutory body, investigating on its own initiative and publishing its advice to the Secretary of State. As to its composition, CRAE believes that its lay members should 'hold the balance between the scientific and animal welfare interests'. Registered laboratories should be open to members of the Advisory Committee and the Home Office Inspectorate should be increased.

All applicants must be trained in, and show competence in, basic care techniques and humane skills such as anaesthesia, euthanasia, and analgesia before being licenced.

Breeders and suppliers of animals must be registered.

Lower organisms and humane alternative techniques must be used where available, and the Government should promote the further development and use of such techniques.

CONCLUSIONS

As this chapter goes to press*, it is unclear what outcome there will be to the vigorous reform movement of the 1970s. Nothing short of total abolition can be accepted by those who believe that speciesism is as unacceptable as racism as a basis for the exploitation of sentient beings, and even most moderate opinion within the welfare movement will require at least the following curbs upon experimentation:

(a) restriction or abolition of pain in all experiments;
(b) restriction to worthwhile medical purposes;
(c) requirements to use humane alternative techniques wherever feasible;
(d) increased public accountability.

REFERENCES

Bellairs, C. (1979). 'Animal welfare', *Politics Today*, no. 13, 17 September.
French, R. D. (1975). *Antivivisection and medical science in Victorian society*, Princeton University Press, Princeton, N. J.
Hampson, J. (1979). 'Animal welfare—a century of conflict', *New Scientist*, **84**, 280–282.
Labour Party (1978). *Living without cruelty: Labour's charter for animal protection*, Labour Party, London.
Paterson, D. and Ryder, R. D. (eds) (1979). *Animals' rights—a symposium*, Centaur Press, Fontwell, Sussex.
Report of the Departmental Committee on Experiments on Animals (1965). *Cmnd 2641*, HMSO, London.
Report of the Royal Commission on the Practice of Subjecting Live Animals to Experiments for Scientific Purposes (1876). *Cmnd 1397*, HMSO, London.
Report of the Royal Commission on Vivisection (1912). *Cmnd 6114*, HMSO, London.
Ryder, R. D. (1975). *Victims of science*, Davis-Poynter, London.
Singer, P. (1976). *Animal liberation*, Cape, London.
Smyth, D. H. (1978). *Alternatives to animal experiments*, Scolar Press, London.
Vyvyan, J. (1969). *In pity and in anger*, Michael Joseph, London.
Vyvyan, J. (1971). *The dark face of science*, Michael Joseph, London.

* On 24 April 1980, the Report of the Select Committee on the Laboratory Animals Protection Bill [H. L.] was published (HMSO 246). This report's radically new approach was to suggest that licensed experiments be controlled by Regulations (i.e. statutory instruments subject to annulment by resolution of either House of Parliament) made by the Secretary of State after consultation with a statutory Advisory Committee which is given wide powers. Clearly the composition of such a Committee becomes a matter of central importance. In 1980 the RSPCA proposed that the Advisory Committee be composed equally of 'users' and welfarists.

Animals in Research
Edited by David Sperlinger
© 1981 John Wiley & Sons Ltd.

Chapter 2

European Animal Experimentation Law

RICHARD W. J. ESLING

INTRODUCTION

This chapter deals with the control of animal experimentation within Europe, and will contrast the legislation existing at the time of writing in the various countries. Legislation controlling the use of live animals for experimental purposes differs greatly from one country to another and indeed in several countries there is virtually no precise control of the practice.

The countries of prime concern, and which are dealt with in greatest detail here, are the countries of the European Economic Community, Norway, Sweden, Austria, and Switzerland. Some brief details of the state of legislation in other countries which participate in the Council of Europe are also given, and reference is made to the countries of Eastern Europe. The latter, however, cannot be dealt with fully since details are understandably hard to ascertain.

There are various moves afoot, through private organizations and at government level, to review and change the laws concerning animal experimentation in several European countries. But the most important of these new developments is the work being undertaken by a Committee of Experts on Animal Protection which sits at the Council of Europe building in Strasbourg, and which is presently occupied with drawing-up a European Convention on the Protection of Laboratory Animals. These promising developments from the animal welfare point of view, will be dealt with towards the end of the chapter.

The legislation in different countries of Europe has evolved over the past century in the direction of better protection for animals, and the idea has gradually developed, and become incorporated in legislative documents, that a close watch must be kept that animals are well treated before, during, and after experiments. Another principle which is developing yearly, is that of necessity and justification for the use of live animals. Since all experiments cannot be performed under general or centrally acting anaesthetic agents, such as with toxicity studies for example, some degree of suffering or discomfort of the animals may occur. It must thus be shown that the experiments are absolutely necessary and, whenever possible, alternative methods which do not involve the use of live animals must be employed. Obviously, some areas of research using live animals are difficult in this respect, since it is not always possible to determine whether the results of the experiment will be of value or not. The term 'of value' is in itself relative, and criteria for judging a procedure as justifiable must differ considerably from person to person, as well as from nation to nation. However, the idea that some form of judgement and thus 'control' must be made at some stage is gaining increasing support, as may be seen later when the various national laws are described.

Animal experimentation which involves physical or psychological interference is a difficult problem for animal protection societies. The benefits derived from the practice, which are undeniable, principally affect various aspects of human welfare, either directly, such as with human medicines and pharmaceutical products, or indirectly. The development of a new pesticide, for example, which is highly effective against the target species and relatively non-toxic to non-target species, may vastly increase the 'welfare' of entire human populations in Third World countries. Some experiments, on the other hand, may be of benefit to other *animals*, as with the development of veterinary products. If people are going to keep pets, they must then accept the responsibility of keeping them in good health, and this requires the development of animal medicines.

Animal protection societies are thus faced with an obvious dilemma. Some societies have solved this by adopting the totalitarian philosophy of aiming to ban all experiments, and these anti-vivisection societies occur throughout western Europe. However, other more responsible societies have adopted a more reasonable approach. They accept that there is a need for a certain amount of animal experimentation, but campaign for the most humane approach possible at any given time. The two foremost societies which operate in Europe are the World Federation for the Protection of Animals and the International Society for the Protection of Animals. Both of these societies appear to have adopted this reasonable approach, and both have observers at the Council of Europe meetings on animal protection, where they have the possibility of voicing an opinion on the matters discussed.

Anti-vivisectionism has never established wide popularity in any country of Europe, and the only country which has a total ban on animal experimentation is the Principality of Lichtenstein. Even this was not brought about by pressure from abolitionists, but by the autocracy. It is generally accepted in virtually all European countries that some degree of usage of live animals in laboratory experiments is an unpleasant necessity. However, even those who may be described as non-'animal-lovers' agree that animals must not be made to suffer needlessly, and the progression of European legislation is directed along these lines.

Since, however desirable it may be, research and medical science cannot yet dispense with the animal experiment, every government is under the moral obligation of ensuring its practice under humane conditions and to develop, promote, and apply alternative methods where possible. The basic attitude of man towards the animal given into his charge has developed continuously over the course of time. However, economic patterns, as well as science and technology, have also undergone developments—particularly in the last few decades—which have resulted in a wide divergence between the interests of man and the needs of the animal. Within this ever-increasing field of tension, the idea of animal welfare has gained importance and topicality both at national and international level. The development and progression of legislation in various European countries shows to a large extent this developing humane approach.

Legislation on animal experimentation in Europe which is in actual operation, spans more than a century, and this is one of the factors accounting for the variablity of control which is offered in different countries. Much of the recent legislation is strongly on the side of animal welfare, and contrasts markedly with earlier laws, as shall be seen later. It is thus of great importance that the Council of Europe is presently drawing up a European Convention, since an attempt will be made to harmonize the different laws and give a basis for equivalent protection throughout the member states.

The variety of legislation on this subject existing within Europe can best be demonstrated by studying the various laws in some detail. The legislation which is decribed in the following sections ranges from very detailed to very vague and represents the principal countries of western Europe.

SCANDINAVIA

Some of the most recent legislative developments in this field have occurred in three principal countries of the northern region of Europe—Denmark, Norway, and Sweden. Finland has a law relating to animal experiments passed in 1971, but little is known of its implications or effect. The other three countries, however, have detailed and effective legislation of a modern nature, which is amongst the most advanced on this subject in the world.

Denmark

In Denmark, an Act was passed in 1891 under which rules were established about the use of animals for experiments that might cause pain. A more detailed Act came into force in March 1953 entitled *On the Use of Animals in Biological Research and Combating Disease* and many of the elements of this Act remain in the present Act of 1977. This Act relates to vertebrates only, and for biological research and the combating of diseases such animals may only be used by licensed persons. The basic principle is that the use of live vertebrates for medical and scientific research is prohibited if the procedure is likely to cause pain or other suffering in the animal. Thus, feeding experiments, injections, the taking of blood, organ, or other samples by puncture or superficial incisions are excluded, and may be done by any person without observing any of the formalities associated with obtaining a licence. These latter experiments, however, may only cause minor, transient pain or suffering.

A licence is required for the use of vertebrates in all other medical and biological research which will cause pain or suffering in the animal. Licences are granted only to persons who are deemed to be qualified to deal with such use of animals. Administration of the Act lies with the Ministry of Justice, although a licensee may, under certain conditions, have experiments carried out by other persons under his responsibility. Licensing, however, is partly automatic and partly decentralized. The heads of government institutes for biological and medical research are automatically licensed, and the medical faculty of a university or the Royal Veterinary and Agricultural School may grant licences to other professionals for periods of up to 5 years.

Supervision of the working of the Act is carried out by a Board appointed by the Minister of Justice, and in this respect is similar to the Advisory Committee in Great Britain appointed by the Home Secretary. The terms of reference of the committee, and its composition, however, are somewhat different. The members are an administrative skilled jurist as chairman, two university professors of medicine and two of veterinary science, as well as three other independent veterinary surgeons. The members are appointed on the recommendation of the Ministry for Public Health, permanent public councils for scientific research, the animal welfare organizations, and the drug industry. The terms of reference of the board are wide-ranging, and, apart from being concerned with overseeing the licensing procedure, include(a) ensuring that licensees observe the law at all times during their work; (b) the inspection of experiments, unannounced when desirable or necessary; and (c) making inquiries into particular experiments and animal facilities.

The Act lays down rules for certain types of experiments. Surgical interventions and physical and chemical effects that might cause pain shall be carried out under anaesthesia. The animals must be killed before the effect of the anaesthesia wears off, unless it may be assumed that the pain has come to an end or unless the experiment requires that the animal is kept alive. If the animal is kept alive, it must be treated with analgesics and special care. If the animal is in any other way brought into a condition that is not its normal state, and if this state might cause suffering, this suffering must be relieved as far as possible. The abnormal condition should be brought to an end as soon as possible, or the animal must be killed. If animals are not killed, the reason must be given, and for dogs, cats, and monkeys, the exact details of disposal must be given.

All experiments must be entered into protocols, which must then be authorized by the supervising Board. In this way the Board is given the possibility of intervening if experiments are not in accordance with the law. Moreover, the

Board periodically inspects laboratories—often unannounced. Records of experiments must also be kept on numbers and species of animals and purposes of experiments.

For any experiment, it is stated that not more than the necessary number of animals must be used. It is also stated that a species of animal of the lowest possible rank must be used—a principle which may create problems. The intention was probably to try to curb the use of animals which are often regarded as the 'higher' mammals—dogs, cats, and especially primates. But the actual wording, however, may be less helpful than intended. The zoological classification of animals cannot be used in such a way, since species are adapted to particular ecological niches. Furthermore, it has never been proved that a bird, for example, is more capable of suffering than, say, an amphibian. This same principle is at present causing problems in the discussions concerning the Council of Europe Convention. Another factor not to be overlooked is that it may be infinitely more difficult to cope with numbers of fish, amphibia, and reptiles, than with numbers of birds or small mammals such as rodents. More suffering may thus be caused unintentionally.

A new concept which appears in the 1977 Act is that of 'alternatives'. Animals must not be used for experiments if the use of cell, tissue, or organ cultures, or other methods, may be employed with equal relevance.

Norway

In Norway, the law governing animal experimentation is contained within a general animal welfare law, *The Welfare of Animals Act* of 20 December 1974. This Act was brought into force on 1 January 1977, by Royal decree, and this same decree delegated the administration of the section dealing with animal experiments to the Ministry of Agriculture. Control is in fact afforded by detailed regulations drawn up by the Ministry, and these Regulations came into force on 1 February 1978.

The Norwegian law affords greater control in some respects than the Danish Act. For example, not only are vertebrates covered by the Norwegian Act, but the order *Decapoda* of the arthropod phylum, i.e. the crustaceans, are also included, and the Ministry of Agriculture is considering extending cover to cephalopods. The general prohibition is stricter than the Danish law also. All biological experiments involving animals are prohibited unless a license is obtained from the Experimental Animals Board. 'Biological experiment' is defined as 'any experiment with live animals of a biological, psychological, ethological, physical, or chemical nature, including those carried out as an essential part of the training given by an institution'. However, marking experiments which do not affect the animal's normal way of living and behaviour, nor cause anything but completely transient and slight pain or unpleasantness, are exempted. Also exempted are experiments concerning feeding, breeding, or environmental changes in animal husbandry and aquaculture of fish and decapods. If, however, an abnormal physiological condition is produced in the animal, the exemption is void.

Licensing is undertaken by the Experimental Animals Board, which grants licences when the purpose of the experiment is to gather knowledge, test a hypothesis, produce or control a product, or register the effect of a certain procedure. As with the Danish law, permission is refused if the experiments can be performed with cell, tissue, or organ culture, or other methods not involving animals. The purpose of the experiment must also be restricted to (a) the diagnosis of disease in animals and man, (b) research, and (c) the production and control of medicines, drugs, preparations, poisons, etc., for use in animals, man, and plants.

Special permission is required to carry out experiments which there is reason to believe may cause pain, but which must be performed without employing anaesthesia. Permission to use monkeys, dogs, or cats is given only when strictly necessary. These special cases need approval by the Experimental Animals Board, which may impose any restrictions it thinks fit.

In Norway there is a two-tier licensing system. Licences may be granted to individuals and also to firms or institutions having an approved experimental animal department. The department must be headed by a graduate of a university or college who has the necessary biological knowledge and experience with animal experiments. Individuals who are licensed must have personal biological expertise and either access to an approved experimental animal department, or have an approved plan for field trials. Supervision by another licence-holder may be required. The head of an experimental animal department is made responsible for all activities carried out in that department, and all licencees—firms, institutions, and individuals—must complete and return an annual report to the Board. The Board may withdraw permission to carry out animal experiments if the laws, regulations, or special conditions have not been complied with, or if the basis for granting permission is no longer present.

Again, as with the Danish law, there is a definite stipulation that the number of animals used must be kept to the minimum. Furthermore, it is stated that if the methods of experimentation are untried, or if there is uncertainty as to the total number of animals required, pilot studies must be undertaken.

Requirements concerning anaesthesia, analgesia, and euthanasia are quite explicit, and the inclusion of regulations requiring analgesics to be used is to be noted, and commended. Experiments where no anaesthesia—local or general—is used, but where pain is likely to be caused, are severely restricted, and animals used for such purposes may not be used in further experiments. Exsanguination may only be performed after the animal has been pre-stunned, or is totally anaesthetized, and any form of euthanasia may not be performed in a way in which there is a risk of the animal being subjected to unnecessary suffering.

Inspection of premises is carried out by either the Experimental Animals

Board, or a person appointed by them, and includes inspection of the animals kept on the site and the actual performance of experiments. The Board also acts in an advisory capacity and must provide information which is necessary for compliance with the laws and regulations. It also presents an annual report to the Ministry of Agriculture. The Board is chaired by an Appeal Court Judge and has four other members consisting of a physician, two veterinarians, and a zoologist.

The 1974 Act, Chapter 6 of which deals with animal experimentation, is a great advance on the previous laws of 1902 and the by-laws of 1905 and 1908. These original laws were of a very elementary nature, and did not afford a great deal of protection to the animal. The *means* of administering the laws were also unsound. The by-laws of 1908 contained a section stating that the species selected must be of the lowest possible rank in the animal kingdom, similar to part of the Danish law described earlier. This section has been dropped, however, from the recent law and the relevant regulations.

The Norwegian Research Council for Science and Humanities (NAVF) undertook a study of animal experimentation in 1974, and they set up a committee for laboratory animals. The NAVF published a report in 1976 (NAVF, 1976) which pointed out several matters for further investigation and action. The main criticisms that they made were concerned with animal supply and housing. It was considered that all institutes should have qualified personnel in charge of animal experimentation, and that facilities for education and training in this topic should be more widely available. The NAVF committee also undertakes the consideration of applications for grants where animal usage is envisaged, with a view to ensuring adequate planning and sufficient justification where animal experiments are to be performed. A similar situation applies in Sweden with the Swedish Medical Research Council (SMFR).

Sweden

In Sweden, present control is afforded by the relevant sections of the Act of 1944 *On Protection of Animals*. This law appears to be kept under continual review, and the most recent amendments at the time of going to print came into effect in July 1979. The governmental body administering the law is the Ministry of Agriculture Veterinary Board, and licensing, as in Norway but *un*like Denmark, is undertaken centrally by this body. The licensing system itself has different factors in common with some principles of both the Norwegian and Danish systems. Thus, as in Norway, permission may be granted both to institutes and to individuals, and, as in Denmark, this system is partly automatic. The categories automatically licensed under the 1944 Act are, however, greater in number than in Denmark, and include universities, research institutes, laboratories, and hospitals which are owned by the government, a county, or a borough.

Responsibility in the case of institutes, as with the other Scandinavian laws described, lies with the head of the institution or department, who must be a dentist, pharmacist, physician, or veterinarian. Qualifications in other fields must be approved by the Ministry of Agriculture Veterinary Board. The Board also grants licences to carry out animal experiments, which are otherwise prohibited under Article 12 of the Act (Articles 12 and 13 relate basically to animal experimentation), to individuals and other institutions which are not automatically licensed. In such cases, the Board will consider the qualifications of the individual responsible, the purpose of the experiment, the species and number of animals to be used, and the standard of the facilities available. The licences are issued for a maximum period of 3 years and may be withdrawn by the Board if necessary.

The principal legal provisions contained in Articles 12 and 13 of the 1944 Act are succinct, more detail being set out in a Royal Edict published later in the same year. Basically, permission must be obtained from the Borad of Agriculture for any use of animals for the purpose of scientific research or education, diagnosis of disease, production of pharmaceutical products, or other similar purpose, when the use is combined with surgical intervention, injection, drawing of blood, or otherwise involves fear or other suffering for the animals. This is the only piece of legislation of this kind where 'fear' is specifically mentioned, and is perhaps of significance. Suffering must be limited to the inevitable, and animals which are expected to suffer considerably following the termination of surgery or other procedures must be killed as soon as possible.

Again, as in Norway and Denmark, a distinction is made between 'higher' and 'lower' vertebrates. Regulations provide that no severe surgical operation may be performed on a *warm*-blooded animal for educational purposes when another method may serve this purpose. When surgical operations are performed it is stated that the animal must be anaesthetized in such a way as to feel no pain. However, the operation may be carried out under incomplete anaesthesia, or none at all, when it is necessary for the particular purpose. In such cases, the animal shall, when possible, be given pain-killers or tranquillizers to limit its suffering.

Other regulations provide that detailed information concerning each animal must be entered in registers concerning their origin and their use for experiments, under the supervision of the Board of Agriculture. However, this procedure only applies to certain species—non-human primates, dogs, cats, horses, and ungulates. Record keeping is not compulsory for other species, which in fact include the majority of animals used.

Supervision of animal experimentation is organized under the central administrative body—the Ministry of Agriculture Veterinary Board. However, unlike the Danish and Norwegian systems, which have one supervising committee for the whole country, supervision in Sweden is highly decentralized. It is the duty of Municipal Health Boards, assisted by County or District Veterinary Officers, to organize local supervision. These bodies report back to the County Councils and through these to the Veterinary Board, which prescribes the terms of reference for the local supervisory bodies.

The latest changes in the Swedish law were brought in on 1 July 1979. Ethical Committees have now been established on a regional basis. Six regional boards, one for each university district, undertake a preliminary examination of all requests for permission to use animals for experiments, and exercise an ethical control. The Boards have a total of at least 15 members, composed of equal numbers of researchers, animal personnel, and laymen (of which at least half the number represent animal welfare organizations).

Another important change, which was introduced in 1979, concerns the supply of animals. The law now states that animals used in experiments must have been bred specifically for the purpose. The breeders must obtain permission from the National Board of Agriculture, which stipulates rules for breeding such animals. The users of animals for experiments must only buy animals from these authorized breeders.

As mentioned briefly earlier, the Swedish Medical Research Council (SMFR) has a Laboratory Animal Committee, and this too exercises ethical control. The Committee includes lay members, one of which is a representative of the National Federation of Animal Protection Societies. Proposals for experiments are evaluated *ethically* as well as scientifically, before funds are granted by the SMFR.

FEDERAL REPUBLIC OF GERMANY

Leaving Scandinavia, but still in the Nordic region of Europe, we pass on to West Germany. A new law passed in 1972, called the *Animal Protection Act*, has recently become effective, and has replaced the law of 1933 on this subject. Parts 5 and 6 of the 1972 Act deal with 'Animal experiments' and 'Operations for training purposes' respectively. The 1972 Act is quite detailed in respect of animal experimentation, and is an advanced piece of animal protection legislation. Further regulations are being drawn up in association with the Act in order to define more clearly certain areas, such as animal care and husbandry, and committees of expert scientists are engaged in drawing up recommendations for these regulations.

In the terms of this new law, experimental procedures or treatments of animals which may be accompanied by pain, suffering or injury must be declared to the competent authority prior to their commencement. The competent authority is the Ministry of Food, Agriculture, and Forestry, which administers the law and from which the necessary authorization must be obtained when these procedures or treatments are carried out on live vertebrates. Authorization may be granted to university or scientific research establishments, or to persons performing scientific research. In the case of licensed establishments, the director holds responsibility. Persons who are licensed must be qualified in medical, veterinary, natural or biological science fields, or have requisite professional knowledge.

Licences are granted only after a dossier has been presented to the Ministry by the director or leader of the proposed experimental project. This must show that the desired results cannot be obtained by other reasonable methods than those of animal experimentation and that the experiments are essential for the prevention, diagnosis, or healing of diseases in man or animals, or that the experiments are otherwise serving scientific purposes. Before authorization is given, the installations and equipment present at the establishment must be seen to be adequate, as well as sufficient staff to ensure proper performance of the experiments and care of the animals. Proper accommodation, care, and veterinary treatment must be guaranteed for the animals.

Embryos or fetuses are excluded from cover under the 1972 legislation, and breeding experiments and behaviour experiments not involving stress or other trauma are beyond the scope of this Act. There are also other areas of work where licences are not required, even though pain, suffering, or injury may be inflicted. Animal experiments which must be carried out on account of statutory provisions or judicial instructions do not require licensing. Likewise, some procedures undertaken in human and animal medicine but involving other animals do not require a licence, although the procedures must serve the prevention, diagnosis, healing, or improving of disease, suffering, physical injury, or ailments of men or animals, or be used for obtaining or testing sera and vaccines, or for pregnancy testing. The procedures, such as vaccination, blood sampling, etc., must be carried out by standard or officially approved techniques.

Certain other principles are laid down in the Act, which must be respected whether or not the experiments require authorization. Thus it is stressed that the experiments must be restricted to the absolutely necessary measure, and that higher vertebrates should only be used if lower vertebrates will not suffice for the intended purpose. Also, warm-blooded animals must only be used when cold-blooded animals would be unsuitable. Similarly, it is stressed that pain, suffering, or injury can only be inflicted when they are inevitable for the intended purpose of the experiments.

Anaesthesia is obligatory for all animals subjected to experiments, except where its use would not be compatible with the object of the experiment, or where the experimental procedure would be less painful than the administration of an anaesthetic. Only one painful procedure may be performed on an unanaesthetized animal, except where the object of the experiment requires otherwise. An experimental animal which has undergone a major procedure whilst under the influence of an anaesthetic may only be used for another experiment when this does not cause pain, suffering, or injury.

Animals of certain species, mainly domestic types, are treated differently with regard to euthanasia. Thus solipeds, ungulates, non-human primates, dogs, cats, and rabbits must be examined by a veterinarian at the end of the experiment, so that a decision may be made as to whether the animal is to be kept alive or humanely destroyed, such as in cases where prolonged suffering will persist. For animals of other species, this decision is taken by the person who has carried out the experiment.

Records of all animal experiments must be kept, stating the purpose of the experiment, the number and species of animal, and the experimental methods used. In particular, if permission has been granted for the use of higher animals, the reasons for their use must be given. The records must be kept for a minimum of 3 years and are available for inspection by the supervising body. There is no requirement for specialized breeding of animals for experiments, but it *is* required that when dogs or cats are obtained for experimental purposes, the records kept must additionally show the name and address of the previous owner.

THE NETHERLANDS

By contrast, in the Netherlands, as in Sweden, experiments may only be performed on animals which are either raised at the establishment where the experiments are to be performed or which come directly from another establishment where animals are raised solely for experiments. This requirement is contained in the recent Dutch Act which was passed in 1976 and became law on 12 January 1977. The law is very detailed and contains no less than 30 articles. Supervision is undertaken by the Ministry of Public Health, from which licenses must be obtained for experiments on vertebrates which are likely to be injurious or cause significant pain. Permission will only be given for experiments that are of importance to the health and feeding of man or animals, or which are for other work of scientific interest.

The purposes for which experiments may be performed are set out in Article 1 of the 1977 Act, and include the *production* of sera, vaccines, and diagnostic products, as well as their testing. The other purposes allowed are for toxicological and pharmacological investigations, to test for pregnancy and disease conditions, to develop knowledge of the human or animal body, and to answer a scientific question.

The licensing system introduced in the Netherlands by the new law of 1977 is distinct in that licenses are not issued to individual scientists, but only to the heads of institutes or establishments requiring to perform animal experiments. This system cuts down considerably the administrative difficulties for the government department concerned—the Ministry of Public Health—but supervision and control of experimental conditions, methods, and facilities appears to be enhanced. The director or head of the establishment who is licensed, is held responsible for all the animal experiments carried out at that establishment, and, although he may not necessarily be an expert in laboratory animal care himself, he is responsible for appointing experts, a system which assures good animal care and conditions of experimentation.

Supervision of animal experimentation under the Dutch legal system is built in throughout all levels. The Ministry has a team of veterinary inspectors under a chief inspector, who, in addition to supervision, have an advisory task. These inspectors work closely and constructively with supervisors *within* establishments, appointed by the licensed director. These supervisors are mainly veterinarians, but may also be biologists and are resposible for the welfare of the laboratory animals, being trained so as to be capable of enforcing the law within their particular establishment. The laboratory animal experts must in their turn work in association with the research workers who actually perform the experiments, guaranteeing the best conditions for the protection of the laboratory animals.

The other form of expert laid down in the Act is the animal technician, and a programme of training exists for these, similar to that of the British Institute of Animal Technicians.

Thus, through co-operation and mutual consultation between these various groups of experts who act at all levels in the system, the welfare and protection of the laboratory animals is greatly enhanced.

Other provisions contained within the Act give the Minister of Public Health the power to specify requirements for laboratory animal housing and to regulate the origin of the animals. With regard to species, it is prohibited to use horses, apes, dogs, or cats for experiments if it can be assumed that the same results can be achieved by experiments on other species. It is also forbidden to perform experiments on *any* animals if the result may be obtained by using existing alternative methods not using live animals. This same principle now figures in all the most recent legislation on this subject.

With regard to anaesthesia, Article 13 of the Act stipulates that the experiment must be performed under partial or total anaesthesia if more than negligible pain will be caused. However, this obligation does not hold if the anaesthetic would frustrate the purpose of the experiment. If the experiment causes pain which is likely to persist for a long time, the animal must be killed immediately after the experiment.

The Act also allows for an Advisory Committee to advise the Minister of Public Health on all matters concerning licensing and statutory provisions. The Committee consists of experts both in laboratory animal science and in animal welfare, but it is an *advisory* committee only and not a supervisory committee such as in Denmark or Norway.

The law of 1977 has also introduced the registration and publication of information concerning animal experiments, with regard to the numbers of animals used and the purposes for which they are used. This principle is also of importance generally, since, with more information available, the public may be in a better position to weigh the importance to the welfare of the community of a given animal experiment.

BELGIUM

Contrasted strongly against the detailed and well-constructed law in Holland on animal experimentation are the legal provisions of neighbouring Belgium. The only control in existence in Belgium is afforded by a single article of the general *Animal Protection Act* of 2 July 1975. This Act replaced that which was previously in operation, dating from 1929. The relevant article is Article 6, which basically prohibits any person from carrying out vivisection experiments with certain exceptions. The out-dated term of vivisection is still employed, and no system of licensing or supervision is prescribed.

The exceptions to the prohibition are experiments performed on animals which are necessary for scientific research, for medical purposes, or for veterinary purposes. These experiments can only take place in the laboratories of universities or other establishments as laid down in a Royal Decree of 13 September 1976. The experiments are performed under the control of the director of the laboratory, who is held to be responsible and, except in cases of strict scientific necessity, must be carried out under anaesthesia appropriate to the species. Adequate post-operative care must also be given.

The only supervision laid down in the Act is provided by state veterinary inspectors, who have free access to the laboratories at all times. The administration is given into the hands jointly of the Ministry of Agriculture and the Ministry of Justice.

Recently, a proposal has been put forward by a group of Belgian parliamentarians to amend the 1975 law. The principal effects of this proposal would be to ensure that experiments may only be performed when it can be shown that they are absolutely necessary, and to introduce the concept of alternative methods being used whenever possible. Various recommendations have also been put forward by the National Council for the Protection of Animals (CNPA) of Belgium, which undertook a study of animal experimentation and published a report (CNPA, 1976) containing several recommendations, many of which would enhance the welfare of laboratory animals. For example, it is suggested that mixed scientific committees should be set up at each research centre performing animal experiments. These committees would be open to people from outside the institute, and would undertake the definition of policy on animal experiments, the vetting of projects, and their surveillance. They would also be held legally responsible.

Other recommendations include adequate training of researchers and animal-house staff, with examination procedures to prove competence; animals—particularly rodents, dogs, and cats—to be specially bred by authorized breeders; veterinary supervision of animal houses; and recommendations on methods of euthanasia. However, it remains to be seen whether any of these valuable suggestions are taken up in law.

FRANCE

Control in France is only marginally more effective, at least in theory, than in Belgium, and lacks definition and precision. The French law of 19 November

1963 states that persons carrying out experiments, scientific or experimental research on animals, must conform to regulations laid down by Order. The Order of 9 February 1968 is the relevant Order, controlling experiments or research carried out on vertebrate animals.

Experiments which cause suffering to live animals, can only be carried out by persons authorized by the different government departments concerned. Certificates are issued, which must be shown on demand. Authorization is only given for experiments for the purposes of scientific research, tests, diagnosis, or the production of substances of benefit to the health of man and animals, or for teaching purposes.

A consultative and administrative committee attached to the Ministry of Agriculture, the government department mainly concerned with supervision, gives advice on all problems concerned with the protection of experimental animals.

The conditions of housing and maintenance of animals in establishments are defined by reference to animal protection in general. Experiments must be performed under local or general anaesthesia, or with equivalent analgesic procedures, with the usual exception of cases where this would be incompatible with the desired experimental results. Animals which are not destined to survive the procedure must be destroyed whilst still under the influence of the anaesthetic. Experiments without anaesthesia must be reduced to the minimum and to cases of strict necessity, the justification for which lies with the person responsible for the experiments. Only one operation may be performed on the animal without anaesthesia, except in cases which must be specially justified by the person responsible for the experiment.

Supervision of the conditions under which experiments may be performed, and of the persons authorized to carry out experiments, is undertaken by veterinarians or pharmacists from the different ministries concerned—respectively the Ministry of Agriculture and the Ministry of Public Health.

ITALY

Moving now to the last country of the EEC to be dealt with (Luxembourg has a very limited law relating to animal experiments, while the Republic of Ireland still uses the British 1876 Act), the Italian law giving protection to laboratory animals dates from 12 June 1931, and was later modified in May 1941. This law prohibits vivisection and all other experiments on live animals except for experimental purposes for the progress of biology. However, the law only applies to warm-blooded vertebrates, and thus is the least restrictive law in terms of animal groups covered.

The experimental procedures, according to Italian law, may only be performed in state institutes or laboratories authorized by the administrative authorities, and must be carried out under the responsibility of the director of the establishment. Experiments may only be carried out by named, authorized persons at these establishments. The experimenter must be a physician, veterinarian, or scientist competent in the field of biological science or natural science. An experimenter may also be a person qualified in another field, but in such a case, must be authorized specially by the administrative authority.

Experiments on live animals must be carried out under general or local anaesthesia, which must be effective throughout the whole experiment, except again in cases of incompatibility with the desired experimental purpose. The animals must be destroyed before the anaesthetic ceases to be effective if pain will persist, except in cases where the experiment necessitates that the animal be kept alive. The decision as to the procedures mentioned above rest with the experimenters, and not with the director of the establishment. Again, animals may only be used once for experiments, except in cases of necessity for the particular experiment.

Supervision of establishments and of animal experimentation is undertaken on a decentralized basis by the provincial health authorities, who appoint animal inspectors qualified in either human or animal medicine.

CENTRAL EUROPE

It is thus evident, from a comparison of the legislation outlined so far, that several countries, particularly in northern Europe, have extensive and detailed laws concerning the use of animals for experiments. Other countries, however, are lacking in their provisions for control of this practice. However, two countries of central Europe which do have very detailed laws on animal experimentation are Austria and Switzerland, which have laws dating from 1974 and 1978 respectively. The Austrian law deals specifically with animal experimentation and is distinct from other animal protection Acts in that country. Control in Switzerland is afforded by Section 6 of the general *Animal Protection Law* and this has yet to be passed by a referendum before becoming operational, although this is expected.

Austria

The Austrian federal law on animal experimentation is one of the most detailed of all the European laws on this subject. The spirit of the Act, and the way in which it is set down, is undeniably on the side of animal welfare and protection. Animal experimentation is divided into three areas and different government departments are responsible for each area. Thus supervision and the issuing of licences for experiments in universities is undertaken by the Federal Minister for Science and Research, whilst experiments in the area of trade and industry are supervised by the Minister of Trade, Commerce, and Industry. The Minister of Health and Environmental Protection is empowered with enforcement of the law

LORETTE WILMOT LIBRARY
NAZARETH COLLEGE

in connection with animal experiments in the third area—affairs of public health, veterinary activities, and nutrition, including the safety control of foodstuffs.

Experiments in the three categories may be carried out for purposes of research and development, for purposes of scientific education and medical diagnosis, and for purposes of testing and investigation of sera, medicines, foodstuffs, luxuries, preservatives, pesticides, and cosmetics. A permit to carry out animal experiments is granted to the named head of an institution by the appropriate authority, but stringent conditions have to be met before the permit will be granted. The permit also specifies the kinds of experiments for which it is granted. Before a permit is granted, it must be shown that the experiments are justified in that they are for prevention, recognition, or cure of disease in man or animal, or for the achievement of scientific knowledge, or for scientific education. It must also be assured that the experimental results or scientific education cannot be achieved by other methods or procedures not involving animals. Other requirements which must be fulfilled before permits are granted involve: the provision and assurance of accommodation and care for the experimental animals, including medical care; proper facilities, equipment, and premises for the conduct of experiments, maintenance, and care of the animals; and persons available before, during and after the experiments, including technicians, for the proper treatment of the experimental animals.

There are, nevertheless, certain exemptions to the licensing procedure and permits are not required for (a) experiments carried out in state testing institutions of the health and veterinary administrations and for state institutions concerned with foodstuffs, (b) experiments carried out under legal or judicial orders, and (c) experiments for the routine testing of sera, vaccines, and diagnostics, when done by standard scientific procedure.

Animal experiments on vertebrates using operative procedures may only be conducted by persons qualified in the fields of veterinary or human medicine, pharmacy, or biology and who also have adequate special knowledge or are supervised by such persons. Other animal experiments may be undertaken by these persons and by persons with qualifications in the other natural sciences with the same requirements for special knowledge.

Regarding the conduct of experiments on vertebrates, it is required that the state of health of the animals be verified as proper for the experiments. Experiments on vertebrates must be performed under anaesthesia, with the usual exception of cases where the desired aim of the experiment precludes it, or it is more injurious to the animal than the experimental procedure itself. The use of muscle-paralysing substances are forbidden without anaesthesia. The animals (vertebrates) may only be used once when operative procedures are undertaken which cause serious injury to their condition, except for their use in acute experiments under general anaesthesia. If animals suffer pain after an experiment, they must be treated by a veterinary surgeon. The experimenter is responsible for the animals' condition after experimentation has terminated and, if prolonged suffering is likely, the animals must be immediately, painlessly destroyed.

A general statement appearing in the Act requires that animal experiments are always limited to the indispensable measure and that they are to be conducted with the avoidance of all pain or suffering which is not necessary to the purpose of the experiment. This last phrase is slightly ambiguous, but, if taken in its true sense, could be significant for animal protection.

Other articles of the Austrian law relate to record keeping, supervision by the authorities, and penalties for breaches of the law and its regulations. Records have to be kept for at least 2 years and must include the purpose of the experiment, the results, the number, species and origin of animals, including the name and address of the previous owner in the case of dogs and cats. The supervisory authorities must employ qualified persons for inspection, and these persons have free entry to establishments during working hours, and access to all records. An examining committee may make an inspection and may compel the director of an establishment or his deputy to accompany them on their inspection.

Two types of offence are recognized in relation to the Austrian animal experimentation law. Direct breaches of the federal law are judicially punishable offences, but breaches of certain sections of the law are regarded as 'administrative violations' and are punishable by fines (or imprisonment) imposed by the district administrative authority.

One further article contained in the 1974 Austrian law, is of interest and concerns the refusal of an employee to conduct an animal experiment as defined by this law. It is stated that such a refusal does not represent a violation of duty when this duty does not result immediately from the employment contract and if the employee has not signed on specifically to perform such tasks. It is also not a violation of duty if the animal experiment is connected with danger to the operator's health. The enforcement of this part of the law is given into the hands of yet another government minister, the Minister of Social Administration. This section of the law appears to be unique amongst European animal experimentation laws.

Switzerland

Many of the provisions of Section 6 of the 1978 Swiss law—the section dealing with experiments on animals—are similar to those of the Austrian legislation, but they only apply, however, to vertebrates. For instance, it is stated that animal experiments requiring a permit must be restricted to the level of what is indispensable, and that competent personnel and appropriate installations for the keeping of the animals concerned must be available. It is also required that the experiments be conducted only under the direction of a trained specialist, and by personnel who have the necessary specialized information at their disposal, as well as required practical training. The animals must be kept, fed, and given medical care in accordance with the most recent developments in the field, before, during, and after the experiments.

Permits have to be obtained for animal experiments which cause pain to the experimental animal, cause it great 'anxiety', or considerably injure its general health, and they are issued by the authorities of the respective cantons. Permits are only granted to the director of an establishment.

The purposes for which experiments may be performed as laid down in the Act differ slightly from those stipulated in the Austrian law. Experiments must serve the purposes of scientific research; the manufacture or testing of sera, vaccines, diagnostic reagents, and medicinal substances; the investigation of physiological and pathological processes and conditions; teaching in colleges, where the experiments are unconditionally necessary for the teaching; and the maintenance or multiplication of living material for medical or other scientific purposes, if this cannot be done by other means. The Act specifically mentions also that experiments in behaviour research are controlled by the regulations.

Pain, suffering, or injury may only be caused when this is unavoidable for the stated objective of the research. Anaesthesia must be used if more than insignificant pain is likely to result, but the usual exceptions apply. The principle of only using higher animals, such as mammals, when lower species will not suffice, found in several countries, is embodied also in the Swiss law. Requirements regarding euthanasia of animals suffering pain, or for their subsequent use in further experiments, are similar to those under the Austrian system, although under the Swiss system there is no exception and animals which have suffered considerable pain or serious anxiety may not be used for *any* other experiments.

A requirement for records to be kept is made, in that a protocol has to be followed for all animal experiments requiring a licence, and these must be kept for a minimum of 2 years and must be made available on inspections.

Supervision is carried out by the cantonal authorities which regulate the issuing of permits and supervise the keeping of experimental animals and the conduct of experiments. For this purpose, each canton has a Commission composed of various specialists, the tasks and powers of which are determined by the canton. Supervision is also aided by an Advisory Commission appointed by the Swiss Federal Council to advise the Swiss Veterinary Administration.

OTHER WESTERN EUROPEAN COUNTRIES

Regarding the remaining countries of the Council of Europe, no formal legislation specific to animal experimentation is in existence. Spain has no legislation governing animal experiments, although the Ministry of Agriculture is at an early stage in the preparation of a Bill based on the regulations existing in other countries. Likewise, Greece has no special regulations on animal experiments, although it appears that it would ratify the European Convention of the Council of Europe when it is completed.

In Cyprus, whilst there is no actual legal control, experiments with animals are carried out in line with the *Cruelty to Animals Act 1876* of the United Kingdom. The Cypriot *Cruelty to Animals Law* of 1959 provides for the protection of animals from unnecessary suffering, but does not deal specifically with experiments on animals. However, the experiments are all carried out by veterinarians and skilled technicians and need the formal permission of the Director of the Department of Veterinary Sciences. The experimental work undertaken includes mainly vaccine testing and control, and the use of animals for other kinds of work, such as toxicology, is rare. It is envisaged that an Advisory Committee will be established in the near future. This Committee will be composed of competent experts and will draft regulations governing the use of live animals for experimental purposes, which will eventually become national legislation. The terms of reference of the Committee will be in line with the principles of the eventual European Convention.

EASTERN EUROPE

Regarding the countries of eastern Europe, information is difficult to obtain. However, it is known that there is no legislation on the subject in Czechoslovakia. In Russia, the protection of laboratory animals is controlled to some degree by the USSR Veterinary Regulations of 1967, on the keeping of animals in captivity. The keeping of animals for experiments is determined in detail by rules adopted by corresponding scientific organizations, under the supervision of the veterinary service of the USSR Ministry of Agriculture.

In Poland, animal experiments are controlled by a Decree of the President of the Republic of 22 March 1928 concerning the Protection of Animals. In this Decree, it is stated that scientific experiments with animals are not considered cruel treatment if the experiment is essential to important scientific research. The Decree requires separate regulations to be issued in order to restrict unjustified suffering and these regulations are presently contained in a Decree of the Minister of Higher Education of 16 November 1959. The Regulations allow experiments to be conducted for scientific research, educational purposes, and for diagnosis, as well as for the control and production of pharmaceuticals, chemicals, and foodstuffs. Experiments for educational purposes may be conducted only if previously programmed by the state. No unnecessary experiments may be performed where the results are already well known and experiments may only be conducted in scientific institutions of higher education, Institutes of the Polish Academy of Sciences, other scientific institutes, and laboratories for the control of medical products, sera, vaccines, and food products.

Permission needs to be obtained to carry out animal experiments, and is granted to individual persons having the 'proper' qualifications. This permission may be withdrawn if the person conducting the experiments does not observe the regulations protecting animals. Permission is granted by the heads of educational institutions, the Polish Academy of Sciences, and the respective state bodies supervising the other institutions at which experiments may be performed.

Requirements regarding care, the use of anaesthesia, the use of animals more than once, and euthanasia are basically the same as in other European legislation which has been described. However, a variation of the principle of using lower animals in preference to higher animals occurs, in that it is required to use animals which are of a low level of *'psychological'* development, and only exceptionally may experiments be performed with animals which are highly developed psychologically. A register has to be kept at all institutions where such animals are used, and the institutions themselves must also be registered by the supervisory authority concerned.

Thus it is evident that at least some measure of control is exercised in certain countries of eastern Europe, although it is not possible to determine how effective these measures are. Nevertheless, thought is being given to the subject in eastern Europe as well as western Europe, and further legislation is presently being drafted in Poland.

EUROPEAN LAW AND THE COUNCIL OF EUROPE

On 1 July 1969, a Resolution was presented to the Consultative Assembly of the Council of Europe at Strasbourg, by a group of parliamentarians. This Resolution requested three things in relation to what has now become popularly known as 'the alternatives question': (a) the establishment of a research institute for the development of new experimental methods to replace the use of live animals; (b) a documentation centre should be created for the purpose of making known these new techniques; and (c) it should be prohibited to carry out experiments on live animals in *all* cases where 'alternative methods' permit of their dispensation.

The Resolution passed before the Consultative Assembly of the Council of Europe who adopted a Recommendation—no. 621—in 1971. Many of the problems of the use of live animals in experiments were broached and several points were accentuated in this Recommendation such as, for example, the disparity of the national regulations existing in Europe—which can be seen from the preceding sections of this chapter. Experiments performed on live animals without anaesthesia were condemned, although this would lead to problems with regard to procedures carried out for the production of biological products. The necessity of doing everything possible to develop alternative methods to the use of animals was highlighted, as well as the obligation of respecting the fundamental rules of human ethics, whilst recognizing the necessities of human and animal health and of biomedical and scientific research.

In order to put the various clauses of the Recommendations into practice, a Committee of Experts was set up, which has now become the Ad Hoc Committee of Experts on the Protection of Animals and which started, in 1978, the elaboration of a European Convention on the use of live animals for experimental purposes, in accordance with the mandate given by the Committee of Ministers of the Council of Europe.

The Committee of Experts has a difficult task in attempting to harmonize the different national regulations and several years of discussions will doubtless be necessary. Several sessions of this Committee have now taken place and although it is not possible to pre-judge the outcome of the final Covention, nor the date on which it will become applicable and open to ratification by the member states, it is interesting to examine the main objectives of the Convention and the areas of the discussions and problems which arise.

The object of the Convention is to limit experiments on live animals to the strict minimum for humanitarian as well as economic reasons. However, certain factors are allowed for in this objective, namely the maintenance of scientific and technical progress demanded by our civilization and the assurance of the needs of man in respect of his food, health and well-being, and the protection of his environment.

However, such a limitation is not aimed at simply through regulations and prohibitions instigated after consultation with the scientific authorities, but more as a result of adherence to a professional code of ethics, the principles of which would be determined by the Convention. It will be left to the member states to determine the necessary methods of control in relation to their individual needs.

It appears that some system of authorization will be stipulated for all persons *and* establishments using animals for experimental purposes. A system of control is also envisaged for the commercial breeders of experimental animals, notably with regard to certain species to which the general public seem to be particularly sensitive, i.e dogs, cats, horses, and non-human primates. The procuring of animals from sources other than specialized laboratory animal breeders may be prohibited, mainly in order to cut out the 'back-door' trade in dogs and cats, dubiously obtained by equally doubtful animal 'dealers'— a practice which is known to occur in several European countries, including Britain, France, and Italy.

The eventual European Convention will require the provision of adequate statistical information, as well as the introduction of some form of consultative or supervisory body, on which all groups interested in the protection of experimental animals would be represented, including the users of the animals and representatives of animal protection societies. This should help to alleviate many of the current controversies caused by lack of *informed* knowledge.

Statistical information is also likely to be required on the breeding, keeping, and trade in animals to be used for experiments, and both this information and the other statistics on species and numbers used, purposes, techniques, etc., may be required to be supplied to the Secretariat of the Council of Europe.

Another question of some difficulty which has been broached in the discussions at Strasbourg, is that of the recognition in one country of experimental results obtained in another. If this could be achieved, it is probable that a considerable amount of repetitive experimentation would be abolished. However, owing principally to the factor of 'biological variability', inherent in all living systems, there are great difficulties to be overcome, since results may differ according to the conditions of experiment, the strain of animal used, the sex of animal, its weight, and a whole host of other factors. In toxicity screening, the main field where repetition occurs, results may differ between laboratories (and even within the same laboratory) due to one small factor being changed. It is thus infinitely more difficult to compare results internationally. However, the problems are not insurmountable and, through greater standardization, significant progress could be made.

Other interesting features of the proposals being put forward for a European Convention include the establishment of a Code of Ethics. This would involve such restrictions as, for example, not carrying out experiments on live animals where a valid non-animal alternative exists. The Code of Ethics would also attempt to restrict the infliction of pain, suffering, distress, or injury on animals. Experiments without anaesthesia would have to be restricted to the absolute minimum, as would the infliction of any pain or suffering. Only one severe procedure would be able to be carried out on any one animal, and analgesics and sufficient treatment and after-care would have to be given after procedures have been terminated. Exceptions to these stipulations would have to be supported by special justification.

Present proposals for the Convention do, however, make a distinction—which can often be found in national European legislation—between higher vertebrates and lower vertebrates. The distinction would give greater protection to birds and mammals—the warm-blooded vertebrates—than to their cold-blooded counterparts—the reptiles, amphibia, and fish. This idea, as has been explained before, may be based more upon a reaction to public opinion than to sound principles of animal protection, since there is no direct evidence that cold-blooded vertebrates are less capable of experiencing pain and suffering than warm-blooded vertebrates.

Throughout the proposals for a European Convention, the accent is firmly upon animal protection, with a great number of safeguards being built into the system of supervision and control. Various individuals are made responsible at different levels in the system and adequate training of all personnel is stressed. More emphasis appears to be put upon safeguarding the animal as opposed to safeguarding the experimental results, and this principle, if achieved, could be of great value for animal protection. For example, animals which were suffering severely would be destroyed immediately, without waiting for the experimental results to be achieved. This type of provision does not appear in any present European legislation controlling live animal experiments.

CONCLUSION

In spite of persistent attempts in some countries, and in spite of some progressive animal laws, it can be seen from the preceding comments that there are a number of problems regarding animal protection in relation to animal experimentation within Europe. It may be that the only solution to many of these problems will be through supranational regulations—such as the Council of Europe Convention.

It appears to be becoming more and more apparent as European animal protection legislation evolves, that the obligation to protect animals is not so much based upon human needs and desires, nor upon human sentiments, but more upon the respect of an animal as a sentient being. Protection is based increasingly on appropriate ethological principles and scientific knowledge of the standards and requirements necessary for different species, and these principles will hopefully develop further in the future and become widely accepted.

The objective of new European regulations concerned with the welfare of animals in experiments, must be to bring about a balanced relationship between ethical, economic, and scientific needs. The eventual European Convention should undoubtedly introduce further constraints on experimenters and result in more stringent control over establishments performing experiments on animals. However, the constraints and controls cannot be too exacting or stringent, since their effective application and enforcement has to take into consideration existing means and practicalities. The interest manifested by the EEC Commission and the USA, both of whom have observers at the Council of Europe Expert Committee meetings on this subject, exemplifies the concern that impractical, unnecessary, and over-stringent control is not established in the Convention.

As has been mentioned earlier in this chapter, apart from the proposed European Convention, various moves are afoot in different countries of Europe to bring about changes in the law so as to give further protection to laboratory animals and to restrict their usage. The basic principle, which is gaining greater and greater acceptance throughout Europe, is that animals must be protected not simply because they represent an economic value, nor because the pain *they* endure offends *our* sensitivity or emotions, nor even because they are a part of our environment, but because they are living, sentient beings, which must be protected for humanitarian reasons. Precise legal obligations must be established at an international level, so as to assure this protection and lead to a greater respect for the interests of animals.

REFERENCES

NAVF (Norges almenvitenskapelige forskningsrad) (1976). *Forsoksdyrsituasjonen i Norge En Kartlegging av forbruk, kvalitet og oppstalling i 1974*, NAVF, Oslo.
CNPA (Conseil National pour la Protection Animale) (1976). *Le livre blanc de la protection animale*, CNPA, Brussels.

Animals in Research
Edited by David Sperlinger
© 1981 John Wiley & Sons Ltd.

Chapter 3

Legislation and Practice in the United States

MARGARET MORRISON

INTRODUCTION

Nearly 100 years following passage of the British *Cruelty to Animals Act*, the United States Congress passed the *Animal Welfare Act* (AWA). The delay in enactment of a federal laboratory animal statute was due more to the nature and evolution of the American political system than to any other factor. Until the middle of the twentieth century, the states were considered the proper level of government to regulate animal research, if there was to be regulation. The Tenth Amendment to the US Constitution reads:

'The powers not delegated to the United States by the Constitution, nor prohibited to it by the states are reserved to the states respectfully or to the people.'

Gradually, states' rights were eroded by an expanded interpretation of the Constitution, and the federal government became directly involved in matters formerly reserved to the states.

The *Laboratory Animal Welfare Act* of 1966 (see Appendix) was based on the premise that the authority of the federal government to regulate interstate commerce could be applied to the sale and use of experimental animals. Appropriately, the original law was passed during the 'Great Society' of the Johnson Administration at a time when the federal government became involved to an extent unprecedented in time of peace and prosperity in such areas as education, jobs, urban development, and other economic and social problems.

State legislative efforts to protect laboratory animals resulted in a few pyrrhic victories and a string of defeats (McRea, 1910). During the second half of the nineteenth century, most states passed basic anti-cruelty laws to prevent the mistreatment of animals. Vivisection was not listed as an offence under these laws. In many states, such as New York (the home of Henry Bergh and the American Society for the Prevention of Cruelty to Animals), vivisection was expressly exempted. Even with no specific exemption, research was beyond the purview of cruelty laws (McRea, 1910). It is believed that only one researcher has been convicted of cruelty. In 1958, the Massachusetts Society for the Prevention of Cruelty to Animals successfully prosecuted a researcher and an animal caretaker for cruelty (Stevens, 1978). However, the charges were based on care and transport abuses outside of the laboratory.

Bills modelled on the British Act or designed to ban vivisection totally met ultimately with defeat or repeal in state after state. These failures can not be blamed on any one factor. It has been suggested that humanitarians were more concerned with establishing societies than stopping painful experiments. Anti-vivisection efforts were further handicapped by the diffusion of their resources throughout the states. Also, the research community was better organized and able to find allies in state legislatures. Anti-vivisectionists' credibility suffered, too, because of their occasional identification with unorthodox medical theories (Bordley and McGehee, 1976), such as the repudiation of the germ theory of disease (Schultz, 1924).

The American medical community has a long tradition of political action. During the era of the greatest anti-vivisectionist agitation, polemical writings were circulated to counter the claims of medical research opponents. Evarts A. Graham, Professor of Surgery, argued:

'In light of the amazing conquests achieved in that war [World War I] by the use of vivisection, how can anyone now cry "stop" and attempt to tie the hands of those investigators who are finding ways to prevent the pain and mutilation of little children that is still caused by disease! As it happens, moreover, in practically all of the work in vivisection, the animal which is used suffers no pain because the experiment is carried out in full anaesthesia from which the animal is not allowed to recover.' (Graham, 1935).

By discrediting their opponents and warning of dire consequences, the research defence interests avoided any infringement on their freedom to conduct research. In the United States, medical research using animals was allowed to grow and gain in prestige without government regulation. Since legislation restricting animal experimentation was not passed during the early years, enactment of such laws was unlikely in later years. Even so, prior to World War II, the amount of biomedical research, compared to today, was relatively small.

After World War II, there was a tremendous explosion in the amount of research sponsored, supported, and conducted by the federal government. In 1945, the Public Health Service gave $700 000 in grants through the Department of Health, Education, and Welfare's National Institute of Health (NIH). NIH now stands for the National Institutes of Health with a budget of several billion dollars per annum. Congress caused this acceleration by providing massive funding for research. Public concern over health matters, especially cancer, also played an important role in research expansion.

Legislative interference into the business of animal experimentation might never have occurred if the demand for research animals had not similarly increased. Researchers had in the past acquired animals from a variety of sources. The professional research animal dealer is a recent phenomenon. Previously, animals were casually procured from pounds, or from individuals who collected strays and sold them to laboratories. Humanitarians were relatively unsuccessful in attempts to block the practice.

'Pound seizure' laws requiring that unclaimed, impounded animals be turned over to laboratories, were passed in several states, largely at the instigation of the National Society for Medical Research. Enacted in a total of 11 states, a few of these laws contained animal care provisions in an attempt to pacify humane concerns, but these were weakly enforced. In a few states, the practice was banned. Mainly it was left to the discretion of the individual pounds and shelters.

In 1979, the New York State pound seizure law, the Metcalf—Hatch Act, was repealed. Several local and national humane groups coalesced in the successful repeal efforts, basing their arguments on two premises: (1) strays are not reliable research subjects and (2) pound seizure deters responsible animal control since many owners prefer to abandon pets rather than place them in a shelter from which they could be sent to research laboratories.

The political acumen of the humane groups in 1979 was not visible 25 years ago. Animal welfare organizations, including anti-vivisection groups, lacked political savvy. The major national humane organization in the early 1950s, the American Humane Association, was not a vocal advocate for laboratory reform. The dissatisfaction of activists within the humane movement led to the formation of more aggressive national organizations such as The Humane Society of the United States (HSUS) and The Animal Welfare Institute (AWI).

Until the late 1950s, the animal protection position was perceived as being sentimental and made up of 'soft argument'. However, it is now more sophisticated and professional, and has attained a certain degree of scientific credibility. The resulting political activity is more effective and is beginning to affect the control by the medical research community on the use of animals in laboratories.

THE FEDERAL LABORATORY ANIMAL WELFARE ACT

The first national laboratory animal bill in the US Congress was introduced in 1916 by Senator Jacob H. Gallinger, a homeopathic physician. The bill would have authorized the Secretary of Agriculture to examine the 'extent and conditions of the practice of experimentation on living animals'. Hearings on this bill were held on 19 June 1916 (Schultz, 1924), however no further action was taken. Throughout his political career, Gallinger had also sponsored bills to ban or restrict vivisection within the District of Columbia (at the time Washington, DC was totally governed by Congress). In 1897, one bill for the District of Columbia almost came to a vote, however it was withdrawn (Bordley and McGehee, 1976). Similar bills continued to be introduced in Congress, and as late as 1946 hearings were held which pitted anti-vivisectionists against the medical community.

Interest in legislation with a national scope coincided with the formation of the Animal Welfare Institute (AWI) and The Humane Society of the United States (HSUS). Reports of cruelty in experimentation and the terrible conditions in some dealers' facilities surfaced, and the public demanded action. From 1959 to 1965, bills were introduced, including several modelled on the British Act. Hearings were held in 1962, but no action was taken until 3 years later. On 30 September 1965, the Health Subcommittee of the House Interstate and Foreign Commerce Committee held additional hearings. Two British medical researchers testified that their work had not been harmed by the constraints of British law (Stevens, 1978). American researchers contradicted this, alleging that British research was badly hamstrung. Anti-vivisectionists opposed the bill as too moderate. Clearly, a bill on the British model was a dead issue in the US Congress at that time, but the desire to protect research animals was not.

It was at this watershed that the course of American legislation was determined, making it very different from the British approach. The British Act applies to actual experimentation, while the American *Animal Welfare Act* applies to care and not to experimentation.

Between 1965 and 1966, two events occurred that may have determined the eventual path of US legislation. First, an allegedly stolen family pet was traced to a laboratory. Congressman Joseph Resnick, who had aided the pet's owners, introduced a bill to regulate dog and cat dealers and laboratories (Stevens, 1978). His bill was attacked by both anti-vivisectionists and researchers. The second event was a raid on a Maryland dealer's facility. The squalid conditions were depicted in graphic detail by a photo essay in *Life* magazine (*Life*, 1965).

On 2 September 1965, Congressman W. R. Poage, Chairman of the Livestock and Grains Subcommittee of the House Agriculture Committee, held hearings on the Resnick bill and 30 others, including one introduced at the request of the National Society for Medical Research that deleted all provisions affecting laboratories. Poage was sufficiently moved to introduce his own bill. A weakened version of his bill, exempting laboratories, was passed by the House of Representatives.

In the Senate, Senator Hugh Scott introduced a bill identical to the original Poage bill that might have had the same fate, except for the intercession of Senator Mike Monroney. Monroney was able to restore the laboratory provisions. At this time, a split developed within the humane movement. The Animal Welfare Institute supported the Monroney restoration, but The Humane Society of the United States preferred a bill on the model of the British Act. In the end, the Monroney amendment was modified, passed the Senate Commerce Committee, and the bill sailed through the full Senate.

A House–Senate conference committee resolved the differences between the two bills, keeping the laboratory section. President Lyndon B. Johnson signed the *Laboratory Animal Welfare Act* on 24 August 1966. The law took effect for dealers on 25 May 1967 and for laboratories on 24 August 1967.

THE *ANIMAL WELFARE ACT*, 7 USC 2131–2156, ENACTED 1966, AMENDED 1970 AND 1976

Today the *Animal Welfare Act* regulates a broad spectrum of animal users, including zoos, circuses, and the wholesale pet trade. However, for discussion purposes here, the narrative will be limited to laboratory animal dealers and laboratories. The goals of the original Act were the humane care and treatment of laboratory animals, and the prevention of pet theft for sale to research facilities.

The Act is enforced by the US Department of Agriculture (USDA) through the Animal and Plant Health Inspection Service (APHIS) whose major mission is livestock disease eradication. By law, the USDA is directed to issue regulations for humane care, handling, treatment, and transportation. Standards for housing, sanitation, feeding, watering, employees, veterinary care, separation of incompatible species/animals, and transportation were also developed and issued.

Under the 1966 Act, laboratories and dealers using *dogs and/or cats* were covered. In these facilities, the care of rabbits, non-human primates, hamsters, and guinea pigs was also regulated. In 1970, Congress dropped this qualifier and authorized coverage of all users of most warm-blooded species to the extent that money and manpower were available for administration and enforcement. The law exempts farm animals. Under the feasibility exclusion, rats, mice, birds, and, until recently, marine mammals were exempted.

Under the Act, a laboratory 'dealer' is any person or entity who in commerce for compensation or profit deals, buys for resale, or sells animals covered under the Act for research. The 1970 amendments added dealers who acquire live animals and prepare them as biological specimens. The 1966 Act limited 'commerce' to that between the states. In 1970, intrastate activities were added.

Under the original Act, 'research facility' was limited to facilities which purchased or transported members of the above-mentioned species or received federal funds. In 1970, 'research facility' was extended to all users of warm-blooded animals as defined by the revised Act. However, elementary and secondary schools, and small-scale diagnostic laboratories which did not use live dogs or cats, were exempted.

The *Animal Welfare Act* requires registration by research facilities and licensing of dealers. Dealers pay fees and are subject to civil and criminal penalties for infractions. Research facilities pay no fees, and since they have no licenses to lose, are not as vulnerable as dealers when they violate the Act.

Although the federal government accounts for approximately one-third of the research animal use in the US, federal facilities are not inspected, and hence there is no real enforcement of the AWA in federal agencies. The Act merely directs federal agencies to comply. While several agencies conducting animal research have incorporated the *Animal Welfare Act* standards into their own intra-agency policies, the issue of US government compliance is a controversy that has not been resolved. For example, in 1979, officials of the National Institutes of Health alleged that two particular animal shipments were not subject to the AWA regulations.

To deter pet theft, dealers and research facilities must keep detailed records and tag dogs and cats. Dealers must hold animals for 5 days before resale.

The *Animal Welfare Act* was never intended to affect the design, protocol, or performance of any research, experiment or test. The 1966 Act contained a broad exclusion that led to a total self-exemption by several research facilities from any adherence to the Act. Animals were declared 'under experimentation' from the moment of their arrival at the facility. In 1970, Congress amended the research exclusion to exempt from the Act only *the design, outlines, guidelines or performance of actual research.* Congress grasped the notion that painful experiments needed special attention and enacted the requirement that

every research facility shall show that professionally acceptable standards governing the care, treatment and use of anaesthetic, analgesic and tranquilizing drugs, during experimentation are being followed by the research facility during actual research or experimentation.' (7. U. S. C. 2142)

The AWA also requires the yearly reporting of numbers of animals under experimentation by every registered research facility and agency of the US government that uses live animals. The numbers are broken down into three categories: no pain, pain and drugs, and pain and no drugs. Pain and no drugs is only permissible when medication would invalidate the research. The National Society for Medical Research has long opposed the publication of data on painful experimentation.

Although the analgesic requirement seems adequate, the mechanism has not worked. The legislative history of the Act shows that USDA was to issue guidelines on analgesics. This has not been done to a sufficient extent. A scan of the Annual Reports shows that research facilities have different interpretations as to what constitutes a painful experiment and what demonstrates that professionally accepted standards are being followed. Often, the veterinarian who signs the reports knows nothing about the experiments, but relies on perfunctory assurances from the researchers. In short, no one is accountable. The reports are inconsistent and incorrect. For example, the 1978 Annual Report of the National Institutes of Health lists only seven animals in the pain and no drugs category, an absurdly low figure considering the number of research animals used by NIH. Some research facilities and government agencies have not filed the required reports.

Attempts are being made by humanitarians to make   research   facilities directly accountable for the screening of analgesic use. Also, the USDA must clarify the issue so that the research facilities will better understand their obligations in this regard.

ENFORCEMENT OF THE ANIMAL WELFARE ACT

In the past, the USDA has been an ineffective administrator of the Act. Many of the abuses that the Act was meant to rectify continue. USDA did not want jurisdiction over the Act, and has been unenthusiastic about its responsibilities. In addition, Congress has not given USDA sufficient funding. When the original Act was passed, it was estimated that $1 million was the minimum appropriation needed for the first year, but Congress provided only $ 300 000. Funding by Congress is often controlled by a few powerful legislators. For several years, the House Appropriations Agriculture Subcommittee Chairman, Congressman Jamie Whitten, opposed funding increases; in 1980 USDA shut down the programme for six weeks because the funds ran out.

No requests for adequate funding have been submitted by the Johnson, Ford, Nixon, or Carter Administrations. USDA bureaucrats have not aggressively pushed for a significantly higher animal welfare budget. Current AWA funding is $ 4 million, but $ 8 million is necessary. Small increases have been wrestled from Congress by concerned legislators such as Senator Birch Bayh. Since the *Animal Welfare Act* never benefited from the largesse of the government during the era when federal funds were poured into other areas, it is unlikely, because of the present trend toward austerity, that the funding will greatly improve.

Manpower has been an especially serious problem. The Act has been a low priority for APHIS. In 1977, 5 per cent of total APHIS man-hours was spent on the Act. Many APHIS employees are neither trained or motivated to work on animal welfare. Many violations go unchecked because the inspectors do not report them, often because of ignorance of the standards. Further, in at least one case, a USDA inspector failed to report his findings. Facilities should be inspected more frequently, especially those with serious deficiencies. Even when violations are noted, cases will drag on for years.

Only one research facility has been subject to legal proceedings. Legal action was taken against the National Institute of Scientific Research in California for mistreatment of animals. The charges included dirty, unsafe housing, inadequate food and water, and failure to provide adequate veterinary care. The institution agreed to a cease and desist order. No fines were levied against the facility.

Usually the Department relies on persuasion, and only threatens legal action. In 1979, the University of Iowa finally corrected several serious violations only when confronted with that threat. USDA has also been overly generous in allotting time in which to make corrections. Reports of poor conditions in research facilities still surface, although it is difficult for humane society investigators to gain access to a site.

By far USDA's toughest assignment is obtaining the compliance of the dealers. Pet theft continues, although not at the pre-1966 level. Many dealers fail to keep accurate and complete records. Records can be forged, and USDA has neither the time nor the money for full investigations of suspected pet thefts.

Many animal dealers still house animals in squalor. Even though a dealer's professional reputation should rest on the quality of the animals he produces, many have flourished despite their disregard for animals. For example, Marshall Research Animals, a prominent dealer in research beagles, was cited for a violation of the transport standards in 1979.

One of the most notorious violators of the Act was Animal Research Center, Inc. of Massachusetts, owned by one Ivan Likar. In 1973, Likar was prosecuted for a serious violation. The resulting cease and desist order was disobeyed. The subsequent court case was lost on technical grounds because of the ineptitude of the USDA legal staff. Another case was developed and in 1978 Likar's license was finally revoked. For more than 5 years, animals were kept in filth, and the USDA never took action to alleviate their suffering. For example, a dog sent from this dealer to a research facility was received with a metal chain embedded in the flesh of its neck. Obviously, the chain had been placed on the animal as a puppy, and never replaced.

Despite this bleak enforcement record, the *Animal Welfare Act* has benefited animals in research. Many facilities find it easier to obtain funds to improve animal housing and care because they can argue that improvements are mandated by federal law. The Act has also sensitized researchers to the right of animals to receive humane care. Many support the Act and are genuinely concerned about animal welfare. Others have found that they can live with the *Animal Welfare Act*, perhaps because it has been poorly enforced.

Animal welfare advocates criticize USDA for its failure to enforce the Act aggressively. APHIS has made changes designed to improve its performance and the public's perception of its record. Several part-time inspectors and veterinarians qualified in animal care have been hired. However, it is somewhat disconcerting that several veterinarians currently employed by research facilities have been contracted by USDA to inspect other research facilities. Although the level of technical expertise may be high, the potential conflict of interest may outweigh the benefits. In another attempt to improve its performance, the agency assigned several regular APHIS employees to work solely on animal welfare matters, and comprehensive training courses were held throughout the country.

Truly, there are some USDA employees who are concerned about animal welfare, and have done a remarkable job considering the funding limitations and the administrative difficulties. As a result, there has been noticeable improvement in enforcement. While some USDA employees have considered animal welfare groups in an unfavourable light, others recognize the assistance which these groups can provide and are responsive to them.

ANIMAL WELFARE ACT REFORM

The Animal Welfare Act should be strengthened. The penalties should be revised to provide a less cumbersome mechanism to force compliance. The analgesic provision should be modified to require more accountability for painful research in which painkillers have been withheld. The regulations are inadequate in many respects. The standards in some respects are too restrictive and, in others, too permissive. Ambiguity is rampant as evidenced by the liberal use of words such as "adequate" and "sufficient" leaving too much discretion to inspectors who are often ill-informed.

The solution to this regulatory nightmare must be the grounding of the standards in scientific data. The USDA has not been able to produce much reliable evidence to date. Funding for this type of research is rarely available and USDA veterinarians lack the specific expertise needed.

In 1980, the USDA embarked on a review of the regulations. Conceivably, the standards could be either strengthened or weakened as a result depending upon the persuasive skills of the special interest groups involved. Animal welfare organizations are advocating a general upgrading of the standards to surpass the so-called survival threshold and arrive at a truly humane set of standards. The regulated parties view the current USDA initiative as an opportunity to "get the government off their backs".

In addition, a bill has been introduced into Congress by Representative Pat Schroeder that would amend the Act to a significant degree. The bill, HR 6847, contains provisions to include *all* vertebrates under the Act, to statutorily require more humane standards, to prevent certain categories of unnecessary painful experimentation and to create actual accountability of research facilities by requiring the creation of research facility animal care committees that will be responsible to review the conduction of animal research. At this time, these groups perform little more than a rubber stamp function. Additional laboratory animal legislation will be discussed later on in this chapter.

Even if the law and regulations were ideal, enforcement problems would remain. Administrative difficulties continue to rise from a bureaucracy in which many employees consider animal welfare to be an inconsequential issue at best. This seemingly insurmountable problem may only be solved by the establishment of an independent enforcement squad responsible only to high officials within the Department of Agriculture.

OTHER FEDERAL LAWS PROTECTING LABORATORY ANIMALS

Both the *Endangered Species Act* and the *Marine Mammal Protection Act* protect animals that are subject to controls under these laws. If one wishes to purchase, acquire, or capture any marine mammal or endangered species, a permit must be obtained from the agencies that enforce these laws. Only scientists and public display facilities may obtain these permits. Proposed research outlines must be submitted along with a description of the care and facilities available. The *Endangered Species Act* goes one step further and requires that experiments may be done only if they directly benefit the species. Thus breeding programmes, dietary need studies, etc., are acceptable but not solely for the benefit of human research.

BEAGLES IN THE PENTAGON

In 1979 the Department of Defense (DOD) was under heavy fire for its use of dogs in testing chemical warfare agents. One of the largest blitzes of mail on a single issue hit federal legislators. In some congressional offices, 'beagle mail' outweighed mail on Watergate. The DOD, in assessing the mail they received, commented that they received more mail on this issue than on any other since President Truman fired General MacArthur (Holden, 1974). Senator Hubert Humphrey and Congressman Les Aspin championed the beagle cause in Congress. The *Military Procurement Authorization Act* of 1974 prohibited research on dogs for the purpose of developing biological or chemical warfare agents'. However, the provision was weak since the Conference Report on the bill indicated that it was not the intent of Congress 'to prohibit research on dogs for other purposes such as establishing immunologic levels, occupational safety hazard levels and other vital medical research to improve and save lives'. The remaining eviscerated prohibition was in effect for only one year. Subsequently, DOD announced its intention to continue to abide by the restriction. Other bills have been introduced into Congress that would place broader constraints on defence research, but none have passed. 'Benefit to human health' is repeatedly invoked by DOD to justify other research such as the radiation experiments on rhesus monkeys which have been connected to neutron bomb evaluations. As a result the Indian Government cut off the export of these animals to the US.

FEDERAL LAWS REQUIRING OR SUPPORTING ANIMAL TESTING

The purpose of this chapter is to discuss US laws protecting laboratory animals. However, this government activity is slight compared to the massive involvement of the US government in support of animal testing, either through direct funding or through legal requirements to protect human health and safety. Roughly one-third of all animal testing is conducted or supported by the government.

In the area of biomedical research, the largest federal supporter of animal research is the National Institutes of Health. However, other federal agencies sponsor biomedical research either through grants and contracts or through research actually conducted by the agency. In 1977, in the area of toxic substances alone, 37 agencies were involved at a budget obligation surpassing $ 640 million. In 1980, the figure will exceed $1 billion.

Research on toxic substances and other areas of biomedical research is largely spurred by the legal commitment of the federal government to test, evaluate, and regulate substances harmful to humans, and to otherwise assure and further the health of its citizens. The federal legal authority over toxic substances is vast and widely scattered throughout the bureaucracy as Table 3.1 illustrates (Council on Environmental Quality, 1979).

Enactment of the *Toxic Substances Control Act* (TOSCA) was a partial consolidation of scattered federal control over the thousands of chemical substances present in the environment. Through TOSCA, the Environmental Protection Agency controls the manufacture and production of chemical substances, and requires the submission of test data before manufacture may commence. During House consideration of the Toxic Substances bill, Congressman Richard Ottinger offered an amendment to give preference to available tests which do not involve animals. The amendment failed to pass. Another amendment sponsored by Congressman Andrew Maguire was accepted. This amendment authorized the Secretary of Health, Education, and Welfare to create grants for the development of inexpensive and efficient tests that later could be used for the development of test data required by TOSCA.

The National Institutes of Health, as the prime US government user and supporter of animal research, became increasingly sensitive about its role in animal research after passage of the *Animal Welfare Act*. The NIH found it advisable to issue its own policy on care and treatment of research animals for grant and contract recipients. First issued in 1971, the NIH guidelines on laboratory animals were updated in 1979 (Institute of Laboratory Animal Resources, 1979). Grant and contract applicants submit 'assurances' to the government that they will comply with the guidelines. An institutional committee within the facility is responsible for reviewing animal care. The principles for use of live animals specify qualified personnel, fruitful research, use of available alternatives, adequate analgesics, and avoidance of unnecessary pain.

A random sampling of federal research grant applications showed that the applications contained insufficient information on animal care and use with which to make an informed judgement. This means that the peer review committees could not have evaluated the applications on animal care requirements (Fox *et al.*, 1979).

In addition to the peer review within the various Institutes of NIH, an NIH employee in the Office for Protection from Research Risks evaluates compliance of the applications with the guidelines, and reports his findings to those responsible for scoring the applications. For the first time, in mid-1978, that employee followed up his earlier reports and found that of the 29 applications he found unacceptable on animal welfare principles, only two were scored high enough to guarantee funding. However, he could not measure the effect of his

Table 3.1 Federal laws and agencies affecting toxic substances control

Statute	Year enacted	Responsible agency	Sources covered
Toxic Substances Control Act	1976	EPA	All new chemicals (other than food additives, drugs, pesticides, alcohol, and tobacco) and existing chemicals not covered by other toxic substances control laws
Clean Air Act	1970, amended 1977	EPA	Hazardous air pollutants
Federal Water Pollution Control Act (now Clean Water Act)	1972, amended 1977	EPA	Toxic water pollutants
Safe Drinking Water Act	1974, amended 1977	EPA	Drinking water contaminants
Federal Insecticide, Fungicide, and Rodenticide Act	1947, amended 1972, 1975, 1978	EPA	Pesticides
Act of 22 July 1954, (codified as Section 346 (a) of the Food, Drug, and Cosmetic Act)	1954, amended 1972	EPA	Tolerances for pesticide residues in food
Resource Conservation and Recovery Act	1976	EPA	Hazardous wastes
Marine Protection, Research, and Sanctuaries Act	1972	EPA	Ocean dumping
Food, Drug, and Cosmetic Act	1938	FDA	Basic coverage of food, drugs, and cosmetics
Food additives amendment	1958	FDA	Food additives
Color additives amendments	1960	FDA	Colour additives
New drug amendments	1962	FDA	Drugs
New animal drug amendments	1968	FDA	Animal drugs and feed additives
Medical device amendments	1976	FDA	Medical devices
Federal Meat Inspection Act	1967	USDA	Food, feed, and colour additives; pesticide residues
Poultry Products Inspection Act	1957	USDA	

Act	Year	Agency	Coverage
Egg Products Inspection Act	1970	USDA	in meat and poultry products
Fair Packaging and Labeling Act	1976	FDA	Packaging and labelling of food and drugs for man or animals, cosmetics, and medical devices
Public Health Service Act	1944	FDA	Sections relating to biological products
Occupational Safety and Health Act	1970	OSHA, NIOSH	Workplace toxic chemicals
Federal Hazardous Substances Act	1960	CPSC	Hazardous (including toxic) household products (equivalent in many instances to consumer products)
Consumer Product Safety Act	1972	CPSC	Hazardous consumer products
Poison Prevention Packaging Act	1970	CPSC	Packaging of hazardous household products
Lead-based Paint Poisoning Prevention Act	1973, amended 1976	CPSC, HEW, HUD	Use of lead paint: on toys or furniture, on cooking, drinking, and eating utensils, in federally assisted housing
Hazardous Materials Transportation Act	1975, amended 1976	DOT (Materials Transportation Bureau)	Transportation of toxic substances generally
Federal Railroad Safety Act	1970	DOT (Federal Railroad Administration)	Railroad safety
Ports and Waterways Safety Act	1972	DOT (Coast Guard)	Shipment of toxic materials by water
Dangerous Cargo Act	1952		
Federal Mine Safety and Health Act	1977	Labor (Mine Safety and Health Administration), NIOSH	Toxic substances and other harmful physical agents in coal or other mines

Source: Environmental Law Institute (1978). *An analysis of past federal efforts to control toxic substances*, Environmental Law Institute, Washington, DC.

negative reports. It is probably impossible to assess correctly the weight that the animal welfare considerations play in the grant award process.

RECENT CONGRESSIONAL PROPOSALS REGARDING ANIMAL EXPERIMENTATION

In 1976, a House Committee held hearings on a bill to establish a Commission on the Humane Treatment of Animals. The use of animals in research was included on the proposed agenda. Government officials *and* private citizens with expertise on the various issues were to serve on the panel. After a 2-year study, findings and legislative recommendations were to be submitted to Congress. The hearings were dominated by pro-trapping opponents and the bill was effectively killed. Although a modified proposal has been repeatedly re-introduced, enthusiasm for this bill appears to have waned, possibly because of the more concrete proposals on animal research now pending in Congress.

In 1979, three significant 'alternatives' bills were pending in Congress. In this sense, 'alternatives' are research methods that do not use live animals, use fewer animals, or cause less pain and stress to those used. Animal welfare groups have seized upon the concept that the best way to alleviate animal suffering is to make animal testing obsolete. Although research using alternatives is entitled to funding through existing programmes, and roughly one-seventh of existing NIH research funds support this type of research, the consensus in the animal welfare community is that the full potential of alternatives has not been realized, and more incentives are needed. The anti-cruelty position has been strengthened by arguments based on the higher costs and relative slowness of animal testing. Considering the economic forecast, these reasons will be persuasive in light of the tremendous task of evaluating tens of thousands of hazardous substances. It is hoped that cost and speed will be the decisive factors necessary to gather the broad support for passage of any of the bills described below.

HR 282 (House Rule), sponsored by Congressman Robert Drinan, authorizes $ 12 million each year for 5 years to the Department of Health, Education, and Welfare, for grants to develop valid and reliable alternatives. HR 4479, sponsored by Congressman Ted Weiss, is similar to the original Humane Commission Bill in that it would establish a commission to study the use and potential of alternatives and also to make recommendations to Congress.

HR 4805, sponsored by Congressman Fred Richmond, is the most ambitious bill. It requires that between 30 and 50 per cent of all federal research money now used for animal research be re-allocated for the development of alternative methods. Acceptable alternative methods would be listed in the Code of Federal Regulations, and all federal agencies would be required to use these methods rather than conduct animal research. As the most restrictive bill, HR 4805 is expected to encounter the strongest opposition.

Any legislation regulating animal experimentation will raise hysteria within a segment of the research community and the federal government. Predictably, the red herring labels of 'overregulation' and 'the end of biomedical research progress' will be dragged out. However, supporters of legislation will be able to cite the cost and speed factors. Reform advocates will also be able to draw on the resurgence in popular concern about laboratory animal welfare.

The animal rights movement is partially responsible for this renewed interest. At least one researcher has ceased his animal experimentation when he concluded that his was an unjustified exploitation of animals (Curtis, 1978). In particular, discoveries regarding the intelligence of non-human primates and marine mammals have raised serious questions regarding the morality of painful and manipulative research. The notion that animals possess intrinsic rights is gaining in popularity. Even though most rights are relative, the absolute right of animals to humane treatment and consideration by man is difficult to deny by reason of man's new understanding of his own status on Earth. For centuries, man tried to subjugate his world and failed. We are now redefining our relationship to Earth and we can no longer irresponsibly exploit whatever or whomever we please.

Recently, the federal government issued regulations regarding the rights of human subjects of biomedical and behavioural research. Special attention was paid to the rights of prisoners and the mentally ill, two groups which have been exploited in the past. A case could be argued that animals need similar protection beyond the present ineffective and largely voluntary guidelines.

Just as anti-vivisection was the popular ethic leading to research animal protection in nineteenth-century Britain, the espousal of the animal rights philosophy may lead to reform in the United States at the end of the twentieth century. Great effort and commitment are required to turn an ethic into political action. However, it is a challenge that animal welfare proponents will eagerly accept.

REFERENCES

Bordley, J. and McGehee, A. (1976). *Two centuries of American medicine 1776–1976*, Saunders, Philadelphia.

Council on Environmental Quality, Toxic Substances Strategy Committee (1979). Draft Report to the President, Washington, DC.

Curtis, P. (1978). 'New debate over experimenting with animals', *New York Times Magazine*, 31 December 1978.

Fox, M. W., Ward, A. M., Rowan, A. N., and Jaffee, B. (1979). *Evaluation of awarded grant applications involving animals*, Institute for the Study of Animal Problems, Washington, DC.

Graham, E. A. (1935). *Animal experimentation, its importance and value to scientific medicine*, American College of Surgeons, Board of Regents, Chicago.

Holden, C. (1974). 'Beagles: Army under attack at Edgewood', *Science, New York*, **185**, 131–132.

Institute of Laboratory Animal Resources (1979). *Guide for the care and use of laboratory animals*, US Department of Health, Education and Welfare, Bethesda, Maryland.

Life (1966). 'Concentration camp for dogs', photographs by Stan Wayman, *Life*, **60**, 22–29.

McRea, R. (1910). *The humane movement, a descriptive survey*, Columbia University, New York.

Schultz, W. J. (1924). *The humane movement in the United States 1910–1922*, reprinted 1968 by AMS Press, New York.

Stevens, C. (1978). 'Laboratory animal welfare', in *Animals and their legal rights* (E. S. Leavitt, ed.), Animal Welfare Institute, Washington, DC.

APPENDIX

Legal Nomenclature and References in the United States

Statutes enacted by the United States Congress are incorporated into the volumes of the *United States Code* (USC). 'Public Law' (PL) is the designation for laws as they are enacted. They are published in the *Statutes at Large* numerically as they are passed. The first number in the citation refers to the Congress that passed the law. For example, PL 95-632 was passed in the 95th Congress which was in session from the beginning of 1977 until the end of 1978. The three federal laws protecting laboratory animals are:

(a) *Animal Welfare Act* 7 USC 2131 *et seq.* (passed 1966, amendment 1970 and 1976). Originally entitled the *Laboratory Animal Welfare Act*, the title was changed in 1970.

(b) *Endangered Species Act* 7 USC 136, 16 USC 1531 *et seq.*, PL 95–632 (passed 1969, amended 1973 and 1978).

(c) *Marine Mammal Protection Act* of 1972 16 USC 1361 *et seq.* (passed 1972, amended 1973 and 1976).

The Code of Federal Regulations (CFR) contains the rules and regulations issued by administrative agencies to implement federal statutory law. The CFR is divided into 50 titles which represent broad areas subject to federal regulation. For example, *Animal Welfare Act* Regulations are contained in 9 CFR, Chapter III. *Endangered Species Act* Regulations are in 50 CFR, Chapter I and *Marine Mammal Protection Act* Regulations are in 50 CFR, Chapters II and V. Recent rules and administrative actions are in the *Federal Register*, a daily publication.

Records of the legislative histories of these laws can be found in the volumes of the *US Code Congressional and Administrative News* which is issued by session of Congress in which the law was enacted. Votes in Congress and record of other congressional activity can be found in the *Congressional Record*, which is published daily. Transcripts of some hearings are kept and published also by the congressional committees that hold the hearings.

These legal publications are generally available in the libraries in the United States, but not so elsewhere. Embassy libraries may be able to offer some assistance.

Animals in Research
Edited by David Sperlinger
© 1981 John Wiley & Sons Ltd.

Chapter 4

Natural Relations—Contemporary Views of the Relationship between Humans and Other Animals

DAVID SPERLINGER

INTRODUCTION

The lives and deaths of millions of experimental animals appear to take place in an isolated, cut-off world. Most research laboratories are physically isolated from any kind of regular scrutiny by members of the public, both by their actual location and by regulations which prevent 'unauthorized' people gaining admission. In addition, the scientists themselves have their own languages for describing their experimental procedures and their experimental 'subjects'. These create a further barrier for those non-scientists who wish to learn more about the fate of experimental animals in our laboratories.

But this picture, of an isolated and pure scientific vacuum uncontaminated by contact with the outside world, is only part of the story. Firstly, science itself is an activity that takes place in a particular social situation and current scientific theories will often offer important reflections of that social situation. Secondly, anyone growing up in a Western (or any other) society will have been assimilating a whole series of complex (and frequently contradictory) ideas and feelings about animals. The scientist cannot discard these ideas as he enters his laboratory and dons his white coat.

Thus, in order to understand animal experimentation as a scientific activity, one also needs to understand the background against which it takes place. In Western societies a major component of that background is the general attitudes which are held toward the non-human animals. These general attitudes are, of course, in their turn influenced by the theories about animals which are being developed by scientists. So one is looking at a very complicated process of interaction between the ideas of scientists and the ideas of the wider society. This will be true of many areas of science. But it applies with particular force in any of the sciences involved with animals, for our views of other animals have a direct impact on views of ourselves. (In common with most other writers, the word 'animals' in this chapter will often be used as a shorthand for 'non-human animals'. Although this usage, in itself, exemplifies the human view of themselves as being 'not-animal', different from other animals.)

This chapter aims, therefore, to explore some of the major contemporary themes in discussions about non-human animals and their relationships to human beings. This is an area of enormous scope and complexity; it will not be possible to do more than touch on some of the more important issues. The chapter will, therefore, focus on three topics: (1) contemporary attitudes to animals in Western societies; (2) human nature and animal nature; (3) the question of the rights of animals. For all of these topics, contemporary views have important historical roots. It is, however, beyond the scope of this chapter to attempt to sketch these roots. Attempts have been made to trace parts of this history in other places—see, for example, Brumbaugh (1978), S. R. L. Clark (1977), Klingender (1971), Passmore (1974), Porter and Russell (1978), and Singer (1976, Chapter 5).

CONTEMPORARY ATTITUDES TO NON-HUMAN ANIMALS IN WESTERN SOCIETY

Children brought up in a modern Western society will be able to observe a bewildering array of attitudes and behaviour towards animals. Often this observation will be indirect—for their direct contact with animals, apart from pets, will be slight—their views of animals having become increasingly moulded by what they see on television, read in books or comics, etc.

Just consider, for example, the different views of animals revealed by the following:

(i)   parental injunctions not to be 'cruel' to individual animals;
(ii)   the fact the animals are killed to provide us with clothes or food;
(iii)   the idealized picture of farm animals described in most children's books (as compared to the factory-like conditions of most modern farms; see, for example, Harrison, 1964; Singer, 1976, Chapter 3);
(iv)   animals in zoos;
(v)   adult attitudes to garden 'pests' or insects such as spiders, moths, or 'daddy long-legs' (crane flies);
(vi)   cartoon films involving animals, such as Tom and Jerry or Yogi Bear; or Disney films 'starring' animals;
(vii)   the traditional glorification of fox hunting;
(viii) the dissection of animals, such as frogs, in biology lessons;
(ix)   pet-keeping, including 'putting animals to sleep' when they get old or ill;
(x)   nursery rhymes or poems involving animals.

This is not intended to be a comprehensive list of the variety of ways in which children may encounter other animals. It merely illustrates the enormous complexity of feelings and behaviour which the child must somehow learn to integrate. ('Integrate' is, however, probably the wrong word here, since it implies bringing together in some unified and consistent way. Whereas, in practice, most of us learn to apply different standards towards animals in different situations. So that lambs are cute and lovable symbols of innocence, but this does not prevent one enjoying braised lambs' hearts next time they appear on the canteen menu.)

It is astonishing how little work has been undertaken into what might be called the 'psychology of human relationships with animals'—that is, the roles which animals play in human thought and behaviour.* The major body of research into human/animal relationships has explored the role of pets in human society (see, for example, Anderson, 1975; Levinson, 1972).

One important study of the development of attitudes to animals was made by Susan Isaacs in her major work on the intellectual growth of young children (Isaacs, 1930). Here she acknowledges the varying standards and ways of adult behaviour with which any child is bound to come into contact. And she notes:

'The extraordinarily confused and conflicting ways in which we adults actually behave towards animals, in the sight of children. What children make of our injunctions to be "kind" and our horror at any impulse of cruelty on their part, in the face of our own deeds, and the everyday facts of animal death for our own uses and pleasures, would be hard to say. There is probably no moral field in which the child sees so many puzzling inconsistencies as here.'

She goes on to explore how these confusions and contradictions in the parents' attitudes are further complicated by contradictory impulses within the child—that is, impulses to cherish alongside the desire to mastery and hurt. Her concern, as an educator, was in directing these latter impulses so that they are taken up in the direction of understanding: 'The living animal became much less an object of power and possession, and much more of an independent creature to be learned about, watched and known for its own sake'. She provides some beautifully detailed observations of the vast range of responses which children have to both real animals and to the animals created by their own fantasies.

Desmond Morris has also collected some information about children's attitudes to animals, from a survey of responses to questions asked on a children's zoo television programme (Morris, 1967). Morris found that the most popular animals were almost entirely mammals and that they tended to be well endowed with anthropomorphic features (e.g. hair, facial expressions, rather vertical postures). These features obviously aid the children in identifying with the animals, so that, as Morris says, the animals are seen as being reflections of

---

* As the current volume was in press Kellert (1980) published results of a large-scale and comprehensive study of American attitudes towards and knowledge of animals; this is a major contribution to research in this area.

ourselves. He also found that the younger children preferred bigger animals and the older ones smaller animals (the children ranged from 4 to 14 years of age). Morris attributes this finding to the fact that smaller children were viewing the animals as parent-substitutes, while the older children were viewing them as child-substitutes. In keeping with his preference for biological explanations, Morris considers that, although the choices of particular animals may be culturally determined, the *reason* the animals were chosen reflects a biological process at work; that is, that favourite animals are selected according to fundamental symbolic needs of humans.

One might want to question to what extent the processes Morris is describing do reflect a *biological* process. But, nonetheless, the issue that he raises is clearly an important one in our relationship with other animals—namely that our view of other animals is often determined by the outlook of the humans involved, rather than by the actual characteristics of the animals. This is not to say that it is purely a matter of *individual* psychology, for clearly our culture provides us with a myriad of different ideas about animals.

Several authors have examined the way in which human needs and perceptions play a crucial role in determining our views of non-human animals. Thus Lopez (1979), in describing the sad and complex history of the relationship between humans and wolves, notes that:

> 'We create wolves. The methodology of science creates a wolf just as surely as does the metaphysical vision of a native American or the enmity of a cattle baron of the nineteenth century. It is only by convention that the first is considered enlightened observation, the second fanciful anthropomorphism, and the third agricultural necessity.'

Willis (1974) examines the way that three different African societies regard animals as a means of understanding the way in which the peoples of these societies conceive of themselves and the ultimate meaning of their lives. Thus, on this view, the way in which animals are regarded, the meaning given to them, can be seen as one way in which the fundamental values of a particular social system are expressed. This argument has been extended to Western societies by others who have noted how our current treatment of animals is an accurate mirror of our current social and economic relationships. Thus, Berger (1977a,b,c), in charting the changing views of animals in Western societies, notes how in the first stages of the industrial revolution animals were used as machines (as also were children), while in so-called post-industrial societies they are treated as raw materials and processed like manufactured commodities. He continues:

> 'This reduction of the animal, which has a theoretical as well as economic history, is part of the same process as that by which men have been reduced to isolated productive and consuming units. Indeed, during this period an approach to animals often prefigures an approach to man. The mechanical view of the animal's work capacity was later applied to that of workers.' (Berger, 1977b)

In the same article, he then goes on to describe the 'cultural marginalisation' of animals in present-day Western society. That is to say that while animals are still important in sayings, games, dreams, stories, etc., 'the category of *animals* has lost its importance'. They have instead been co-opted mostly into two other categories—firstly, *the family*—whereby 'the pettiness of current social practices is *universalised* by being projected on to the animal kingdom' (e.g. Beatrix Potter's books); secondly, *the spectacle*—through films, photographs, etc., where animals are always observed and the 'fact that they can observe us has lost all significance'.

As well as offering reflections on these wider social relations, animals are also used by humans as symbolic expressions of their own individual needs. Thus some animals are seen as representing what are considered to be unacceptable aspects of human behaviour. Even the word 'animal' is frequently used in this sense as a form of abuse—'You're worse than an animal'. These animals of our imagination are, of course, almost pure myth. Wolves, for example, which are often represented as the epitomy of ruthlessness and savagery, rarely kill more animals than they require for their immediate needs. They are also highly social animals, who devote considerable care and attention to the rearing of their young and show great loyalty to their pack (see, for example, Hall and Sharp, 1978; Lopez, 1979; Mech, 1970). (Russell and Russell (1978), however, have argued that the fear and hatred of the wolf is not entirely without justification and that for many centuries the wolf was the most dangerous animal for humans and their livestock.)

This whole topic, of the way in which we project onto animals our own unacknowledged feelings and impulses, has been sensitively and comprehensively examined by Midgley (1978, particularly Chapter 2). She argues there that, for example, man has always been unwilling to admit his own aggression and has tried to deflect attention from it by making animals out to be more aggressive than they are. Thus animals are not only used as symbols of evil but actually become seen as instances of evil. (This process was aided until fairly recently by the lack of substantial observations of how non-human animals did behave in the wild, as opposed to how they behaved in various forms of human captivity, such as zoos.) Midgley explores the ramifications of this idea of what she calls a 'Beast Within'—that is a lawless monster inside us who is responsible for all our 'bestial' actions, which could not be attributed to a true 'human' feeling. This idea of a Beast Within has distorted man's view both of himself and of the other animals. As Midgley puts it:

'If the Beast Within was capable of every iniquity, people reasoned, then beasts without probably were too. This notion made man anxious to exaggerate his difference from all other species and to ground all activities he valued in capacities unshared by the animals, whether the evidence warranted it or no. In a way this evasion does the species credit, because it reflects our horror at the things we do. Man fears his own guilt and insists on fixing it on something evidently alien and external. Beasts Within solve the problem of evil.' (Midgley, 1978)

The point that is being developed here is that attitudes to animals in Western society are highly elaborate and contain many (often contradictory) ideas and feelings within them. This set of ideas and feelings will inevitably influence reactions to animal experimentation (both by scientists and the general public), as it influences reactions to other areas of animal use or abuse. Take, for example, the enormous furore that arose in Great Britain recently over the case of the 'smoking beagles'. The public reaction to this story of dogs being required to inhale cigarette smoke appears to have been importantly motivated by the fact that the animals being experimented upon were *dogs*. It seems clear that the particular experiments were no more cruel and no more trivial in purpose than thousands of experiments carried out on animals each year.

Another significant indication of the influence of general attitudes to animals on animal experimentation is the fact the British *Cruelty to Animals Act*, which governs the practice of animal experimentation, requires special certificates if the experiments are to involve cats, dogs, or horses. There is no evidence to suggest that these species are more liable to suffer under experimental procedures than many other species. This regulation clearly reflects the contradictory feelings of the general public for certain categories of animals. These double standards towards animals are also revealed by the fact that, for example, a dog can be kept, quite legally, in conditions in a laboratory which would result in a prosecution for cruelty if the dog was a pet being kept in identical conditions. This point has also been made, with a rather different emphasis, by Harrison (1964):

'If one person is unkind to an animal it is considered to be cruelty, but where a lot of people are unkind to a lot of animals, especially in the name of commerce, the cruelty is condoned, and, once large sums of money are at stake, will be defended to the last by otherwise intelligent people.'

Scientists share these contradictory attitudes to animals too. It is surely not a coincidence that animals which are often seen as frightening or unpleasant by the general public, mice and rats, form the bulk of experimental animals. (In Great Britain, in 1978, out of a total of nearly 5 200 000 animals which were experimented on, over 4 million were mice or rats.)

It is difficult to summarize our main ways of seeing other animals, since these are so diverse and overlapping. But there appear to be at least five important strands:

(1) *Concern for individual animals* Humans often form close, personal ties to individual animals. This is seen particularly in relation to pets in Western societies and often their deaths can lead to a bereavement reaction similar to that found with the loss of a loved person (see, for example, Keddie, 1977; Levinson, 1972). Such attachments may at times seem to be pathological, in the sense of meeting some defensive needs of the person (see, for example, Rynearson, 1978), but in general they would appear to be a normal and healthy extension of other kinds of caring relationships. It is this concern for particular animals (of particular kinds) which tends to be the main theme of much newspaper and

television coverage concerning animals—paralleling the 'human interest' stories about people.

The converse side of this type of relationship to animals is that there is less concern with animals not met face to face or with large groups of animals. (This, again, is paralleled by a typical human response to be concerned with known individuals or groups, but to show less concern for unfamiliar groups.) Thus, the fate of a racehorse that breaks its leg and is shot becomes a focus for concern, in a way that the millions of cows who are slaughtered to provide us with food are not.

This point has some relevance for animal experimenters, since they will often be faced by having to experiment on a particular, known animal. In order to overcome their natural ways of reacting to these animals some kind of learning to inhibit or devalue these natural responses has to take place. Thus part of learning to be 'a scientist' consists of learning not to feel concerned about the effects of the experimental procedures on the animals involved. (This does not make these scientists sadists, who enjoy any pain or discomfort they are inflicting. No more than doctors are sadists because they learn to inhibit their natural reactions to cutting up other humans.) Part of this learning involves avoiding the language of individual feelings and relationships. (The development of such 'scientific' attitudes in children is further explored in the chapters by Paterson and by Fox and McGiffin in this volume.) As Heim (1970) notes in relation to psychological experiments with animals:

'The work on "Animal behaviour" is always expressed in scientific, hygienic sounding terminology, which enables the indoctrination of the normal, non-sadistic young psychology student to proceed without his anxiety being aroused. Thus techniques of "extinction" are used for what is in fact torturing by thirst or near-starvation or electric-shocking; "partial reinforcement" is the term for frustrating an animal by only occasionally fulfilling the expectations which the experimenter has aroused in the animal by previous training; "negative stimulus" is the term used for subjecting an animal to a stimulus which he avoids, if possible. The term "avoidance" is O.K. because it is an observable activity. The terms "painful" or "frightening" stimulus are less O.K. since they are anthropomorphic, they imply that the animals have feelings—and that these may be similar to human feelings.'

Similarly a Working Party of the British Psychological Society (1979) which looked at animal experimentation commented that 'it is likely that repeated references to animal experiments without comment desensitizes students to the ethical and scientific issues involved'.

It is also surely significant that in her sensitive observations of chimpanzees van Lawick-Goodall (1971) chose to give her chimps *names*. The same is true of other writers who have established some kind of *relationship* with the animals whom they are observing (see, for example, Schaller, 1965). Most animals who are experimented upon never achieve this mark of individuality—they merely have numbers and become statistics in the research data.

(2) *Animals as projections of human feelings*   This aspect of our relationship

to animals has already been examined above. It may be seen in the relationship of a particular person to a particular animal and it may involve either positive or negative feelings. It can also be seen at a wider, cultural level in the way in which particular groups of animals are viewed. Compare, as a simple example, the associations of 'lamb' and 'pig'. Leach (1972) has examined some of the anthropological aspects of the English language classification of animals and writes that it is 'not just a list of names, but a complex pattern of identifications subtly discriminated not only in kind but in psychological tone'. Again the feelings that are projected on to the animals may be either positive or negative. Menninger (1951) provides examples, at both the cultural and individual case history level, of the ways in which feelings towards humans may be displaced on to animals.

(3) *Animals as objects for human use*   This is one of the most significant ways in which non-human animals are viewed by humans. The purpose of the animals' existence is seen as primarily that of serving various human interests. Thus, cows, pigs, and chickens are seen as 'animal machines' (Harrison, 1964); while experimental animals are seen as pieces of laboratory equipment, 'tools of research' (Singer, 1976). The uses to which animals are put by humans are, of course, vast—for food, clothing, sport, entertainment, etc.

The implications of this view of animals have been thoroughly examined by Godlovitch (1971), who writes:

'We have come to regard animals, like houses and chairs, as constituted by their utility-value to us. We have in our eagerness to classify the world, taken it for granted that one may ask of a natural process: what is the reason for this entity's having come to exist? and, assuming the meaningfulness of such a query, answered it with: In order that we may benefit (. . .). Simply because we can use a seal pelt does not entail that a seal is essentially "the bearer of a (usable) pelt". Without men there could not exist such things as chairs. The seal, on the contrary, goes its own way with or without us. Problems arise about seals and not about stones because seals, like us, are purposive beings.'

Not many scientists who experiment on animals would want to subscribe to the blunt view of animals stated by Godlovitch. But, nonetheless, it is clear that some such view—that animals may be freely used for various human purposes—must underlie practices such as animal experimentation.

(4) *Observation*   Another important strand of our relationship to the other animals is that of standing back and observing, without interference, the patterns of their lives. This is seen in an everyday way in practices such as bird-watching, but takes a more developed scientific form in the work of some ethologists. The most extreme form of this kind of relationship is the work of people like Jane van Lawick-Goodall (1971) and George Schaller (1965), who have gone to *share* the animals' environment with them. It says something for the openness of these workers to learn *from* the animals they were observing that, for example, van Lawick-Goodall can write about being so impressed by the way the chimpanzee

mothers coped with their infants 'that we made a deliberate resolve to apply some of these to raising our own child'.

Another aspect of the observing relationship to animals would include the vast range of aesthetic and religious responses evoked by animals. Animals from the time of the earliest humans have often been considered with a sense of wonder and have provided rich inspiration for much painting, poetry and spiritual thought (see, for example, K. Clark, 1977, Klingender, 1971).

(5) *Ecological approaches*   There has recently been an increased concern with the idea of the Earth as an integrated and balanced whole. This has resulted in a recognition of our interdependence with all the other animals and plants with whom we share this planet and an awareness of how our actions may have widespread and unanticipated consequences for the other parts of 'the natural community' (as S. R. L. Clark, 1977, calls it). For, as Passmore (1974) has noted, in his examination of ecological approaches to nature,

> 'when men act so as to transform their environment, they never do only what they want to do. (. . .) Nature, in other words, does not simply "give way" to their [human] efforts; adjustments occur in its modes of operation, and as a result their actions have consequences which may be as harmful as they are unexpected. That is the force of the dictum, now so popular among ecologists, "it is impossible to do one thing only".'

Such ecological concerns, it is true, are often centred on considerations about the effects on other humans (in this or future generations) if we abuse the natural resources of this planet. Nonetheless, part of such an ecological attitude does involve an appreciation of the importance of the other animals, a respect for their right to continue to exist and an acknowledgment that we *need* them if we are not to become 'unbalanced'.

I have been examining in this section some of the backcloth against which much experimental work with animals takes place. But the general ideas about animals that are held in most Western societies have also been crucially influenced by ideas that have originated with scientists themselves. The most central of these ideas is of course that associated with Darwin: evolution through natural selection. These ideas have not only radically affected our views of the other animals but have also produced an irrevocable change in humanity's view of itself.

HUMAN NATURE AND ANIMAL NATURE

The 'Darwinian revolution' (Mayr, 1972) was one of the most fundamental of all intellectual revolutions. For not only did it challenge many established scientific theories, but it also brought into question many major metaphysical, ethical, and religious ideas. In particular, Darwinian ideas challenged an anthropocentric view of nature—that is the view that human beings were the central fact of the universe, to which all surrounding facts had reference and that humans had evolved/been created in a manner that was completely separate from the other animals. This challenge to human uniqueness still reverberates today. The difficulties it produces can be seen at work in many scientific writings, as well as in many popular writings about the origins of humanity.

There are two quite distinct and opposed lines of argument confronting one another over the relationship between humans and the other animals. Firstly, it is argued that, in spite of evolution, nonetheless humans are *unique*. That humans have various attributes which make them qualitatively different from the other animals. This view would tend to reject biological explanations for human behaviour as being reductionist and deterministic. The second view aims to place humanity in its *evolutionary context*. This view would see humans as being essentially biological animals and would seek primarily to explain their behaviour by reference to the same concepts as are used in the explanation of the behaviour of the other animals. Many who take this view would wish to reject explanations of human behaviour which involve the use of mental concepts.

This question about the relation of humans and the other animals is not only of theoretical interest to many scientists but also seems to be of great concern for many non-scientists as well. The popularity of, and controversy around, the books by people such as Morris (1967), Ardrey (1976), and Lorenz (1966) indicates that questions about humanity and its origins have a significance beyond their mere scientific interest.

It would be impossible in a chapter of this length to make any attempt to review all the work and argument that there has been in this area. Midgley's (1978) book *Beast and man* provides an excellent and comprehensive examination of the issues involved. But an attempt will be made to outline the major arguments that have been put forward by the proponents of the two camps in this debate.

Unique Humanity?

Human life unquestionably has a richness and diversity that can be found in no other animal species. This is not in dispute. The core issue is whether this richness and diversity is a result of qualities that are *unique* to humans. Brown (1958) has, humorously, drawn attention to some of the concerns behind the scientific arguments:

> 'I grant a mind to every human being, To each a full stock of feelings, thoughts, motives and meanings. I hope they grant as much to me. How much of this mentality that we allow one another ought we to allow the monkey, the sparrow, the goldfish, the ant? Hadn't we better reserve something for ourselves alone, perhaps consciousness or self-consciousness, possibly linguistic reference?
>
> Most people are determined to hold the line against animals. Grant them the ability to make linguistic reference and they will be putting in a claim for minds and souls. The whole phyletic scale will come trooping into Heaven demanding

immortality for every tadpole and hippopotamus. Better be firm now and make it clear that man alone can use language and make reference. There is a qualitative difference of mentality separating us from the animals.'

What criteria could be used to show that humans are qualitatively different from the other animals? What evidence is there to support these criteria? The main criteria which have been proposed are: (i) tool-making, (ii) language, (iii) rationality intelligence, and (iv) culture. The evidence to support each of these attempts to draw *the* line between humans and animals will now be briefly examined.

(i) *Tool-making*   Again there is no dispute that humans produce a greater variety of more complex tools than any other animal. But is there any evidence that other animals use tools of any kind, i.e. that they manipulate particular objects in order to attain a particular end? Work suggesting that apes were capable of using tools was conducted in the 1920s by Köhler (1957). But the most striking evidence for animal tool use has come from van Lawick-Goodall's observations on chimpanzees in the wild. Her initial observations were of a chimp using grass stems to fish for termites from a mound. She also later observed them, for example, using leaves to sop up water or to wipe dirt from their bodies. Thorpe (1974) has reviewed the evidence for tool use by other mammals and by birds. It seems quite clear from this evidence that some animals, particularly primates, can show some elementary kinds of tool use—that is, selecting a particular stone or stick for a particular use, with clear evidence of some degree of planning and foresight.

(ii) *Language*   Language is an ability that many would think of as being particularly fundamental for and distinctive of human beings. Language forms the basis for much of our complex interactions with other humans. It is also the foundation on which much of our culture rests and on which its transmission from one generation to another depends. Chomsky (1968) has argued that there are no evolutionary continuities and no similarities between human language and the various animal systems of communication. He argues that human language is *not* simply a more complex example of processes to be found in other animals, but represents a qualitatively distinct phenomenon.

However, this view has been questioned by those who have closely examined animal communication systems. Thorpe (1974), for example, lists 16 'design features' that are shared by all human languages and examines to what extent these can be found in the various animal communication systems. Thorpe's view is that human language is made up of many features which are found in the various communicative systems of animals—fish, insects, mammals, and birds. He also concludes (from work with chimpanzees to be discussed below) that if chimpanzees had the necessary equipment in the larynx and pharynx, then they could learn to talk at least as well as children of 3 years of age. Thus, for Thorpe, there is no single characteristic which can be used as an infallible criterion to distinguish human and animal communication systems; what is unique about human language is the way it combines and extends attributes which can be found in other groups of animals. A similar conclusion is accepted by Griffin (1976), who points out that it often seems to be an *assumption* of much research into animal communication that language is unique to human beings, and that, no matter how complex animal communication turns out to be, it cannot possibly be continuous with human language. While Stephen Clark (1977) has suggested that

> 'it is not that we have *discovered* them [other animals] to lack a language but rather that we define, and redefine, what language is by discovering what beasts do not have. If they should turn out to have the very thing we have hitherto supposed language to be, we will simply conclude that language is something else again.'

The most significant development for these arguments has been the work teaching chimpanzees to use a human sign language to communicate both with humans and with each other (much of the early work in this area is described in Linden, 1976; see also Hill, 1978). This work has shown that chimpanzees are able to utilize many of the features of language that were previously thought to be the unique preserve of humans. Thus Washoe, the first chimpanzee to be initiated into the pleasures of American sign language (ASL), could invent new words for objects for which she had not been taught the ASL sign and showed the beginnings of the development of syntax.

Linden (1976) was over-stating the case when he remarked that 'Washoe poses the greatest threat to the integrity of the Western vision of reality since Darwin'. Nonetheless, this work is a very clear example of how our assumptions about human uniqueness may prevent us from examining the continuities which we share with other animals.

(iii) *Rationality and intelligence*   Some of the views which would argue that humans are unique amongst the animals in their rationality and intelligence, would tie these qualities up intimately with our ability to use language. But if this definition is not to be circular, so that only beings who could use a human language could be said to be acting rationally or intelligently, then we will have to look for other definitions of these terms. This does not seem unreasonable. For it is surely not crucial to our calling an act 'rational' that the person should be able to explain (to themselves or others) what they did and why they did it.

But once one begins to look at wider definitions of these terms, it becomes increasingly difficult to see how they can be confined to humans. For example, Midgley (1978) defines rationality in terms of having a definite structure of deep, lasting preferences, which are linked to character traits. This definition does not deny the distinctive human forms of rationality involving conceptual thought and language, but does indicate a structure which can be used to compare what is rational or irrational about both human and animal acts.

What is rational or intelligent has to be taken in the context of the behaviour that is natural for each species. It surely makes perfectly good sense, for example, to talk about an animal's having acted intelligently, when it has not just reacted in a mechanical, blind way when faced by a threatening or puzzling situation. That animals can act 'instinctively' and 'unintelligently' at times—i.e. in ways that appear to be against their own interest—seems to be clear, but then the same could be said of many human actions. Indeed the *possibility* of its acting unintelligently is central to the idea of being prepared to describe an organism as having acted intelligently. (A fuller discussion of these issues can be found in S. R. L. Clark (1977) and Midgley (1978).)

(iv) *Culture*   Humanity, it has been argued, has its own unique form of evolution, alongside the Darwinian evolution which it shares with the other animals. Medawar (1976) has suggested that this distinctively human form of evolution has three main characteristics: (a) that it is Lamarckian in style—i.e. that what is learned in one generation can be passed on to the next; (b) that 'cultural heredity' is mediated through non-genetic channels; and (c) that it is potentially reversible—i.e. that should any catastrophe strike most of the human race, the remaining humans could return to a Stone Age level. It is this system of evolution Medawar suggests which is the most important characteristic which has allowed us to gain biological supremacy over all other life. (Medawar sees this cultural evolution as being dependent on humans having a high selective premium for the capacities of teachability and imitativeness; however, this part of his argument is not crucial to the point being developed here.)

But while it is clear that cultural evolution is unique in many major respects when compared to Darwinian evolution—for example, in the rate at which it proceeds and in the ease with which complex traits can be modified—it is, nonetheless, far from clear that many animal species do not have cultures. If one takes 'having a culture' to mean having a distinctive set of *learned* and flexible behaviours, which are communicated amongst the members of the species and which vary in the particular form that they take between social groups within a species—then there is some evidence of cultural behaviour in several groups of animals. Thus, for example, Hall and Sharp (1978) have argued that wolves 'are cultural animals with a learned—and hence variable—social organisation. Their social organisation in different places, even under similar ecological conditions, is neither identical or entirely predictable'. They have also drawn attention to the way in which, in response to similar ecological relationships to their prey (caribou), artic wolves and the Chipewyan Indians have produced remarkably *similar* social systems. Thorpe (1974) also describes various 'primitive proto-cultures' in different primate groups.

This brief overview of the major criteria which have been proposed as differentiating humans from the other animals shows very clearly that it is difficult, if not impossible, to find *a* criterion which will serve the purpose for which it is intended. What evidence is there, then, to support the views of those who wish to emphasize the similarities between human and animal behaviour?

Humanity's Animal Roots?

The authors in this camp are a motley group. Nonetheless, what they share in common is a belief that much of human behaviour can be explained by using the same concepts as can be used in the explanation of animal behaviour. They emphasize, therefore, concepts deriving from biology and the sciences of animal behaviour and have a distrust of the traditional language for explaining human behaviour (in terms of motives, intentions, etc.).

In terms of his popular influence, Desmond Morris (1967) can be taken as representative of this approach. Other writers who follow this general view would include Ardrey (1976), Bleibtreu (1968), and Tiger and Fox (1972), ethologists such as Lorenz (1966), and the sociobiologists such as Wilson (1975, 1978).

Morris starts his book off by boldly proclaiming the creed of those holding this view. He states that *Homo sapiens* spends far too much time 'examining his higher motives', while Morris is instead going to look at his more fundamental motives, 'his basic behaviour'. He also dismisses, in one paragraph, what can be learnt from anthropology by the extraordinary argument that simple tribal groups are examples of biological failures which can safely be ignored as having no relevance to 'the ordinary successful members of the major cultures'. His basic strategy appears to be to re-label human behaviour with terms derived from studies of animal behaviour, with the assumption that this provides some kind of (scientific/biological) explanation for the behaviour. One small example will illustrate this point. In relation to 'feeding behaviour' Morris asserts that the patterns developed at the time our ancestors were hunters have 'become deep-seated biological characteristics of our species'. As a result, since the majority of adult males in our society no longer have a hunting role, they

> 'compensate for this by going out to "work". Working has replaced hunting, but has retained many of its basic characteristics. It involves a regular trip from the home base to the "hunting" grounds. It is a predominantly masculine pursuit and provides opportunities for male-to-male interaction and group activity. It involves taking risks and planning strategies.' (Morris, 1967)

It is not clear what one gains from such an explanation of why people go to work. The re-labelling appears to be of dubious validity, to be based on no evidence and to leave out many of the major factors (social, economic, and political, as well as personal) that *are* involved in such a complex human activity.

At a rather more sophisticated level is the work of people such as Lorenz (1966) and, in a rather different field, Wilson (1975, 1978). These authors both provide a clearer idea of the empirical data from which their theories of human behaviour have been developed. They provide detailed and sophisticated analyses of various aspects of animal behaviour (aggression in Lorenz's case, and social behaviour in Wilson's). And yet, the doubts still remain when their theories become transferred to the human level. Thus Lorenz can write,

'During and shortly after puberty human beings have an indubitable tendency to loosen their allegiance to all traditional rites and social norms of their culture, allowing conceptual thought to cast doubt on their value and to look around for new and perhaps more worthy ideals. There probably is, at that time of life, a definite sensitive period for a new object-fixation, much as in the case of object-fixation found in animals and called imprinting. (. . .) At the post-pubertal age some human beings seem to be driven by an over-powering urge to espouse a cause, and, failing to find a worthy one, may become fixated on astonishingly inferior substitutes.' (Lorenz, 1966)

Similarly, doubts arise with Wilson when he attempts to describe genes for, amongst other things, flexibility in social behaviour, a capacity for culture, indoctrinability, innate moral pluralism, and aesthetics. There seem to be at least two major problems facing such theories:

(1) The difficulty of producing any evidence which could substantiate such views about the origins of human behaviour. (See, for example, the detailed methodological criticisms of sociobiology made by Burian (1978).)

(2) A second major difficulty is to do with the richness and complexity of much human behaviour. It is not adequate merely to find a particular piece of human behaviour that appears to fit one's case, without analysing the behaviour in the same depth as one has applied to the animal behaviour from which one is generalizing (see, for example, Ambrose, 1976, for an examination of some of the problems in this area). This is particularly striking in Lorenz's (1966) book on aggression, where he explores in elaborate detail the function and meaning of small pieces of behaviour by the greylag goose, but makes no comparable analysis of the human behaviour he relates it to.

It is not without reason that ethology has been defined as the science that pretends humans cannot speak (Washburn, 1978). There really appears to be a major danger of 'beastomorphism' in many writers. Because their theories are founded on the assumption that humans will behave like the other animals that they observe, they transfer their analysis straight from these animals to humans, with no attempt to make a reasoned translation to meet the human case.

The two views of human nature and animal nature explored above—that humans are unique amongst the animals and that human behaviour can be directly explained in terms derived from work with animals—appear to be untenable in their present form. Future work in this area will need to take account of at least the following issues:

(1) Clearly humans and the other animals *are* related. Humanity did not arrive on Earth as an alien being from another planet. There will, therefore, be areas of continuity with some other animal species. There will also be important areas of difference and discontinuity. Since, as Midgley (1978) has eloquently pointed out, there is no such thing as 'an animal nature', each species having its own particular nature, then it only makes sense to compare humans with a

particular species of animal in relation to a *particular* purpose. Thus, if one is examining human social life it may make more sense to make comparisons with one of the other highly developed social animals (such as wolves), rather than with some groups of primates (who are closer biological relations to humans, but may not have such a rich social life).

(2) In considering any animal/human comparison one needs to take account of the potential richness of *meaning* of human behaviour (see, for example, Sahlins' (1977) criticisms of sociobiology) and also of the *context* in which the behaviour takes place. One also has to take account of the flexibility of human behaviour and not to assume that because a particular behaviour is adaptive that, therefore, it is genetic. In our society biological explanations of behaviour are often felt to be more scientific than explanations involving individual motives or social processes. But it may turn out that a careful extension of the latter types of explanation to other animal species may ultimately prove more fruitful (for our understanding of *both* humans and other animals) than the generalization of concepts derived purely from animal behaviour to humans.

(3) For many aspects of animal behaviour, in order thoroughly to understand various animal species and their different natures, it will be necessary to undertake careful observation in the field. The value of laboratory work, of the kind where complex procedures are devised and then imposed upon experimental animals, would appear to be extremely limited.

(4) Humans are animals, with their own particular biology and their own particular nature. In practice this does not mean that we have a set of innate, fixed behaviour patterns. But, rather, this biology acts as a set of (wide) natural limits within which human functioning takes place. These genetic constraints on human behaviour

'do not define the specific social acts that men must perform; they are, rather, limitations on the range of their alternatives. Stating this differently, there is nothing that man by reason of his genes must do; but there is much man by reason of his genes can do. (. . .) Evolution does not govern behaviour; it makes behaviour possible.' (Jaynes and Bressler, 1971).

(5) Humans do excel at many skills when compared with the other animals. The areas in which our skills are particularly excellent may well be related to characteristics which we share with other species. This point has been put succinctly by Midgley (1978) in discussing structural properties—that is properties that affect the whole organization of the life of a species. She writes:

'Structural properties, then, do not have to be exclusive (to a species) or necessarily excellent. Nor do they have to be black-or-white, yes-or-no matters. And certainly no one of them is enough alone to define or explain a species. We commonly employ a cluster of them, whose arrangement as *more* or *less essential* can be altered from time to time for many reasons. And what is really characteristic is the shape of the whole cluster. (. . .) What is special about each creature is not a single, unique quality but a rich and complex arrangement of powers and qualities, some of which it will certainly share with its neighbours. And the more complex the species, the more true this is. To expect a single differentia is absurd.'

THE QUESTION OF ANIMAL RIGHTS

The last section examined some of the ways in which human attitudes towards other animals have developed during the time since Darwin. This section looks at the implications of these changing attitudes for our *treatment* of the other animals. There is nothing new in individuals, or indeed whole societies, showing compassion for other animals and being concerned to minimize the suffering which is inflicted by humans on animals. (Ryder (1975) provides a brief 'history of compassion'; while Leavitt (1978) traces the considerable history of legal rights which have been given to animals.) But it is really only since 1970 that a consistent movement has developed which has taken account of the changed scientific view of other animals and has attempted to explore the implications of this for their relationship with humans. To put it rather simply, this movement has tried to look at the implications of seeing animals as being our natural relations, rather than being beasts who were created for human use.

The first significant landmark in this movement for animal rights was the publication of a collection of essays, *Animals, men and morals*, edited by Godlovitch, Godlovitch, and Harris (1971). This volume examined and criticized many of our current practices towards animals—in farming, science, and fashion. The importance and the urgency that the contributors attributed to their arguments is well caught in these sentences from the editors' introduction:

'Once the full force of moral assessment has been made explicit there can be no rational excuse for killing animals, be they killed for food, science, or sheer personal indulgence. We have not assembled this book to provide the reader with yet another manual on how to make the brutalities less brutal. Compromise, in the traditional sense of the term, is simple unthinking weakness when one considers the actual reasons for our crude relationships with other animals.' (Godlovitch *et al.*, 1971)

And also by the closing sentences of Corbett's postscript:

'Our conviction, for reasons we have given, is that *we* require *now* to extend the great principles of liberty, equality and fraternity over the lives of animals. Let animal slavery join human slavery in the graveyard of the past.' (Godlovitch et al., 1971)

It is this latter parallel, between the abuse of various human groups and abuse of other animals, that has been particularly taken up in more recent writing. A significant development from this point of view was Ryder's coining of the term 'speciesism' in his book on animal experimentation (Ryder, 1975). Ryder argues that, since humans are biologically related to the other animals and share with them the important qualities of life and sentience, there are no morally relevant grounds for continuing to exploit animals. He sees a clear parallel between discrimination practised by humans against the other species and discrimination against different racial groups. Thus, he writes:

'Racism today is condemned by most intelligent and compassionate people and it seems only logical that people should extend their concern for other races to other species also. Speciesism and racism both overlook or underestimate the similarities between the discriminator and those discriminated against and both forms of prejudice show a selfish disregard for the interest of others, and their sufferings.' (Ryder, 1975)

This idea that our treatment of other animals was a form of discrimination, parallel to discrimination on grounds of race or sex, has proved to have had a crucial influence on the way the argument has proceeded since that time. Singer (1976), in his book *Animal liberation*, produced what has turned out to be the central statement in a campaign for animal rights.

Singer's book is in part philosophy and in part polemic, based on his description of two of the major examples of 'speciesism in practice'—the use of animals in agriculture and in science. The major part of his argument is to extend and develop Ryder's thesis about speciesism. Thus, he states that speciesism

'is a prejudice or attitude of bias towards the interests of members of one's own species and against those of members of other species. It should be obvious that the fundamental objections to racism and sexism made by Thomas Jefferson and Sojourner Truth apply equally to speciesism. If possessing a higher degree of intelligence does not entitle one human to use another for his own ends, how can it entitle humans to exploit non-humans for the same purpose?' (Singer, 1976)

Singer's central ethical principle is that equal consideration should be given to the interests of different groups. This does not imply that these groups will necessarily have equal treatment or equal rights. It also does not imply an actual equality of the beings involved—just as the assertion of the equality of humans does not depend on what qualities they have or what abilities they possess. What gives all animals the right to an equal consideration of their interests, according to Singer, is their sentience, their ability to suffer or enjoy. Singer restricts his argument largely to a concern with animal *suffering* and leaves on one side the issue of killing other animals.

On this view, then, some animals would have an equal capacity for suffering to some humans (for example, infants or people who are severely mentally handicapped). Therefore, if we are prepared to condone experiments on these animals but not on these groups of humans, this is likely to be based purely on a discriminatory and morally indefensible preference for members of our own species.

Singer's book has provided the central argument and concepts for subsequent debate on this issue. Partly as a result of it there has been an enormous volume of writing produced which is concerned with animal rights. See, for example, the articles collected in Morris and Fox (1978), Paterson and Ryder (1979), and Regan and Singer (1976). The topic has become a highly controversial one in philosophical circles (see, for example, the October 1978 issue of *Philosophy*); but much of this debate centres around the rather technical issue of the definition of 'rights', which it is beyond the scope of this chapter to explore. In addition to Singer's book there have been other examinations of humanity's treatment of animals, which have come at the issues involved from a rather different angle (see, for example, S. R. L. Clark, 1977; Linzey, 1976). Benson (1978) has made a useful critique and comparison of the books by Singer and Clark.

These issues have, however, also had considerable reverberations outside the walls of universities and publishing houses (as Singer clearly intended that they should). Thus, for example, in Great Britain an 'Animal Liberation Front' has been formed. Their activities have included 'liberating' beagle puppies from a company that breeds animals for animal experimentation and attacking the office of a Home Office inspector employed to oversee animal experiments. There has also been growing militancy on behalf of animals in many other areas— e.g. the Greenpeace campaigns on behalf of whales and seals; the activities of the Hunt Saboteurs Association; organized protests against circuses; etc. (Interestingly much of this activity was anticipated in a novel written by one of the contributors to the book by Godlovitch et al., (Duffy, 1973).)

The essential core of the arguments of those who advocate an animal rights position is: Any argument which would show that all human beings had a right, e.g. to life or to not be caused pain, would also show that many animals also possess such rights. While attempts to exclude other animals from having rights will also inevitably show that certain humans (such as the severely mentally handicapped) do not have these rights either.

The argument about speciesism has been helpful in drawing attention to significant parallels between our exploitation of animals and the ways in which groups of humans have exploited one another. However, it also ultimately leads one to morally unacceptable conclusions about human life, *even if* one is in agreement about its conclusions about our treatment of other animals. (Significantly several major critics of Singer's particular arguments have also acknowledged the validity of his criticisms of current *practices* towards animals (see, for example, Francis and Norman, 1978; Vandeveer, 1976).)

There is much that is of value, and which raises major challenges to contemporary practices towards animals, in the writings of Singer and those who stress the need for 'animal rights'. Nonetheless, the argument is based on assumptions which seem to be contrary to what we know of human nature. Benson (1978) has put this point clearly.

'By biological inheritance reinforced by tradition most human beings are pro-grammed to treat their young and the young of con-specifics with special tenderness, and there is a widespread, if not universal recognition of special duties of compassion and protection towards the weak and handicapped. Whether a child with grossly impaired mental capacities should be preserved or not is a question that most people would think should be answered by reference to the interests of the child.'

Benson further distinguishes between an objectionable sort of speciesism— that only human life and suffering matter—and the sort of speciesism which begins with our natural affections towards kin, friends, colleagues, and so on, and which then becomes extended to other humans and to other animals. This latter form of partiality towards our own species, Benson suggests, *may* be a form of speciesism but Singer has not shown that its abandonment is necessary to produce the practical changes in our treatment of animals that he wants to secure. For once we have recognized our kinship to the other animals, our continuity with them, then it will be more natural than not for us to extend our concern outside the human family to include the whole natural community.

Thus, a change in our attitude to, and our treatment of, other animals will require a major shift in our view of them as being *objects* for our use. It will require a recognition of them as *subjects*—with their own needs, patterns of life, views of the world, etc. Such a change will be particularly difficult for much scientific work involving animals. For such work is commonly based on the assumption that animals can be treated as objects to be measured, experimented upon, and disposed of. Most scientific research would find it impossible to reconcile their own procedures with the idea that their experimental animals were subjects, who had their own outlook on the experimental situation.

This changed attitude to animals will also require a recognition that, in situations where there are important conflicts of interest between humans and the other animals, that these other animals *do* have interests of their own. These interests will include being able to fulfil the capacities of their own species' particular nature, not being made to suffer, and not having their life ended prematurely. There may be situations (as there are sometimes with other humans) where the interests of the animals may have to be over-ridden in order to meet a major human interest. But it is crucial to recognize that there *is* a conflict of interests, which must be taken into account when decisions affecting animals' lives are being taken. Once the interests of other animals are acknowledged, then it will be evident that not just any human interest, no matter how trivial, can be used as a justification for sacrificing the interests of the animals involved. It will also be evident that much of our current use of animals, including much animal experimentation, is not of such central importance to human life as to justify these practices continuing.

CONCLUSION

This chapter has attempted to trace some of the major contemporary attitudes and feelings towards non-human animals which form a background against which animal experimentation takes place. These attitudes and feelings cannot be ignored as irrelevant by the scientists who experiment on animals, for they share many of them with the rest of Western society. Indeed, much of animal experimentation rests on one of the most pervasive of these attitudes to animals—namely that animals are objects which may be used as means to serve human ends.

Many of our reactions to animals remain muddled and contradictory. But, if there is one major theme emerging from this chapter, it is that we human beings should recognize our continuity with the other animals, as well as our distinctiveness; that we recognize ourselves as being, as well as members of human societies, part of a wider natural community, which we share with our natural relations; that we recognize that these natural relations are all distinct individuals too, with their own needs and interests.

Such changes in our views of the other animals will inevitably result in changes in the way in which we treat them, as we begin to extend to them the concern which we show for some of our human relations. It will also inevitably lead to a re-evaluation of much animal experimentation and to an examination of how much of this experimentation involves such major human interests as to justify the deaths and suffering of millions of animals in our scientific laboratories.

REFERENCES

Ambrose, A.(1976). 'Methodological and conceptual problems in the comparison of developmental findings across species', in *Methods of inference from animal to human behaviour* (M. Von Cranach, ed.), Aldine, Chicago, pp. 269–317.

Anderson, R. S. (ed.) (1975). *Pet animals and society*, Ballière Tindall, London.

Ardrey, R. (1976). *The hunting hypothesis*, Collins, London.

Benson, J. (1978). 'Duty and the beast', *Philosophy*, **53**, 529–549.

Berger, J. (1977*a*). 'Animals as metaphor', *New Society*, **39**, 504–505.

Berger, J. (1977*b*). 'Vanishing animals', *New Society*, **39**, 664–665.

Berger, J. (1977*c*). 'Why zoos disappoint', *New Society*, **40**, 122–123.

Bleibtreu, J. (1968). *The parable of the beast*, Gollancz, London.

British Psychological Society (1979). Report of the Working Party on animal experimentation, *Bulletin of the British Psychological Society*, **32**, 44–52.

Brown, R. (1958). *Words and things*, Free Press, New York.

Brumbaugh, R. S. (1978). 'Of man, animals and morals: a brief history', in *On the fifth day* (R. K. Morris and M. W. Fox, eds), Acropolis Books, Washington, DC, pp. 6–25.

Burian, R. M. (1978). 'A methodological critique of sociobiology', in *The sociobiology debate* (A. L. Caplan, ed.), Harper and Row, New York, pp. 376–395.

Chomsky, N. (1968). *Language and mind*, Harcourt, Brace and World, New York.

Clark, K. (1977). *Animals and men*, Thames and Hudson, London.

Clark, S. R. L. (1977). *The moral status of animals*, Clarendon Press, Oxford.

Duffy, M. (1973). *I want to go to Moscow*, Hodder and Stoughton, London.

Francis, L. P. and Norman, R. (1978). 'Some animals are more equal than others', *Philosophy*, **53**, 507–527.

Godlovitch, S. (1971). 'Utilities', in *Animals, men and morals* (S. Godlovitch, R. Godlovitch, and J. Harris, eds), Gollancz, London, pp. 173–190.

Godlovitch, S., Godlovitch, R. and Harris, J. (eds) (1971). *Animals, men and morals*, Gollancz, London.

Griffin, D. R.(1976). *The question of animal awareness*, Rockefeller University Press, New York.

Hall, R. L. and Sharp, H. S. (eds) (1978). *Wolf and man*, Academic Press, New York.

Harrison, R. (1964). *Animal machines*, Stuart, London.

Heim, A. (1970). *Intelligence and personality*, Penguin Books, Harmondsworth.

Hill, J. H. (1978). 'Apes and language', *Annual Review of Anthropology*, **7**, 89–112.

Isaacs, S. (1930). *Intellectual growth in young children*, Routledge, London.

Jaynes, J. and Bressler, M. (1971). 'Evolutionary universals, continuities, and alternatives', in *Man and beast: comparative social behaviour* (J. F. Eisenberg and W. S. Dillon, eds), Smithsonian Institution Press, Washington, DC, pp. 333–344.

Keddie, K. M. G. (1977). 'Pathological mourning after the death of a domestic pet', *British Journal of Psychiatry*, **131**, 21–25.

Kellert, S. R. (1980). 'American attitudes toward and knowledge of animals: an update', *International Journal for the Study of Animal Problems*, **1**, 87–119.

Klingender, F. (1971). *Animals in art and thought*, Routledge and Kegan Paul, London.

Köhler, W. (1957). *The mentality of apes*, Penguin Books, Harmondsworth.

Lawick-Goodall, J. van (1971). *In the shadow of man*, Collins, London.

Leach, E. (1972). 'Anthropological aspects of language: animal categories and verbal abuse', in *Mythology* (P. Maranda, ed.), Penguin Books, Harmondsworth, pp. 39–67.

Leavitt, E. S. (1978). *Animals and their legal rights*, 3rd edn, Animal Welfare Institute, Washington, DC.

Levinson, B. M. (1972). *Pets and human development*, Charles C. Thomas, Springfield, Ill.

Linden, E. (1976). *Apes, men and language*, Penguin Books, Harmondsworth.

Linzey, A. (1976). *Animal rights*, SCM Press, London.

Lopez, B. H. (1979). *Of wolves and men*, Dent, London.

Lorenz, K. (1966). *On aggression*, Methuen, London.

Mayr, E. (1972). 'The nature of the Darwinian revolution', *Science, New York*, **176**, 981–989.

Mech, D. (1970). *The wolf*, Natural History Press, New York.

Medawar, P. B. (1976). 'Does ethology throw any light on human behaviour?', in *Growing points in ethology* (P. P. G. Bateson and R. A. Hinde, eds), Cambridge University Press, Cambridge, pp. 497–506.

Menninger, K. A. (1951). 'Totemic aspects of contemporary attitudes to animals', in *Psychoanalysis and culture* (G. W. Wilbur and W. Muensterberger, eds), International Universities Press, New York, pp. 42–74.

Midgley, M. (1978). *Beast and man*, Harvester Press, Hassocks, Sussex.

Morris, D. (1967). *The naked ape*, Jonathan Cape, London.

Morris, R. K. and Fox, M. W. (eds) (1978). *On the fifth day*, Acropolis Books, Washington, DC.

Passmore, J. (1974). *Man's responsibility for nature*, Duckworth, London.

Paterson, D. and Ryder, R. (eds) (1979). *Animals' rights—a symposium*, Centaur Press, Fontwell, Sussex.

Porter, J. M. S. and Russell, W. M. S. (eds) (1978). *Animals in folklore*, D. S. Brewer, Ipswich.

Regan, T. and Singer, P. (eds) (1976). *Animal rights and human obligations*, Prentice-Hall, Englewood Cliffs, N. J.

Russell, W. M. S. and Russell, C. (1978). 'The social biology of werewolves', in *Animals in folklore* (J. M. S. Porter and W. M. S. Russell, eds), D. S. Brewer, Ipswich, pp. 143–182.

Ryder, R. D. (1975). *Victims of science*, Davis-Poynter, London.

Rynearson, E. K. (1978). 'Humans and pets and attachment', *British Journal of Psychiatry*, **133**, 550–555.

Sahlins, M. (1977). *The use and abuse of biology*, Tavistock, London.

Schaller, G. B. (1965). *The year of the gorilla*, Collins, London.

Singer, P. (1976). *Animal liberation*, Jonathan Cape, London.

Thorpe, W. H. (1974). *Animal nature and human nature*, Metheun, London.

Tiger, L. and Fox, R. (1972). *The imperial animal*, Secker and Warburg, London.

Vandeveer, D. (1976). 'Defending animals by appeal to rights', in *Animal rights and human obligations* (T. Regan and P. Singer, eds), Prentice-Hall, Englewood Cliffs, NJ, pp. 224–229.

Washburn, S. L. (1978). 'Human behaviour and the behaviour of other animals', *American Psychologist*, **33**, 405–418.

Willis, R. (1974). *Man and beast*, Hart-Davis, MacGibbon, London.

Wilson, E. O. (1975). *Sociobiology*, Bellknapp Press, Cambridge, Mass.

Wilson, E. O. (1978). *On human nature*, Harvard University Press, Cambridge, Mass.

Part II

Animal Experimentation: Major Areas of Research

Animals in Research
Edited by David Sperlinger
© 1981 John Wiley & Sons Ltd.

Chapter 5

The Medical Sciences

LOUIS GOLDMAN

INTRODUCTION

Scientific medicine rests on the premise that diagnosis must precede treatment. To diagnose what is wrong, to detect the abnormal functioning of a part of the body, requires that we know what constitutes the normal. And to determine what is normal in turn implies an understanding of structure (anatomy) and function (physiology). Learning this from the human being is not always practical. What better alternative then than to experiment on animals in the belief that there are close similarities in the way that animals function and that man functions?

Animal experiments go back a long way in time, certainly at least to Claudius Galen (AD 130–200), who dissected living animals in an attempt to discover how various organs functioned. Many of the world's most famous scientists since Galen's time have followed his example. William Harvey (1578–1657) sought to clarify the function of the heart by experiments on living animals. Another Englishman, John Hunter (1728–93), used what he somewhat callously referred to as 'animal machines' to determine the function of muscle tendons and bones.

The greatest impetus to animal experiments as the route to scientific knowledge came in the nineteenth century with the work of Claude Bernard and the French school of investigators. Bernard regarded clinical observation in the hospital as only the antechamber to scientific medicine. The chamber itself was the experimental laboratory in which planned experiments were done in a controlled way to show how the 'living machine' could almost literally be taken to pieces to demonstrate how it worked. Taking human beings to pieces obviously presented problems; nor did the human model seem altogether necessary since Bernard's work on animals turned out to be so productive.

He described several of the functions of the liver, the control of blood vessels by nerves, how red cells carried oxygen to the tissues, and new facts about the brain and spinal cord. More important still, Bernard put forward the fundamental concept of a regulated and stable internal environment in the body (the *milieu interieur*)—in contrast to the constantly changing environment in the world about us. Also from his laboratory came observations about the way certain structures produce 'internal secretions' (nowadays known as hormones), which pass into the circulation to exert their effects elsewhere in the body. Bernard noted that the pancreas was involved in the digestive process, but it took later research to reveal that the organ really comprised two different types of cells: those that manufactured the digestive pancreatic juice, which flowed along a duct into the small bowel, and another group of cells scattered in small islands throughout the organ. These island cells were thought to be responsible for an 'internal secretion' involved in the metabolism of sugar.

Attempts to extract and use this substance to treat diabetes over the next 30 years met with only limited success—until Banting and Best, working in Canada in the early 1920s, speculated that the attempts had failed because the digestive pancreatic juice came into contact with the 'internal secretion' of the island cells during the extraction process and destroyed it. Animal experiments showed that tying off the duct between the pancreas and the small bowel resulted in the destruction of most of the organ, but that the island cells survived. Banting decided that the logical approach towards extracting the 'internal secretion' was to tie off the duct, allow the digestive part of the organ to atrophy, and later seek to extract the substance from the surviving island cells.

Within a year, the two Canadians had not only succeeded in their plan, but they also showed that the extract reduced the high blood sugar in another dog suffering from experimentally-induced diabetes and, moving from animal to clinical experiment, successfully treated the first human case. So insulin first became available to treat the millions of people suffering from what could at times be a dreadful wasting disease.

CONTEMPORARY EXPERIMENTS

This pattern of initial animal experiment and later clinical application continues to the present day—with results that, at least on occasions, have proved as rewarding as those reported by Claude Bernard 100 years ago. A relevant contemporary example is described by a well-known Australian scientist, G. J. V. Nossal, in a book that provides a vivid insight into the aims and methods of medical research (Nossal, 1975).

For about 100 years, we have known that the thymus gland, which lies in the front of the chest behind the breast-bone, consists predominantly of a type of white blood cell, the lymphocyte. We also knew that the lymphocyte was involved in the body's immune defence system. Yet the function of the gland remained a mystery; in adult mice it could even be removed surgically without causing any harm. Another mystery arose from the observation that certain strains of mice suffer an appallingly high incidence of lymphatic leukaemia: 95 per cent or more die from this form of white blood cell cancer. But, oddly enough, surgical removal of the thymus in early adult life prevented leukaemia in these mice.

An immunologist, Dr J.F.A.P. Miller, began to study mouse leukaemia in the late 1950s. As part of his experimental work he began to remove the thymus from progressively younger animals. Eventually, he developed a remarkably skilled technique which allowed him to perform the operation on new-born mice, weighing only $\frac{1}{30}$ oz and measuring a mere 0.5 in in length. These infant mice were also protected against leukaemia when they grew up. But something else usually happened to them: they stopped growing after weaning, became extremely ill, developed widespread infections, and died. Yet the same operation in adult mice proved quite harmless.

Dr Miller realized that he had stumbled on the real function of the thymus. The gland in the very young animal is a school in which young lymphocytes are trained in their role of immune defenders. They subsequently pass from the thymus into the blood stream, where they are available to carry out their task of seeking out and destroying foreign substances. Removing the thymus from the very young thus deprived them of trained lymphocyte defenders against infection. The original purpose of the experiment was put aside as Miller and his immunological colleagues realized the implications of this new insight into the body's defence mechanisms. For the work led directly to a better understanding of the immune response and opened the way towards new clinical applications—in particular, controlling the rejection of the graft in organ transplantation.

Recounting this story, Dr Nossal emphasized that the development of modern transplantation rested on three related strands of research: immunology, which permitted the scientist to keep the lymphocytes at bay; genetics, which helped him to provide matching tests so that donor and host resembled one another, immunologically speaking, as closely as possible; and surgery, which by the development of ingenious skills and techniques advanced the discipline to the point where, in Nossal's words, we now have 'quite fantastic procedures for grafting of virtually any vital organ'. The surgical advance also derives from animal experiment. No reputable surgical pioneer would dream of experimenting with a new operation in man until he had first tried it in animals to learn the technical, metabolic, biochemical, and other problems he is likely to meet.

Another highly-experimental modern surgical technique is stereotactic brain surgery. Under X-ray control, thin probes are introduced into the brain and placed in certain small centres believed to be involved in key processes. If these centres are diseased or functionally abnormal, they can be destroyed by passing an electric current through the probe. Sometimes, it appears that these centres are not fully active. In these circumstances, chronic stimulation electrodes can be placed in the appropriate structure and stimulated daily at frequencies varying from 1 to 100 Hz. These techniques have been used to treat certain types of epilepsy, spastic cerebral palsy, and patients suffering from intractable pain not susceptible to other forms of treatment. The same techniques have been applied, much more controversially, to the management of patients with psychiatric disorders—notably those suffering from aggressive mental states, emotional disorders, and severe phobias.

Whatever the pros and cons of psychosurgery in these latter conditions, many neurosurgeons firmly believe that more precise location of various brain centres will improve the management of some diseases otherwise notoriously difficult to treat. In the devising of better procedures, preliminary animal experiments are vital—a point that was made at the fourth meeting of the European Society for Stereotactic and Functional Neurosurgery held in Paris in July 1979.

Yet the use of living animals to try and advance clinical knowledge every now and then provokes unease and concern. Consider the problem posed when a bone fracture fails to unite. For some reason, the healing process appears to be inadequate or inefficient. Twenty years ago, various scientists began work which showed the existence of electromechanical activity in bone: new bone growth could be induced around the cathode by electrical stimulation of bone in experimental animals. Many problems had to be solved: the type and size of electrical stimulation, the current supplied, the field density, exactly where the anodes and cathodes should be placed, the duration of treatment, the calibre of bone formation so induced—and so on. There were even questions about the validity of the claims made for electrical bone stimulation in healing fractures.

Paterson and a group of his paediatric colleagues in Australia maintained that such an experimental method should not be tried in man until it had been validated under experimental conditions in animals (Paterson et al., 1977). Their experiments involved fracturing the leg bones (tibia) in 69 dogs, preventing the fractured ends from uniting by interposing a silicone block between them, and ensuring that union had not taken place when the silicone block was removed 8 weeks later. An electrical stimulator was then inserted into the fractured bone and stimulation done under 'blind' conditions (so that the investigators did not know whether the particular animal was receiving true or stimulated electrical stimulation). The results clearly showed much better bone growth in those fractures receiving electrical stimulation, compared with those that did not. Paterson and his colleagues concluded that the technique had now been validated and that clinical trial in man was justified.

This report in an internationally respected medical journal drew protests from two senior physicians. Dr Eileen Hill of Birmingham wrote that she was filled with horror to learn that human beings—especially doctors and even more especially those dealing with children—could do experiments which entailed so much suffering (Hill, 1977). No wonder, she added, that the anti-vivisectionists had such an appeal. Dr Gerald A. MacGregor of Guildford in Surrey also thought that the experiment was quite unjustifiably cruel, because the stimulator could have been tried more sensibly on road accident victims (MacGregor, 1977).

The *Lancet* published a brief footnote in which it observed that the experiments had been approved by the Animal Research Committee of the University of Adelaide, one of whose members was an official of the Royal Society for the Prevention of Cruelty to Animals.

Had they chosen to respond to the criticism, Dr Paterson and his colleagues might perhaps have added that their work was directly relevant to clinical medicine, even if it did involve causing animals pain. Or they could have argued that only by such an animal experiment was it possible to control all the factors which can influence the outcome of a study. Rigorous control of these factors is almost impossible in a human study. But if Drs Hill and MacGregor were dismayed by the Adelaide work and appalled to see it given such wide currency in the *Lancet*, how would they have reacted to some of the work reported a few years ago from Cambridge by Dr Colin Blakemore and his colleagues (Blakemore and Van Sluyters, 1975; Blakemore *et al.*, 1975; Blakemore, 1976)?

Dr Blakemore is an experimental physiologist and Director of Studies in Medicine at Downing College. His research field primarily concerns the mechanism of sight. In some of his experiments, kittens were housed for several weeks in a totally darkened room. Others had their eyelids stitched together. Another group of cats were made to wear goggles that permitted the separate control of the visual input to the eye. For example, they might wear goggles fitted with powerful biconvex lenses for as long as 9 weeks. Later, Blakemore and his colleagues took three kittens less than 3 weeks old, another kitten reared in total darkness for more than 5 weeks, and an adult cat. Each of these animals was submitted to an operation in which the muscles which control the movement of the right eyeball were cut and the eyeball rotated. The animals were then housed in a colony room for several weeks or longer before undergoing behavioural and neurophysiological tests.

The investigators reported that the rotated eye was able to mediate behaviour to visual input, although the animal's responses were not very accurately directed in space. This was not particularly surprising, as the investigators themselves recorded, in view of the rotation of one eyeball, the relative immobility of the rotated eye, and the conflicting information which the animal was receiving from the other eye. But if the results were so predictable, why do the experiment at all?

Dr Blakemore's general purpose in following this line of research is clear enough, for in his book (based on the Reith lectures he broadcast in 1976 for the British Broadcasting Corporation) he wrote about the promise that research on the brain would provide a genuine basis for the treatment of mental disorder (Blakemore, 1977). Much more than this, he hoped that we would achieve a better understanding of the nature of man himself. While no one doubts the genuine spirit of scientific enquiry that lies behind the research, it breeds a degree of uneasiness (to put it no higher) to read about experiments in which animals are imprisoned for weeks in an abnormal sensory environment and in which the apparatus of sight is so distorted. The research continues, for Blakemore has recently applied his technique of 'monocular deprivation' to monkeys (Blakemore *et al.*, 1978).

Nor are these experiments so extraordinary or so exceptional as all that— because there appears to be no discernible limit to the amount of damage that can be inflicted on an animal in the pursuit of knowledge. In 1977, headlines in American newspapers accused a research scientist of carrying out 'vicious experiments' and of indulging in 'sex sadism'. The controversy centred on Dr Lester R. Aronson, Chairman of the Department of Animal Behaviour at the American Museum of Natural History in New York City. Aronson planned to investigate various aspects of the cat's sexual behaviour—including the site of action of the gonadal hormones, the role of sensory stimuli, and the function of the limbic structures in the brain. (In passing, we should note that stereotactic neurosurgeons, too, are especially interested in the limbic system, for it appears that certain lesions in this area can not only produce epileptic discharges, but also emotional reactions like terror, screaming, and aggression.)

According to Dr Aronson's application for a research grant from the United States' Department of Health, Education, and Welfare, the work would involve blinding the cats, making them deaf, destroying their sense of smell, severing the nerves in the penis, and removing the animals' testes. In truth, it seemed that they were being used merely as an 'animal machine' (in John Hunter's expressive eighteenth century phrase), parts of which could be removed or destroyed without much consideration for its living state. Scientists always seek to justify experiments like these on the grounds that, however remote they seem from a practical advance in medicine or psychology, they contribute to the general sum of knowledge and so improve the chances of a later discovery.

Such a claim is neither easy to defend nor attack. One apparently reasonable method of judging the value of research is to assess how often the particular work has been quoted or cited by other research workers in the field. The original controversy about Aronson's proposed study arose because a writer for the journal, *Science*, Nicholas Wade, decided to examine his previous research on the sexual behaviour of cats by using the *Science Citation Index* published by the Institute for Scientific Information in Philadelphia (Wade, 1976). Wade concluded that Aronson's previous work was relatively unimportant because other scientists had made scant reference to it.

The founder of the Institute for Scientific Information, Dr Eugene Garfield, carried out his own assessment of Aronson's work and, basing his findings on citation analysis, came to a rather different conclusion: Aronson's experimental work on cats had been reasonably valuable to the scientific community (Garfield, 1977). But he also expressed some doubts. He was perplexed by the assertion of some of the acknowledged leaders in the field that Aronson's work was important when their own citation of it was minimal. Dr Garfield further criticized, as a general defence of all scientific research, the claim that we can never know in advance what value so-called basic research could have in the future:

'In the days when there were just a few thousand people in the world doing basic research, such assertions were acceptable. But when the world's scientific population exceeds one million persons, we need something more than the bland assertions by established investigators or their peers that basic research pays off.'

THE SCOPE OF MEDICAL EXPERIMENTS

So far we have discussed only single experiments or experiments carried out by individual groups of investigators. How large is the problem posed by animal experiments in the medical field? Detailed figures are impossible to find. All that is available in Britain from the figures published by the Home Office are relatively crude totals. Thus 11 700 workers licensed to do these experiments performed 5.4 million experiments on living animals in 1977 (Annotation, 1978). The annual figure has remained at about 5 million since 1968. The world total has been estimated to be in the order of 60 million. As regards the 1977 British figures, nearly 3 million animals were said to have been used for experiments concerned with medical, dental, and veterinary products and appliances, and 1.3 million for studies on the normal or abnormal functions of the body.

These broad categories are not really very informative, because almost any kind of animal experiment could be included in the latter category. A considerable overlap also exists between one supposed category and another. It is even difficult from the published records to estimate with any degree of accuracy how many animals are being used for medical research and how many for testing cosmetics, toilet preparations, pesticides, weed-killers, and the like. Another broad group of experiments, difficult to classify, involves the kind of behavioural research in which animals are submitted to various forms of stress: deprivation of food, reward and punishment (the latter often administered by electric shocks to the cage in which the animal stands), isolation in sound-proof chambers, the administration of alcohol or habit-forming drugs to induce a state of addiction, and depriving the animal of tactile or auditory sensation.

Dr Colin Blakemore's experiments on the cat's eye could well be classified either as medical in intention, since his basic concern is the mechanism of sight in relation to brain function, or as behavioural, since his techniques involve various forms of sensory deprivation or distortion. So, depending on the classification used, some experiments could be considered as medical or non-medical. The confusion led directly to claims that only a minority of animal experiments could be seen to be medical (Ryder, 1975). This prompted Professor Sam Shuster of Newcastle-upon-Tyne to launch a rousing attack on the anti-vivisectionists and other critics of animal experiment in a lecture to the Research Defence Society given in London in October, 1977 and annotated in the *Lancet* (Annotation, 1977).

Shuster argued that most animal experiments could be classified as medical; he estimated the figure to be 75 per cent. Another observer questioned whether this was a reliable estimate, since it was derived from a pilot study by the Home Office (Britten, 1977). Whatever the true figure, there is little doubt that pharmaceutical research organizations use a large number of animals for testing new drugs. Pharmaceutical research may well comprise the largest single group of animal experiments. Before a drug is given to man it undergoes a series of long and complex evaluations in animals during which it is first given in an acute toxicity test (to establish the dose likely to be lethal) and later in sub-acute or chronic (non-lethal) doses for long periods of time. Other tests include the feeding of the drug to pregnant animals to determine whether, like thalidomide, it might damage the growing embryo or young fetus.

The acute toxicity tests, usually in the form known as the LD_{50} test, have attracted much criticism over the last few years. It has been argued that, while they must inevitably cause pain and distress to the animals, the information they provide is often of limited value. The agitation prompted the Home Secretary in 1977 to ask the Advisory Committee on the administration of the *Cruelty to Animals Act 1876* to investigate the extent of usage of the LD_{50} test, as well as the scientific necessity and justification for tests of this kind. The report of the Advisory Committee is a highly informative and up-to-date review of an important aspect of animal experiment and deserves a detailed appraisal (Home Office, 1979).

The LD_{50} (which stands for lethal dose, 50 per cent) is the quantity of a substance which will kill half the number of animals exposed to it. The value is determined by administering graduated doses of the substance, expressed in milligrams of the substance per kilogram of the body weight of the animal (mg/kg), to a group of animals. By using large numbers of animals, the LD_{50} can be determined with great accuracy.

The test was originally devised in 1927 as a more accurate method than any then available for measuring the potency of highly toxic, yet valuable, medicines—such as diphtheria antitoxin, digitalis, and insulin. Where the difference between a clinically beneficial and a dangerous dose is small, the strength of the preparation to be used must be known precisely. In the last 50 years, the development of biochemical and other assay methods has superseded this use of the LD_{50} for most pharmacological purposes. But the LD_{50} has been allocated a new role in the first estimation of the safety (or toxicity) of new substances—including medicines, food additives, detergents, pesticides, and weed-killers.

How many LD_{50} tests are being done each year? The Advisory Committee sought information from the Home Office on the point—only to receive an answer that was not as precise as it obviously wanted, because the test is merely one of a number of acute toxicity experiments recorded under a general heading. The records showed that in 1977 a total of 557 917 acute experiments were done on animals (about 10 per cent of the total). Of these, the Committee estimated that 229 500 referred to the LD_{50} type of experiment, although it acknowledged that this was an approximate figure subject to a substantial margin of error. Of this total, 60 per cent referred to work on mice and 27 per cent to rats. The remainder included studies on fish and birds. Fish are used mainly to study the toxicity of substances likely to find their way into rivers and reservoirs, while birds are used mainly to study pesticides, seed dressings, and similar products. If we assume an average of 60 animals for each LD_{50} experiment, the figures for mice and rats represent about 3300 tests. In other words, 3300 tests—each involving about 60 animals—were done to establish the LD_{50} value for a particular substance or drug. Most of these probably related to the testing of pharmaceutical or medical substances. This, it must be emphasized, is only an assumption, because an applicant to the Home Office for a certificate to perform an animal experiment does not have to specify the particulars of what he proposes to do. The description usually given appears to be simply in the general form of 'administration of substances', followed by a similar, somewhat general, justification in such terms as: 'To evaluate the toxic and other properties of substances likely to be of benefit in the treatment or prevention of disease'.

Are acute toxicity tests cruel or painful? The Committee considered this point at length, especially as the nature of the test is such that it must be done without an anaesthetic. The initial administration of the substance causes little more than temporary discomfort. To give an oral dose, a tube is inserted down the animal's throat: done by a competent technician, the procedure takes only a minute or so. But the larger, toxic doses of the substance, after a varying period of time, may produce the reactions associated with acute poisoning. The Committee received a check list used by one research organization which recorded the possible reactions as aggressiveness, vocalization, salivation, prostration, and coma.

(To enter a personal caveat at this point, I can only object, as I have done before, to euphemistic language which seems to lay a soft, verbal unction on what actually happened. 'Vocalization' sounds so bland—if it really means that the animal screamed or squealed or whined or yelped. American scientists in particular take eager refuge behind jargon of this kind. None of the animals at the end of an American experiment is ever killed. No—the animal is 'sacrificed', a manner of description which induces a degree of self-deception on the part of both writer and reader.)

While the Committee was told that the commonest reaction to the LD_{50} test was a slowing down of activity and a gradual descent into irreversible coma, occasionally accompanied by convulsions, it felt obliged to conclude that the test must cause appreciable pain—at least to a proportion of the animals involved. Can this pain be construed as cruelty? The Committee refused to take an absolutist view on what easily becomes a difficult ethical or philosophical argument. It accepted that those people who believe that any experiments which cause animals pain should be banned because they are cruel are entitled to say that the experimenters who do them are cruel men. Even so, a valid, or at least useful, distinction should be drawn between inflicting pain and cruelty; the two expressions should not be used as if they were synonymous.

Inflicting pain could be justified if the pain was, in philosophical terms at any rate, compensated for by what the Committee termed 'the consequential good'. The criterion implied a presumption in favour of humans over animals, so that research done on animals for the benefit of man could be regarded as justifiable. But the human good envisaged 'must be a serious and necessary good, not a frivolous or dispensable one, if the infliction of pain on animals is to be ethically acceptable'. Even then, the degree of pain must be carefully weighed, because only the minimum necessary for the end was justified. Of course, if one judged the issue strictly by the Committee's own criteria for justification, the number of animals involved in a painful experiment was irrelevant: one animal's pain was not made better or worse merely because other animals were suffering in the same way.

The Committee acknowledged the difficulty, but insisted that a human being had a duty to protect the inarticulate and the helpless and to limit the total amount of suffering inflicted on those he called upon to serve his aims:

'The experimenter will degrade himself and the society of which he forms a part, if he gives no thought to the aggregate of pain he imposes on others for his own, human, ends even though the sufferer has no concept of the quantities involved. A balance must be struck, and thought given not only to the degree of pain which can be justified, but also to the numbers of animals which will experience the pain.'

This ethical question of the number of animals to be included in an LD_{50} test has a crucial relevance to current drug evaluation practice. For by using a large number (between 50 and 140 animals per test), the LD_{50} figure can be calculated very accurately. But this figure looks much more precise than it really is, because it is influenced by so many factors—among them the sex of the animal, species, strain, age, the number of animals per cage, diet, route of administration of the substance, the vehicle in which it is dissolved, the concentration, the prevailing temperature and humidity, and even the time of day or year (Harper, 1976).

Most of the scientific experts who presented evidence to the Committee pointed out that precise LD_{50} determinations involving the use of large numbers of animals were of little use in assessing the safety of the particular substance. What the scientist needed from the test was an idea of the order of magnitude of toxicity. He needed to know if the LD_{50} was in the order of, say, a few mg/kg, 100 mg/kg, several hundred mg/kg, or amounts still larger than this. This kind of information, as the Medical Research Council indicated in its submission to the Committee, could be obtained by using a simple test and small numbers of animals—perhaps only 15 or 20. Sir Eric Scowen, who gave evidence on behalf of the Committee on Safety of Medicines, agreed with this view.

In its general conclusions, the Committee took the same view. What was needed was the fostering of a climate of opinion among scientists the world over which rejected the pursuit of a spurious accuracy based on the killing of a large number of animals. Nor was it necessary to increase dosage in toxicity testing to extraordinary lengths. If a substance was only slightly toxic, the amount needed to kill would be much greater than that which a human being was likely to consume, even by accident. In these cases, there was no need to increase dosing beyond this amount. This procedure is generally referred to as a 'limit test'.

Thus the Committee arrived at its main recommendations. Those who carried out LD_{50} tests should always bear in mind that for safety evaluation of a substance a degree of precision which required the use of many animals was not necessary. And, wherever practicable, a limit test should be done in preference to a formal LD_{50}. So far, it might seem—so good, if Sir Eric Scowen of the Committee on Safety of Medicines had not pin-pointed an almost certain snag. In his evidence, he drew attention to the Norms and Protocols Directive of the European Economic Community (EEC), which appeared to lay down that, where possible, the LD_{50} should be determined with its 95 per cent fiducial limits. Such a requirement means an accurate figure must be established, a result that can be achieved only by using large numbers of animals.

The Committee subsequently returned to the problem which may well arise when EEC Directives take on the force of law in the United Kingdom—although it failed to explain how its own recommendations, which stigmatized the search for a spurious accuracy, could be reconciled with a directive seeming to call precisely for this. To make matters worse, there was an inevitable tendency for the regulatory agency with the strictest requirements to be regarded as the norm. If the United States or France or Germany demand that an LD_{50} be determined to a 95 per cent degree of accuracy, the United Kingdom will have little alternative but to fall in line.

Does this mean that scientists in other countries are less concerned about inflicting pain on animals than the British? In a sentence which seems to be compounded of diplomacy and caution in equal parts, the Committee delicately observed that there was 'reason to think that hitherto some scientists abroad have been less concerned with the welfare of animals than their British colleagues'. One Committee witness asserted that his Continental colleagues often expressed the view that the British had an unduly obsessive concern for animals. Other witnesses, however, thought that there was growing sympathy with our attitude. Evidence for this change of heart did not appear in the Committee's report. The truth seems to be that laws to control animal experimentation similar to those in Britain are found in only a few other countries and that experiments that would cause a public outcry here are done with little general concern abroad.

Two illustrations will suffice. Some years ago, an American scientist, Robert J. White, experimented with the (literally) disembodied brains of dogs and monkeys. He showed that an isolated brain, maintained by artificial means, retained its inherent electrical activity—an observation that prompted Dr White to speculate that some semblance of 'consciousness' was retained as well. Perhaps the most notorious of all experiments has been reported to have been done in Russia and in South Africa: the head of one animal was joined to that of another to create a two-headed dog. Two reactions to this experiment are possible: admiration at an extraordinary degree of technical and scientific skill—and horror that anyone should want to carry out such a pointless experiment.

Even those who most vigorously defend the right of scientists to experiment on animals without hindrance from the law (or from vehement anti-vivisectionists) do not pretend that all experiments are justified. According to the *Lancet* annotation cited above, Professor Shuster admitted that many experiments are scientifically awful and others (including many of those required for new drug testing) futile. Nor did he deny that some experiments were distasteful because they caused undue suffering to the animals. Much the same attitude was adopted a few years earlier in a paper read to the Royal Society of Medicine by Dr W. Lane-Petter, a man who had spent much of his career in the care and breeding of laboratory animals (Lane-Petter, 1972). Some scientific problems, he observed, were really not worth solving; some of them, indeed, seemed to be motivated by little more than curiosity. As for experiments that caused suffering, Dr Lane-Petter argued that they could be justified only if there was a good reason. Without it, the experiment must be considered cruel. Almost exactly the same distinction was of course made by the Home Office Advisory Committee.

Dr Lane-Petter further suggested some practical questions the scientist should ask himself before embarking on an animal experiment. Was an animal the best experimental model? Must the animal be conscious at any time during the work? Could pain or discomfort be lessened or eliminated? Was it possible to reduce the number of animals used? Perhaps most important of all, was the problem the scientist had in mind worth solving anyway?

The number of scientists experimenting with animals who really ask themselves questions like these is, itself, an open question.

ALTERNATIVES

Tests which originally required living animals have been replaced as better procedures became available. To take the obvious example, the LD_{50} test introduced in 1927 for the assay of potentially toxic medicines is now rarely used for this purpose, having been superseded by such sophisticated laboratory procedures as gas chromatography and mass spectrometry. These permit the detection of minute amounts of drugs and other substances in the body. The same techniques are used to study the absorption, distribution, metabolism, and excretion of drugs both in animals and man, with a minimal risk of causing either pain or harm. Alternatively, chemical substances can be labelled with minute amounts of a radioactive material before administration to man or an animal.

This also permits the identification of a substance in the living organism with a minimal risk.

Tissue cultures provide yet another possible solution. The use of cultures of human cells has indeed largely replaced older, live animal techniques in the preparation of vaccines against poliomyelitis and rubella. It seems only logical to believe that the use of tissue cultures could be extended to other fields, including the testing of drugs and chemicals. Progress has also been made in methods which permit the *in vitro* assay of antitoxins, although more work needs to be done to convince national and international statutory bodies that these methods provide an accurate and safe alternative to traditional techniques (Gowans, 1974).

Then there are mathematical models and computer simulation methods. It has been suggested that mathematical models can not only predict the efficacy of certain drugs but may indicate ways of increasing the therapeutic benefits. The Home Office Advisory Committee was intrigued by claims that the oral LD_{50} in the rat could be determined for any compound by a computer provided with data on the molecular formula of the substance. The Committee asked a group of expert assessors to judge the value of this approach. In their report, the experts commented that, for this kind of computer prediction, information on the chemical structures, physico-chemical properties, and pharmacological charac-teristics of a large number of compounds was needed to prepare the initial computer program. The results obtained for a new substance appeared to be only of qualitative importance and rarely provided a reliable index of its potency or other activity.

When measuring acute toxicity, as in the LD_{50} test, quantitative rather than qualitative information was needed—if only because death could not necessarily be related to any individual biological property. Small changes in the chemical structure of a substance often led to profound changes in its pharmacological action and, as a result, to corresponding changes in acute toxicity. Computer-aided prediction, in the experts' view, might therefore provide useful additional information but not enough basic data to satisfy drug regulatory authorities. An even greater drawback would arise with a new class of chemical substance, for which no previous information could obviously be made available.

None of this is to suggest that alternatives to animal experiments should be dismissed or disregarded. Some of the anti-vivisection organizations in Britain have taken a positive attitude in seeking to promote the use of these new techniques and of encouraging scientists to develop new ones. The Fund for the Replacement of Animals in Medical Experiments (usually referred to by the acronym FRAME) not only maintains an information centre aimed at keeping research workers in touch with developments, but urges that government funds should be specifically allocated to the search for new alternatives.

Allegations have also been made that scientists are reluctant to bother with alternative methods as long as animals are freely available to them. In the paper cited above, J. L. Gowans rejected this view as a misconception. There were very strong incentives to replace animals, for the simple reason that *in vitro* techniques were usually more rapid, more accurate and considerably cheaper. What stood in the way of a wider adoption of alternatives, at least in regard to the safety evaluation of drugs, was summed up neatly by Dr A. Spinks, Research and Technology Director of Imperial Chemical Industries, Ltd, in his evidence to the Home Office Advisory Committee. He pointed out that a new compound may affect any of 100 organs, 1000 cell types, or 10 000 enzymes—or any interaction among them.

Thus there was no escape from using the living animal to establish the probable safety (or toxicity) of a compound. Dr Spinks nevertheless believed that non-animal testing could prove useful once the type of toxicity had been established. But clearly this is a prescription for the future—and possibly a remote future—rather than for the present.

A PERSPECTIVE

So far in the appraisal of the use of animals in medical experimentation, the evidence has been presented and largely been allowed to speak for itself. In assessing whether the present usage of animals in Britain is beyond reasonable criticism, however, judgments must inevitably be personal. Let me therefore make my own position clear. I am not an anti-vivisectionist. No reasonable alternative exists, so far as I can see, to animal experiments if the medical science of the future is to make parallel advances to those achieved during the last 50 years. Nor can I see any important alternative, in the present state of knowledge, in tissue cultures, sophisticated biochemical tests, or computer simulation. While these techniques should be developed and refined, they are unlikely to replace more than a small fraction of live animal experiments.

Even so, no-one other than a man wearing rigid scientific blinkers could read some of the experiments done on animals without a deep sense of unease and distress. The most shocking experiments, let it be freely admitted, tend to be reported from other countries. Which is not to say that those done in Britain are always beyond question. As evidence, I need only cite Colin Blakemore's work on cats' eyes at Cambridge. Yet with monotonous regularity, medical scientists over the years have reiterated their firm conviction that the British system offers the best protection for animals and that changes in the law are unnecessary. Even the *Cruelty to Animals Act 1876* is derided as unnecessary, because research workers insist that the true safety of the animal lies in the conscience of the worker, the supervision provided by his seniors in the department, and the critical appraisal of his colleagues.

This was certainly the view put forward in a leading article in the *British Medical Journal* (Leading article, 1974). We could rely on the experimenter being the best judge in his own right, said the journal, because of two vital safeguards. The first was the opinion of the experimenter's own colleagues; the second was the existence of the Home Office inspectorate. (The Journal failed to specify exactly how a small number of inspectors, 13 at the time, were supposed to supervise the work being done by more than 10 000 investigators.) But just to guard against the faint possibility of abuse, the journal went on to advocate a code of conduct in the use of animals.

Exactly the same, surely naive, prescription for a code of conduct came from Professor Shuster in 1977. But to what extent do these reassurances (that the best protection lies in the conscience of the worker, supervision by his peers, etc.) accord with reality? The likely answer may be gleaned from the changes that have taken place in another field of research. Scientists no longer believe that the best protection for the patient who participates in a clinical experiment lies in the conscience of the worker. Almost everybody now accepts that ethical committees in hospital are needed to judge the morality and ethics of any such experiment. If the conscience of the doctor dealing with a patient is not enough to determine what is right and wrong, how odd that the same type of scientist should insist that conscience is enough when it comes to a mere animal?

As for the legal restraints on cruel experiments, the safeguards imposed by the *Cruelty to Animals Act 1876* offer little opposition to the insensitive and ruthless enquirer after knowledge. The reality, as contrasted with the theory, was noted in a letter by the late Lord Platt, President of the Royal College of Physicians of London from 1957 to 1965 (Platt, 1974). He observed that as a professor of medicine he had signed many applications for licences and certificates; as president of a royal college he had countersigned many more.

Most of those he refused to sign were promptly signed by the head of another royal college. As a member of the Home Office Advisory Committee he had disagreed with his colleagues on many occasions but was usually outnumbered. The experiments Lord Platt found particularly objectionable were those which, for purely academic knowledge, subjected animals to physical and psychological stress. He accepted that the control of animal experiments in Britain was better than elsewhere; he was also sure, though, that it could be improved.

Perhaps the most damning criticism of the British system is the fact that in more than 100 years since the *Cruelty to Animals Act* became law, not a single medical scientist has been prosecuted under the Act for cruelty to an animal — even though so resolute a defender of the *status quo* as Professor Shuster conceded that some experiments were 'scientifically awful'. Nor is this all. Early in 1965 a departmental committee of enquiry set up under the late Sir Sydney Littlewood reported that animal experiment was 'a moral and social problem of the first magnitude and one that does not exclusively concern the expert'. The report advocated major changes in the law designed at improving the control of animal experiment. No government since then has paid more than token regard to the Littlewood recommendations.

What, then, are the prospects for the future? The medical establishment remains firmly opposed to any but relatively minor changes in the law. So powerful is its influence that significant changes will be brought about only by constant and unremitting effort. Unfortunately, the anti-vivisectionists, though vocal and committed and genuinely concerned, tend to destroy their case for reform by an absolutist attitude. They seek to ban all animal experiments. No-one, of course, questions their right to advocate such a course. But they face the dilemma that banning all experiments on animals would damage medical research. And most people, however concerned they may feel about animal suffering, believe that our own species has a higher priority. Anti-vivisectionists try to resolve the dilemma by arguing that very little has in fact been achieved by animal experiments—or, alternatively, they try to persuade scientists that alternative methods would do as well.

When scientists claim that virtually all animal experiments can be justified and anti-vivisectionists counter by insisting that none is justified, the middle ground (where the truth probably lies) appears depressingly empty and desolate. So far as I can judge, a number of scientists agree that cruel and unnecessary experiments are sometimes done, that not all is for the best in the best of all possible (experimental animal) worlds, and that the pursuit of academic knowledge without considering the suffering inflicted on animals is improper for a scientist with a moral conscience. There should be a meeting ground somewhere in the middle for people like this and for less-than-absolute defenders of animal rights. Given enough flexibility on both sides, reform will become possible.

And one or two signs indicate that it could happen. At a meeting of the Physiological Society a year or two ago, a foreign visitor presented a paper in which he reported animal experiments which many of those present found repugnant. At the end of the presentation, the chairman of the meeting posed the routine question of whether the paper should be accepted for publication in the Society's journal. He made clear his own strong opposition to this, because of the cruelty inflicted on the animals. The meeting of these scientists, most of them from academic institutions, voted with only one dissentient to reject the paper.

REFERENCES

Annotation (1977). 'Animal torture for vanity?', *Lancet*, ii, 913.

Annotation (1978). 'Studies on animals', *British Medical Journal*, ii, 1797.

Blakemore, C. (1976). 'The conditions required for the maintenance of binocularity in the kitten's visual cortex', *Journal of Physiology, London*, **261**, 423–444.

Blakemore, C. (1977). *Mechanics of the mind*, Cambridge University Press, Cambridge.

Blakemore, C. and Van Sluyters, R. C. (1975). 'Innate and environmental factors in the development of the kitten's visual cortex', *Journal of Physiology, London*, **248**, 663–716.

Blakemore, C., Garey, L. J., and Vital-Durand, F. (1978). 'Reversal of physiological effects of monocular deprivation in monkeys', *Journal of Physiology, London*, **276**, 47p–49p.

Blakemore, C., Van Slutyers, R. C., Peck, C. K., and Hein, A. (1975). 'Development of cat visual cortex following rotation of one eye', *Nature, London,* **257**, 584–586.

Britten, S. (1977). 'Animal experiments'. *Lancet, ii,* 1032.

Garfield, E. (1977). 'Citation analysis and the anti-vivisection controversy', *Current Contents,* (17), 25 April, 5–10.

Gowans, J. L. (1974). 'Alternatives to animal experiments in medical research'. *British Medical Journal, i,* 557–559.

Harper, K. H. (1976). Paper read to a symposium in London on 'Toxicology for the Industrial Chemist', cited in the evidence submitted to the Home Office Advisory Committee by the Committee for the Reform of Animal Experimentation (August, 1977).

Hill, E. (1977). 'Experiments on animals', *Lancet, ii,* 147.

Home Office (1979). *Report on the LD50 test presented to the Secretary of State by the Advisory Committee on the administration of the Cruelty to Animals Act 1876,* HMSO, London.

Lane-Petter, W. (1972). 'The place and importance of the experimental animal in research', *Proceedings of the Royal Society of Medicine,* **65**, 343–344.

Leading Article (1974). 'Animal experiments', *British Medical Journal, i,* 528–529.

MacGregor, G. A. (1977). 'Experiments on animals', *Lancet, ii,* 147.

Nossal, G. J. V. (1975). *Medical science and human goals,* Edward Arnold, London.

Paterson, D. C., Carter, R. F., Maxwell, G. M., Hillier, T. M., Ludbrook, J., and Savage, J. P. (1977). 'Electrical bone-growth stimulation in an experimental model of delayed union, *Lancet, i,* 1278–1281.

Platt, Lord (1974). 'Animal experiments', *British Medical Journal, ii,* 220.

Ryder, R. D. (1975). *Victims of science,* Davis-Poynter, London.

Wade, N. (1976). 'Animal rights: NIH cat study brings grief to a New York museum', *Science, New York,* **194**, 162–164.

Animals in Research
Edited by David Sperlinger
© 1981 John Wiley & Sons Ltd.

Chapter 6

The Biological Sciences

JENNY REMFRY

INTRODUCTION

'Give me to learn each secret cause;
Let number's, figure's, motion's laws
Revealed before me stand;
These to great Nature's scenes apply
And round the globe, and through the sky,
Disclose her working hand.'
Hymn to Science, MARK AKENSIDE (1739)

True scientists, those involved in original research, are driven by the intellectual need to understand the world they live in (or in the case of space scientists, the worlds they do not live in). Sometimes their discoveries have useful applications; sometimes they are employed by others in the hope that they will discover something of commercial or medical value; but essentially they are interested in knowledge for knowledge's sake. Biological scientists find a particular fascination in life and its many and varied forms.

Practical knowledge of the world about us is, of course, older than science. Prehistoric man needed to learn a great deal about animals, plants, topography, and weather in order to survive in competition against animals larger, stronger, and faster than himself. The cave paintings found in France and Spain show that man, 30 000 years ago, had highly developed powers of observation and recording so that the animals depicted can still be clearly identified even though some of them are now extinct. This knowledge must have been of value in hunting and in protecting himself against predators. Knowledge of plants was just as important, in order to recognize those good to eat and those herbs whose infusions were beneficial in sickness.

THE SCOPE OF BIOLOGY

Observations of the 'nature study' type are not necessarily scientific. Science involves using observed data to produce a hypothesis, then by collecting further data, by observation, or experiment, to test the hypothesis. Modern biology is largely experimental.

Classical zoology was, and still is, concerned with the enormous task of finding, describing, identifying, naming, and classifying each different animal found. The task started in the Garden of Eden (*Genesis*, 2) and continues to the present day with new creatures being discovered every time a curious traveller turns over a stone! The task was made more orderly when taxonomists grouped animals according to their fundamental design rather than their superficial appearance and when the binomial nomenclature introduced by Linnaeus was adopted. While protozoa are considered to be the smallest animals, zoologists must also include bacteria which, whilst certainly living organisms, are not necessarily animals or plants. A virus is incapable of life outside a living cell, so is not considered to be a living organism but rather 'simply a piece of bad news wrapped up in protein' (Medawar and Medawar, 1978).

The classical science of zoology was based on observations of the habits, appearance, and behaviour of live animals and the dissection of dead animals. Experimentation on live animals was not much required except for simple behavioural experiments. The modern science of ethology is still essentially observational, requiring little manipulation of the animals. Behavioural science and experimental psychology on the other hand often involves observing the reactions of animals under highly artificial conditions. These will be considered in a separate chapter.

Another new biological science is ecology, which studies the relationships between an animal and its environment. This has great application today as conservation of habitat is seen to be necessary in order to protect some species from extinction, such as the otter in Britain and the great apes of South East Asia. Experimental work in ecology is unlikely to involve surgical interference with the animal. One technique is to trap and kill a representative sample of a wild animal population and examine the stomach contents to identify what the animals have eaten.

EVOLUTION AND ONWARDS

The Theory of Evolution

The science of zoology was revolutionized in the nineteenth century by Charles Darwin's theory of evolution. His voyage in HMS *Beagle* between 1831 and 1836 provided the masses of data on plants, animals, fossils, and geological events which substantiated the hypothesis already put forward tentatively by others, including his own grandfather, Erasmus Darwin. The hypothesis was that individuals of a species tend to vary from one another; those individuals best fitted to the environment will survive and reproduce. Those variations which are unfavourable will die out, so a species may gradually change and eventually a new one may be produced. Since this process has been going on since the beginning of life, all the species could have evolved from one original living creature. Darwin published his ideas in 1859 in his book *On the origin of species by means of natural selection, or the preservation of favoured races in the struggle for life.*

The finches of the Galapagos Archipelago provided one of the main pieces of evidence by showing how one species can merge with another.

'The remaining land-birds form a most singular group of finches related to each other in the structure of their beaks, short tails, form of body and plumage. . . . The most curious fact is the perfect gradation in the size of the beaks in the different species of Geospiza, from one as large as that of a hawfinch to that of a chaffinch, and even to that of a warbler. The largest beak in the genus Geospiza is shown in figure 1 and the smallest in figure 3; but instead of there being only one intermediate species, with a beak of the size shown in figure 2, there are no less than 6 species with insensibly graduated beaks. . . . Seeing this gradation and diversity of structure in one small, intimately related group of birds, one might really fancy that from an original paucity of birds in the archipelago one species had been taken and modified for different ends'. (From the *Voyage of the Beagle*, 1845)

Palaeontology

The theory of evolution, which was so hotly contested at the time, was of great relevance to the science of palaeontology. It opened up the possibility that the curious extinct species found fossilized in rocks had not all met a watery end in the Flood but were earlier forms of life which had been superseded by later more successful forms.

Genetics

An explanation of the mechanism by which species underwent change, or new species emerged, was provided later by the science of genetics. Mendel performed the pioneering work on inheritance in his monastery garden. By cross-pollinating different strains of peas he was able to show that inherited characters retain their individuality from generation to generation and that some characters are dominant and some recessive. Later workers showed his principles to be true for animals as well as for plants. Mendel's work was published in 1866 but was not recognized until 1900, by which time chromosomes had been discovered in the nuclei of cells at the time of cell division. The concept of a 'gene', a piece of chromosome which determines a Mendelian inherited character, followed naturally.

In sexual reproduction the genetic material of the chromosomes is slightly jumbled to give new combinations of genes, so that offspring are always slightly different from their parents. Rarely, alterations in chromosome or gene structure can occur spontaneously, or in response to certain drugs or X-irradiation, and this can lead to altered characteristics in the cells affected or in the offspring, often deleterious but occasionally advantageous. This is known as mutation. A few years ago a litter of kittens was born in Devon with sparse, curly hair. This was probably due to a mutation and certainly constituted a change in the genetic material because the kittens, in due course, 'bred true'. (For a description see Raleigh *et al.*, 1976.)

Genetics is taught today by breeding fruit-flies and mice to produce characteristics that are observable in the intact animal. The applications are mostly in animal production where genetic theories are used to plan the production of new highly efficient breeds of livestock, particularly poultry, and to eliminate undesirable traits which are due to genetic defects, for example progressive retinal atrophy in dogs.

In man, eugenics has not progressed very far, but there are now a number of diseases known to be of genetic origin. Some are caused by chromosomal abnormalities, such as Down's syndrome. Others are caused by particular genes inherited in Mendelian fashion, such as haemophilia and Huntington's chorea. This knowledge has made genetic counselling possible, for older women who are more likely to bear children with chromosomal abnormalities and for prospective parents known to be carrying genes for hereditary diseases.

Molecular Biology

Increased knowledge about chromosomes, genes, and proteins led eventually to the elucidation by Watson and Crick in 1953 of the double helical structure of deoxyribonucleic acid (DNA), which has led to a great surge of interest in molecular biology and the systems by which information coded in DNA is communicated to the rest of the cell. The experimental work depends on animal tissues as sources of the biological material analysed but probably involves very little experimentation on live animals.

Genetic Engineering

Enough is now known about the structure and function of chromosomes and the genetic information contained therein for experiments to be possible in which bits of genetic material are transferred from one organism to another (genetic recombination). So far, this has been done most successfully with bacteria, but there is the possibility that animal genes might be transferrable from the cell nucleus of one species to another. Indeed, the possibilities of the techniques are endless if the technical problems can be overcome. So far the applications have been mainly in the field of agriculture, e.g. producing plants which have the characteristics of nitrogen fixation, which is normally found only in leguminous plants, thus reducing the need for the farmer to apply nitrogenous fertilizers. Other applications are in medicine, for instance in the production of bacteria which can synthesize insulin.

There has been public alarm (as well as many jokes and cartoons about the monsters that could be produced) about recombination and the risks of dangerous organisms escaping from the laboratory after, for example, a pathogenic gene had been tagged on to a ubiquitous and harmless bacterium such as that living in the normal gut. Politicians took the alarm seriously and British scientists have agreed to a code of practice and submission of their plans to outside scrutiny, via the Genetic Manipulation Advisory Group (1978).

Immunology

Immunology is an example of a science with two roots; there is the medical science of immunity to disease which was the basis of vaccine production, and there is also the biological science of immunology developed by geneticists and zoologists, notably MacFarlane Burnet and Medawar who investigated the mechanisms by which an organism recognizes material foreign to itself. This work provided the theoretical basis for transplant surgery. Medawar's early work was done largely with mice, transplanting pieces of skin from a white mouse to a black one and observing the rates and mechanisms of sloughing. Before transplant surgery could be performed in man the techniques were tested and perfected in animals, mostly pigs and dogs. Kidney transplants are already commonplace, and the success rate of heart and bone-marrow transplants are high enough for patients to be willing to undergo these operations. Liver and lung transplants are now being performed. The drugs required to suppress the normal immunological responses which would reject the transplanted organ also have to be tested on animals.

PHYSIOLOGY

History of Physiology

The two essential core subjects of biology are anatomy and physiology. Anatomy is the study of structure; physiology is the study of the function of the various organs of the animal (or plant). Anatomy can be studied on dead specimens but physiology is dependent on experiments performed on living animals. For example, the structure of the mammalian heart was known in antiquity, but not until 1628 was William Harvey able to show by experimentation that its function is to pump blood round a circuit so that it returns again and again to the same point. Similarly, the structure of the spinal cord and its major nerves were well known by early anatomists and some idea of their function was understood, but it was not until the nineteenth century that the conduction of nerve impulses was studied. The functions of the sympathetic nervous system were recognized by Claude Bernard and Brown Sequard in about 1852 and the methods were developed for the quantitative study of spinal reflexes by Sherrington (1906).

The nineteenth century saw a great growth of experimental work using living animals. Lister's ideas about antiseptic surgery were beginning to gain ground in 1867 and these made the survival of experimental animals after surgery more likely and less painful. The greatest problem was anaesthesia. Until the mid-nineteenth century the anaesthetic in general use for man was a mixture of alcohol and narcotics such as tincture of opium. Ether began to come into use following the work of Long in the USA in 1842. Ether was succeeded for a time by chloroform following the pioneer work of the Scottish surgeon Simpson, and became very popular after Snow was brought to London to induce chloroform anaesthesia for the birth of Queen Victoria's son Edward in 1853.

In theory ether and chloroform were satisfactory anaesthetics for animals as well as man. In practice, ether was found to cause excitation of the animals during induction of anaesthesia and chloroform was found to be rather toxic and to cause too high an incidence of mortality. Experimental physiologists therefore preferred to operate without anaesthetics where possible, particularly on the continent of Europe. British scientists of the time who had observed experiments performed without anaesthetics were highly distressed by what they saw and a petition signed by Charles Darwin, Thomas Huxley, and other leading scientists led to the British Government setting up a Royal Commission 'to enquire into the practice of subjecting live animals to experimentation for scientific purposes'. The Commission's recommendations led to the *Cruelty to Animals Act 1876*, which allowed experiments on live animals only under the control of the Secretary of State, only under anaesthesia, and only for the pursuit of useful knowledge.

The situation was made easier by the discovery of chloral hydrate as an anaesthetic for horses in 1886 and improvements in the method of administering ether, but the biggest step forward was taken in the 1930s with the development of the injectable barbiturate anaesthetics. Knowledge of anaesthetics and the ability to produce balanced, reversible anaesthesia in experimental animals is now recognized as being essential for anyone undertaking animal exerimentation which involves surgery. What is still uncertain is the role of analgesics to relieve post-operative pain.

Pain

The drug most widely used to control severe pain is still morphine, an alkaloid derived from opium poppies. In the last few years great progress has been made in the understanding of pain and how it is relieved. Peptide hormones have been discovered within the brain which bind to special receptors. It is thought that the receptors are the same ones that bind morphine, so the peptides have been called endorphins or enkephalins. They blunt the sensation of pain, as does morphine, and their production appears to be partly under the control of the brain. The effects of morphine and its endogenous analogues on the nervous system have been studied in various ways. For example, the guinea-pig ileum with its associated nerve plexuses is a good model for the action of opiates on smooth muscle. Brain and nerve fragments removed from animals are used to study the mechanisms whereby the opiates are bound to nerve endings. Tissue cultures of brain cells are used for studying the effects of opiates on nervous transmission. Man is used to study the effects of morphine antagonists on perception of pain (see Kosterlitz, 1977). It is hoped that the relationship between painful stimuli, the perception of pain, and the level of consciousness may soon be better understood and that rational methods of preventing and controlling pain in animals and man will be developed.

Alternatives to Whole Animals in Physiology

Most lines of physiological enquiry started off using whole animals. Cats and dogs were popular because of their convenient size and their toughness. (Rabbits do not survive anaesthesia and surgery so well.) Now rats and mice are used more frequently. For teaching the basic principles of physiology to undergraduates a few of the larger animals are still used. For example, to demonstrate the nervous control of cardiac output and blood pressure, the nerves and major arteries of the neck are exposed in the anaesthetized cat or ferret, then the nerves are stimulated to record the effects on blood pressure. The animal is killed at the end of the experiment before it recovers consciousness.

Current research is often carried out using tissues removed from animals during deep anaesthesia or immediately after death. For instance, the properties of nerve impulses and the passage of impulses across nerve–muscle junctions can be studied in nerve–muscle preparations taken from frogs or rats. The absorption of nutrients from the small intestine can be studied *in vitro* by removing a length of intestine from a rat and placing it in an oxygenated salt solution resembling tissue fluids. If the length of gut is inverted before placing it in the tissue-bath it is easier to study the movement of nutrients than it would be *in vivo*. The effects of a hormone on its target organ, e.g. the effect of oxytocin on the uterus, is most easily studied *in vitro*, but a whole animal is necessary to discover what other effects the hormone may exert. *In vitro* methods may be more precise than *in vivo* methods but they cannot be used until quite a lot is already known about the phenomenon in question (see Smyth, 1978).

In some cases sophisticated recording methods have made the use of surgical interference unnecessary for recording blood pressure, muscular contractions, and so on, thus permitting the use of a human experimental subject.

Physiologists have always been dependent on the theories and methods developed in chemistry and physics. Now biochemistry and biophysics are sciences in their own right, in which animals are used mainly as sources of material for study.

Human and Animal Physiology

Human physiology traditionally has a close relationship with medicine because of its importance in the training of medical students and also because human disease conditions can throw light on physiological problems. Progress in endocrinology, for example, is greatly assisted by the clinical findings in patients with hormone deficiencies or over-production.

Animal physiology is studied using many of the same techniques as in human physiology. The ways in which insects, fish, birds, and mammals solve the problems of respiration, circulation, locomotion, and reproduction are of great intrinsic interest, and some of the knowledge gained has unexpected applications. For example, it has been known for a long time that bats can 'see in the dark' using an echo-location system involving the production of ultrasonic squeaks. Methods were developed for recording and analysing these high frequencies and using them it was discovered that several species of rodents also produce sounds which are outside the human range of hearing. These are often alarm calls and by tape-recording and playing them back it is possible to influence the behaviour of wild rodents (Ministry of Agriculture, Fisheries, and Food, 1978).

Other mammals given special study are the diving mammals such as seals and otters which have adaptations which allow them to spend much longer below water without breathing than land mammals. Desert mammals have specially adapted kidneys which allow them to survive on a much lower water intake then mammals in temperate zones. The systems by which dolphins, whales, and fish communicate under water are still little understood but could be of great practical interest.

Applications in Disease and Pest Control

Some species are studied because of their medical importance to man, particularly the parasites and disease-carrying insects. The knowledge of the life cycle of the *Anopheles* mosquito and its dependence on still water for reproduction is just as important for the control of malaria as is the study of the effects of the anti-malarial drugs on the protozoan *Plasmodium* in the blood system of the patient.

Other species are studied because of their importance in agriculture, for example, plant pests. It is important to understand their life cycles, food preferences, and their natural predators, so that pest control schemes can be accurately aimed at them. Rats and mice are an age-old problem. The ancient Egyptians domesticated the cat to patrol their grain stores for them. The burghers of Hamelin employed the Pied Piper to rid the town of rats. Their control became even more urgent when it was realized that rat fleas carry *Pasteurella pestis*, the causative organism of plague. In more recent times a wide variety of poisons has been used, from strychine through to the anti-coagulants such as dicoumarol. New substances are investigated as the animals develop resistance to the old ones. The testing of these compounds is bound to be distressing to the experimental animals involved as well as to the biologists observing their reactions and poses a particularly difficult ethical problem. A possible way out has been studied by the Ministry of Agriculture, Fisheries, and Food: the use of chemosterilants, that is substances which, when ingested in a bait, have no painful effects but render the male or female, or both, permanently or temporarily sterile. Several have been tested in field trials with some success. Unfortunately, the more effective the substance is, the more hazardous it is to other animals or birds which mistakenly eat the bait (Ministry of Agriculture, Fisheries, and Food, 1978).

Physiology of Farm Animals

The physiology of domesticated livestock is of practical interest and application. Farmers can benefit from knowing, for example, what the factors are in the growth of wool, or the critical periods in the incubation of hens' eggs. One subject intensively studied has been the production of milk; the mechanism by which the constituents of milk are secreted from the blood into the milk tubules of the udder is a challenge in itself. It is also of practical significance in programmes to increase milk yields, or increase the proportions of fat or protein in it. An ingenious system was devised to study the mechanisms by Linzell (1963). One half of the udder of a goat was transplanted from its normal position to an area under the neck. New blood vessels developed and milk was secreted normally, but the nerve supply was not re-established. There is a mechanism whereby milk is released or 'let-down' when the udder is stimulated by sucking or milking. This mechanism fails in the transplanted udder, thus making measurements possible which are difficult in the normal udder. The goats survived the operation well, and although they looked rather grotesque to human eyes the goats themselves did not appear to mind. The technique is still being used and useful results are still being obtained. In deciding what experiments should be permitted should one respond to human emotional reactions or try to understand the reaction of the animal?

Cell Biology

The life of biologists was made easier by the invention of the microscope and indeed bacteriology would have been impossible without it. The nature of the animal cell as perceived by zoologists has altered as the means of studying it has improved. At first it was seen as a bag of protoplasm, but now as a highly structured package of organelles with even the structure of the membranes made visible by electron-microscopy. The study of cell function has also changed as smaller fractions of the cell can be extracted and analysed. Cell biology is now a science in its own right and has been greatly aided by techniques of tissue culture by which single cells can be encouraged to multiply in nutrient media. Tissue culture has had important applications in cancer research and in virology. Viruses will grow in cells in culture and can be harvested for the production of vaccines—the most notable achievement in this field has been the use of cultures derived from monkey kidney tissue for the production of polio virus vaccine, and the use of human cells in culture for the production of rabies vaccine. By and large tissue culture cannot be a substitute for a whole animal for experimental biology. The technique has opened up opportunities in new fields but has rarely made experimental animals obsolete in old fields.

A microbiologist who wanted to know about all the species of bacteria and yeasts would have an enormous task. They are ubiquitous and life as we know it would be impossible without them, from the yeasts which ferment grape-juice, to the bacteria which we hope will cause the decomposition of spilled oil on sea-beds; but most microbiologists are employed to study just the few which cause disease in man or his domestic animals.

VETERINARY MEDICINE

History

Just as human medicine is heavily dependent on observations and tests made on animals, so veterinary medicine benefits from the remedies and appliances developed for human use. This interdependence of man and his animals was well understood by the Ancient Egyptians, whose physicians were expected to treat both classes of patients, but became forgotten during the Dark Ages when medical studies declined. In Europe animals were treated by farriers and marshals until veterinary schools were established. In the eighteenth and nineteenth centuries the first veterinary surgeons were employed by the military to control contagious diseases amongst the large numbers of horses and mules kept for transport and in the cavalry. They then turned their attention to cattle, sheep, and pigs and later on to companion, zoo, wild, and laboratory animals. Veterinary science was greatly helped by the French chemist Louis Pasteur who, apart from his revolutionary contribution to the germ theory of disease, also made particular studies of fowl cholera and rabies. A little later Robert Koch, via the Koch postulates, put the relationship between micro-organisms and disease on to a scientific footing and also made particular studies of anthrax and tuberculosis, showing them to be zoonoses, that is, diseases which can be transmitted from one species to another, including man (see Reid, 1974).

Veterinary research into animal disease and its treatment is carried out partly on the target species, that is the species which will benefit from the research, and partly on small laboratory animals. Clinical research can be carried out on sick or injured animals with fewer ethical complications than in man. The ethics of veterinary research are particularly perplexing to people who believe that animals should not be exploited to benefit man. Anti-vivisectionists who avoid taking medicines themselves will often present their sick animals to a veterinary surgeon and ask for treatment that has been developed as a result of animal research. It is unfortunately true that remedies for particular conditions can only be developed at the cost of some disease or pain in animals. A safeguard should be that the veterinary surgeons concerned are, by reason of their training, more aware of the signs of pain in animals and the means of relieving it, than scientists trained in other disciplines.

Research in Reproduction

A part of veterinary medicine is concerned with increasing the productivity of domestic animals and thus is directed to the welfare of the farmer and the community rather than the animals. An example of this is artificial insemination, a technique by which semen from particularly valuable bulls can be collected, diluted, stored, and then inseminated into a far greater number of cows than the bull could have covered by natural service. Recently the development of the technique of ovum transplantation and its commercial applications have caused concern amongst those responsible for veterinary ethics. In the 1950s Rowson and his colleagues at Cambridge developed the techniques necessary to remove recently fertilized eggs from the oviduct of one animal and transfer them to another (for a review see Rowson, 1971). Laboratory animals were used first, then sheep. It was found that the ova would develop in their new environment and that normal offspring were produced. If, for example, an embryo from a large breed of sheep was transferred to the uterus of a ewe of a small breed then the lamb might be born small but would soon show the growth characteristics of the larger breed. The value of this research tool for studying implantation, pregnancy, and uterine behaviour was enormous. Its applications in livestock improvement were seen to be interesting although technically difficult.

Several problems had to be overcome. Firstly, the ewe had to be made to superovulate to produce enough ova to make it worthwhile to collect them. The necessary gonadotrophic hormone was produced first from the pituitary glands removed from cattle, sheep, and horses at slaughter. Later they were obtained from horse blood. At a certain stage of pregnancy mare's blood contains a high level of gonadotrophic hormone and by regular bleeding from the jugular vein sufficient quantities can be extracted from the serum (pregnant mare's serum, PMS). Secondly, the fertilized ova had to be flushed out of the oviduct and collected and this required that the uterus was exposed by laparatomy and that a fine cannula was introduced into the ovarian end of the oviduct so that the flushing fluid could be injected and then sucked back into collecting cups. Then the ova had to be stored for transport, or until the recipient ewes were ready to receive them. A good tissue culture medium was developed for this purpose, but it was found early on that a convenient method of storage was to place them in the oviduct of a living rabbit. The embryos of Dorset Horn sheep were exported to South Africa in rabbits, then subsequently recovered and transplanted into native ewes with great success. For the transplanting to be successful the recipient ewe must be in the correct hormonal state, i.e. the same state as the donor ewe. This may be achieved by synchronizing the donor and recipient by the use of drugs; progesterone was used first, but more recently prostaglandins. The ova are transplanted into the uterus either via the cervix, as in artificial insemination, or via a surgical incision in the flanks; the surgical method gives a better success rate. The method was soon developed in cattle, pigs, and horses and fertilized ova were shipped halfway round the world to be transplanted. The economic advantage to farmers wishing to improve their stock by the introduction of an exotic breed or strain was soon obvious and units were set up in North America, Britain, Australia, and New Zealand to perform the necessary operations (Dawson, 1976). In October 1979 some English heifers pregnant with Canadian Holstein calves implanted by embryo transfer were auctioned at Chester market and averaged over £1000 each (Anon., 1979a). The techniques are now established but can lead to abuse if surgery is not aseptic, or if the same animals are used repeatedly as either donors or recipients. The same scientists who developed the techniques became worried about the possibility of commercial exploitation of the animals and asked that embryo transfer should be subject to statutory control. The situation was studied by the Royal College of Veterinary Surgeons and the British Veterinary Association and they concluded that embryo transfer was an acceptable procedure aimed at the better utilization of resources, but recognized that it could be used mainly as a commercial expedient and that the animals' welfare needed to be safeguarded by adequate controls. In particular it was suggested that a recipient animal should be used only once and that a donor animal (which is likely to be very valuable as a source of high quality ova) should be used only at certain time intervals (British Veterinary Association, 1976). Since controls would have to be enforced by the Ministry of Agriculture, Fisheries, and Food, the Ministry was asked in 1976 to make the necessary regulations; in 1979 they began drawing up a code of practice (Anon., 1979b).

Veterinary science is forever extending its boundaries. Much is now known about the diseases of wild animals and birds and this knowledge is utilized in game parks and in conservation projects. More is now being learned about fish, and this knowledge is being applied as fresh and sea-water fishing becomes a matter of husbandry rather than hunting.

Laboratory Animal Science

Rats, mice, guinea-pigs, and other laboratory animals, which are so commonly used as models for man, have also been studied in their own right. A great deal is now known about their anatomy, physiology, genetics, nutritional requirements, and the environmental conditions which are optimal for their health and well-being. Techniques have been developed to produce these animals free from their major diseases and even free from all other forms of life such as skin- or gut-associated bacteria. The pioneering publication in this field was *The UFAW handbook on the care and management of laboratory animals* (Worden, 1947), which is now in its fifth edition (UFAW, 1976).

At about the same time the Medical Research Council's Laboratory Animals Centre at Carshalton was set up to improve the quality of the animals available for medical research in Britain. Disease control was the most urgent problem and this was achieved through the Accreditation Scheme which is discussed in another chapter. The study of the naturally-occurring diseases of laboratory animals is a branch of veterinary medicine. The veterinary profession has been slow to shoulder its responsibilities in this field except in the United States of America where laboratory animal medicine has become an important speciality. These diseases are still not so well understood as those of domestic animals, and the definitive text has still to be written.

CURRENT BIOLOGICAL RESEARCH

What of current biological research in the non-medical and non-veterinary fields? An analysis of papers published in recent journals shows that there are 80 subject headings under Biology, about 30 of which are likely to involve experiments on living animals (*Biological Abstracts*). Under these 30 headings papers are most numerous in pharmacology, neoplasms, cardiovascular research, nervous system, toxicology, blood and blood-forming organs, developmental biology, behavioural biology, endocrines, immunology.

Pharmacology is of obvious importance and application to medicine and the pharmaceutical industry. Neoplasms, i.e. cancer research, is discussed in a separate chapter. Cardiovascular research and the nervous system are important components of physiology. Toxicology was the boom science of the seventies.

Toxicology

Toxicology was originally the study of the pathogenic effects on man and his domestic animals of poisons of plant origin, such as curare and digitalis. Now it is largely concerned with the effect on experimental animals of a variety of substances ranging from new medicinal substances to industrial and agricultural chemicals. Most of the papers published in the journals come from medical schools and research institutes and describe the results of epidemiological surveys of disease, new techniques, or results of general scientific interest. The animals most commonly used, apart from man, are species of fish, such as trout.

The work reflects the academic and political importance of industrial disease and the pollution of the environment.

A comparison between the number of papers published in the field of toxicology and the statistics issued by the Home Office (1978 and 1979) shows a great discrepancy. This is because commercial companies carry out a large number of experiments on the toxic effects of their products and most of their results are not published. Many of these will be tests using standard techniques or on substances of limited interest. But some may be using techniques which could be of interest to others, or testing substances which are of interest to others, but whose properties are a commercial secret. By the same token, wasteful or trivial work could continue without other scientists being aware of it. The science of toxicology is taught at very few universities and there is a danger of its practical applications running ahead of its theoretical basis. One of the standard tests carried out by the toxicologists is the LD_{50}, and this is discussed in a separate chapter. Tests performed on medicines are described by Jolly and Somerville (1973).

One of the basic problems in toxicology is that laboratory animals are used as substitutes for man although they differ from him in size, structure, metabolism, and behaviour, and in their methods of detoxicating and excreting drugs and medicines. No one animal can be a good model for man except in one or two respects, e.g. the immunological responses of the guinea-pig and the female reproductive system of the marmoset. For example, the rat's stomach is less acid than man's and so substances taken by mouth are liable to react differently. Smoking tests on beagles suggested that the dog lung responds differently from that of man when exposed to tobacco smoke, and this could have been predicted from the differences in their anatomy. The behavioural responses of cats are distinct and useful for studying the effects of psycho-active drugs, but the metabolic response of cats to drugs is different from that of man, and indeed of most other mammals. For discussion of these problems see Spiegel (1973).

In some areas the need to find new test systems which will give useful results is recognized as pressing. One of them is cosmetic testing where skin tests on thin-skinned hairy laboratory animals cannot truly predict the likely response in the hairless, but relatively thick, human skin. *In vitro* systems using human skin in tissue culture are being investigated.

Agricultural chemicals are tested as far as possible on the types of animals most likely to be affected in the field: birds, fish, and insects. Even here there are complications because the well-fed creature in the laboratory may react very differently from its half-starved counterpart outdoors during the English winter.

Alternatives to Whole Animals in Toxicity Testing

In vitro systems are already in widespread use for preliminary screening of new substances. Those carcinogens which act by causing genetic damage can be detected in cell systems, e.g. the well publicised Ames test which uses bacteria, and mammalian cell systems using fibroblasts and macrophages. Substances failing these tests can then be eliminated from further testing programmes on animals. Other systems are being developed using heart tissue or whole embryos to test the effects of substances on cardiac action and the development of the fetus. *In vitro* systems have the great advantage that human tissue can be used, obtained from aborted fetuses, placenta, material removed at surgery, or donated material. They do not cut out the use of animals because animal serum is still required to support their growth, there not being enough human serum available in blood banks for this purpose. In any case, human serum is too often contaminated with medicants, alcohol, nicotine, and caffeine (see Dawson, 1977).

Toxicology is of particular interest to the public because of the conflicting demands of safety to the consumer and welfare of animals. It appears likely that the substances most dangerous to the consumer will be those which cause most distress to the animals. In the past, man was his own experimental animal, and boy chimney-sweeps developed scrotal cancer and miners died of pneumoconiosis as a result. Now we value human life more highly and are not prepared to allow our fellow-men to take such risks for our benefit. A certain level of testing in order to safeguard the public is seen as necessary; but where should it end? The present trend appears to be towards a no-risk society where the main hope for the animals will be the recognition that animal experiments are themselves too risky for scientists to undertake!

A FEW QUESTIONS ABOUT EXPERIMENTAL BIOLOGY

Is Biological Research Necessary?

Do we need to know more about the world we live in? Has scientific curiosity taken us far enough already? This question is unanswerable because we cannot know what new facts are waiting to be discovered. Do we need the new products being developed for medical and household use? The collective answer of present society is yes, because the western world is progress-orientated, but grant-aiding agencies are becoming very selective about the biomedical projects they support as the money runs short; and shoppers could say no by being more critical of the goods they buy, since they have the choice of either established products or the new preparations which require testing on animals. Clearer labelling would help the consumers in making their decision.

Some Specific Questions

Are the right animals being used? Animal species need to be carefully selected, whether for study in their own right or as models for human disease or for safety testing. Rational selection depends on a great deal being known about the available species, and this itself requires a lot of experimentation. Special care must be taken to avoid species which are endangered in the wild, or cannot be bred in the laboratory, or do not adapt well to the restrictive life of the animal house. The trend away from wild-caught primates is to be welcomed. Mice and rats have adapted particularly well to captivity and the new inbred strains being developed are useful models for an increasing number of purposes.

Are too many animals being used? The ways in which the number of animals used in a particular experiment can be reduced is discussed in another chapter. Overall, and for the foreseeable future, the number of animals used will depend on the number of biologists employed in research and in commerce, and will thus be pegged to the scientific and commercial activity in the nation as a whole. While the number of animals required for some purposes will fall, for example diagnosis of disease and pregnancy, the numbers used in the safety testing of industrial chemicals will rise. Alternatives to live animals are being sought in many areas, and funds provided by charitable societies have aided the search, but there is no prospect of a rapid switch from animals to *in vitro* research (see Smyth, 1978).

Are there experiments which should never be performed? Scientists in Britain work within the confines of the *Cruelty to Animals Act 1876* and so are prevented from performing experiments which would cause unjustifiable pain. Defining 'justifiable' can be a problem. Should primates be used for studying the withdrawal symptoms from morphine addiction? In Britain this work would probably not be permitted, but it is carried out in other countries. Unnecessary safety tests prescribed in regulations, but causing pain, are infringements of the 1876 Act, but no case has yet been tried in the courts. Experiments performed in the name of the Crown, e.g. weapons research, are not bound by the 1876 Act; the results are not published either, so the scientific community is not able to assess them.

Are experimental animals looked after properly? Legislation alone cannot ensure that the people responsible for the care of animals possess the necessary knowledge and skills. The ideal experimental animal is well-fed, comfortably housed, free from disease, and used to being handled. The last requirement becomes more difficult as manning levels fall. Greater awareness of the needs of laboratory animals is necessary amongst those responsible for allocating resources so that they will see the sense of investing a greater proportion of their finances and manpower in the animal house and experimental laboratory.

Are scientists adequately trained for animal work? Veterinary graduates have some knowledge of an animal's needs. Medical graduates know about fragility of tissue and the prerequisites for health and healing, but do not necessarily have a sympathy with animals. Biologists may be presumed to have a respect for life but will not necessarily have handled mammals in their undergraduate courses. Scientists of other disciplines coming to animal work may regard their experimental subjects as essentially 'test-tubes with whiskers' which, unfortunately, have to be fed. Some sort of basic training in animal handling and experimental procedures could benefit all categories.

Can scientists discuss ethics? The case against the use of animals in biological research is largely a philosophical one (see Clark, 1977). Scientists are rarely taught philosophy or ethics and thus they may feel at a disadvantage when discussing the 'rights' of animals. This is a pity. We need scientists who are prepared to look at experiments from an ethical as well as an academic or practical viewpoint.

REFERENCES

Anon. (1979a). *Veterinary Record*, **105**, 358.

Anon. (1979b). *Veterinary Record*, **105**, 449.

Biological Abstracts, Biosciences Information Services, Philadelphia, published monthly.

British Veterinary Association (1976). *Annual Report*, **1975–6**, 21.

Clark, S. R. L. (1977). *The moral status of animals*, Clarendon Press, Oxford.

Dawson, F. L. M. (1976). 'Review of reproduction and infertility', in *The veterinary annual*, 16th issue, Scientechnica, Bristol, p. 273.

Dawson, M. (1977). 'In vitro systems in basic biomedical research', in *The future of animals, cells, models and systems in research, development, education and testing*, National Academy of Sciences, Washington, DC.

Genetic Manipulation Advisory Group (1978). *Nature, London*, **276**, 103–108.

Home Office (1978 and 1979). *Experiments on living animals: statistics 1977 and 1978*, HMSO, London.

Jolly, D. W. and Somerville, J. M. (1973). *Medicinal tests on animals*, Association of Veterinarians in Industry, available from Huntingdon Research Centre, Huntingdon, Cambridgeshire.

Kosterlitz, H. W. (1977). 'Pharmacological advances in analgesics', in *Pain: new perspectives in measurement and management* (A. W. Marcus, R. Smith, and B. Whittle, eds), Churchill Livingstone, Edinburgh.

Linzell, J. L. (1963). 'Some effects of denervating and transplanting mammary glands', *Quarterly Journal of Experimental Physiology*, **48**, 34.

Medawar, P. B. and Medawar, J. S. (1978). *The life science: current ideas of biology*, Granada Publishing, St Albans.

Ministry of Agriculture, Fisheries, and Food (1978). 'Biology and control of rodents', in *Pest Infestation Control Laboratory report 1974–76*, HMSO, London.

Raleigh, I., Scott, P., Jackson, E., and Jackson, O. (1976). *Practical guide to cats*, Hamlyn, London.

Reid, R. (1974). *Microbes and men*, British Broadcasting Corporation, London.

Rowson, L. E. A. (1971). 'Egg transfer in domestic animals', *Nature, London*, **233**, 379–381.

Sherrington, C. (1906). *The integrative action of the nervous system*, Constable, London.

Smyth, D. H. (1978). *Alternatives to animal experiments*, Scolar Press, London.

Spiegel, A. (ed.) (1973). *The laboratory animal in drug testing*, Gustav Fischer Verlag, Stuttgart.

UFAW (eds.) (1976). *The UFAW handbook on the care and management of laboratory animals* (5th edition), Churchill Livingstone, Edinburgh.

Watson, J. D. and Crick, F. H. C. (1953). 'Molecular structure of nucleic acids', *Nature, London*, **171**, 737–738.

Worden, A. N. (ed.) (1947). *The UFAW handbook on the care and management of laboratory animals*, Baillière. Tindall, and Cox, London.

Animals in Research
Edited by David Sperlinger
© 1981 John Wiley & Sons Ltd.

Chapter 7

The Use of Animals in Experimental Cancer Research

HAROLD B. HEWITT

INTRODUCTION

Human reluctance to accept the certainty of death has fostered an unrealistic notion that the prevention or total control of all fatal diseases will in due course emerge as a product of medical research. This may well be too sanguine a hope in respect of those diseases to which we become progressively more prone as we get older. Among these, heart disease, cancer, and strokes predominate as causes of death. All may be crippling; all may prove to be fatal. Yet, granted a choice between them, it is evident that cancer would be least chosen. This is reflected in the disproportionately large amount of public and freely donated money contributed to the support of cancer research. It is estimated that over $7000 million has been spent on cancer research in the United States alone. Thus, the pursuit of cancer research meets a large public demand. Cancer is regarded as a scourge which can, and should, be removed; and it is unfortunate that this unrealistic presentation of the problem has not always been discouraged by those closest to it.

Some centuries ago, epidemic infectious diseases predominated as causes of death and resulted in an expectation of life which was, in fact, below what is called 'the cancer age'. In those times, the idea of contagion, of invisible agents of disease spreading from the stricken to the healthy, was borne in upon the casual observer long before such agents were proved to exist; but the hunch was irresistible and was soon confirmed by diligent enquiry. The control of infectious diseases by the successive introduction of hygiene regulations, mass immunization, and, most recently, antibiotics represents the golden era of medical science. But it is important to realize that infectious diseases present special features which held out high promise of their eventual control: the causative agents were specific living entities susceptible to destruction; prevention of their access to healthy persons was a conceivable project; the frequency of natural recovery implied that the body possessed natural defence mechanisms which might be artificially evoked or stimulated; and the wide species difference between agent and host made it probable that the two were chemically different and might be expected to display different susceptibilities to selected toxic agents. All of these promising features have been confirmed and successfully exploited, with the result that, in advanced societies, primary infectious disease has become only a minor cause of death. The expectation of life at birth has been dramatically increased; but, inevitably, other causes of death predominate; cancer is among them.

Scientists are not so devoid of optimistic sentiment that they have been able to resist the temptation to force cancer into the biological context which allowed rational prediction that infectious diseases might be susceptible to control. This is to say that very large resources have been devoted during this century to attempts to demonstrate that all or some forms of human cancer are caused by infective agents—in particular, viruses. In the wake of this notion have followed promises of preventive immunization by vaccines and treatment of established cancer cases by 'immunotherapy'. The dominance of this emotionally sustained concept, harking back to the golden era of practically successful medical science, is well displayed in the history of one of the principal cancer research institutions in Great Britain—the Imperial Cancer Research Fund. Three successive directors of the Fund's laboratories were appointed from the field of virology; and in the period 1930–50 a general concept of the virus causation of cancer was pursued and promoted with missionary zeal. It is true that a very small proportion of animal tumours have been shown to be associated with cancer-causing viruses and that a spin-off of valuable information about cell biology has come from the intensive study of such cancers. But a frank evaluation of the influence of the infection/immunity approach at the present time, however subtle its insertion into our present thinking, fails to reveal any indication of direct practical application: no form of human cancer has been proved to be caused by viruses, although circumstantial evidence continues to be strongly asserted; natural recovery from cancer is of extreme rarity (Everson and Cole, 1966); no vaccine has ever been known to prevent the disease; and world-wide clinical trials of 'immunotherapy' have given no encouragement to general adoption of this form of treatment. Thus, the very intensive efforts which have been made during this century to evolve effective preventive or curative measures based on hopeful analogies between cancer and infective disease have been unrewarding. It should be added that the significant metabolic differences between body cells and bacteria which underlie the development of effective chemical or antibiotic treatment of bacterial diseases have not been paralleled by the demonstration of consistent metabolic differences between normal and cancer cells; hence, we have no toxic agents which will kill cancer cells and leave normal tissue cells undamaged. What has become clear is that our fundamental biological knowledge of the orderly normal tissues is insufficient to provide an understanding of their disorder in the shape of cancer. The question to be asked is not: 'How do cancers form and why do they behave as they do?' It is: 'How do the normal cells, many of which are proliferating quite as fast as cancer cells, keep their restrained place in maintaining the normal form and function of the body?' It follows that a large part of modern biological research eligible to rank as 'cancer research' consists of studies of normal growth mechanisms and normal cell structure and function. Clearly, the content of this chapter cannot encompass such an unlimited field of enquiry. I have therefore confined attention to experiments involving the growth of cancers in animals and to effects on animals of the great variety of toxic agents which are in use or under investigation for treatment of the disease. It has to be understood that because we have no method of treating cancer which is not damaging to normal tissues, many modern treatments have a ring of 'kill or cure' about them and animal tests of putative therapeutic agents are commonly used to explore the criticality of the end-point.

The fostering of ever higher standards of humanity in the usage of animals for scientific research requires the development of sympathetic understanding between those who conduct the experiments, who must acknowledge their vested interest, and those whose concern is to preserve and advance ethical standards independently of exigencies claimed for scientific progress. Inevitably this requires communication between the laboratory scientist and lay monitors of his conduct. To assist this communication I have, in the sections which follow, devoted attention to a description of the scientific context in which particular animal experiments are done. The endeavour has been to make them at least meaningful: their justification requires a much wider social reference.

THE NATURE OF CANCER

It is now generally conceded that the critical event which leads to the production of a progressively growing cancer is very rare and occurs within a single cell. The departure from normality to which the event gives rise is irreversible, is inherited by all descendants of that cell over an indefinite number of generations, and (apart from one or two very rare conditions maintained in laboratories) is not transmissible by contagion to normal cells in contact with the transformed cell. Inheritable characteristics are known to depend on unique molecules in the genetic material of a cell. Hereditary diseases such as haemophilia are the manifestation of an abnormal gene which persists in the germ cells and is passed from one generation to the next. It is also known that abnormal genes may arise *de novo* in the germ cells and give rise to new breeds of animals or varieties of plant. Such genetic transformations are known as *genetic mutations*. By analogy, we can conceive that mutations can arise in the cells of the body tissues; they would be called *somatic mutations*. It is a matter of disputation whether cancers are usually initiated from somatic mutations, but the concept has endured for many decades and receives continuing powerful support from the observation that practically all chemical or physical agents which induce genetic mutations are also *carcinogens*. It is this strong correlation which forms the basis of the *Ames test*, whereby the carcinogenic potential of a substance can be inferred from its capacity to induce genetic mutations in bacteria (Ames, 1972).

The one form of somatic cell mutation which could manifest itself is one that induces the cell to undergo proliferation at a greater rate than the normal cells of the tissue in which it arises; and where this is so, some form of new growth, or cancer, is the inevitable outcome.

The chance that a normal cell will transform to a cancer cell is rather smaller than one in one million million, and the deficiency which makes it a cancer cell is loss of 'awareness' or of the ability to 'respond' to that awareness (of its part in the whole organism). Apart from this 'social' deficiency, akin to delinquency, the cancer cell retains all other normal functions: it utilizes nutrients and oxygen, it grows and divides, and as it gives rise to a nodule of cancer tissue it evokes from the surrounding tissue its own blood supply; in many cases the arrangement of cells within the cancer closely mimics that of the normal tissue in which it has arisen. It excretes similar products to those of normal cells and does not secrete any kind of *toxin*, as do many bacteria.

A population of cells which is known to have arisen from one single cell is called a *clone*. Thus, a cancer is an aberrant clone superimposed on the normal body and redundant to its requirements. The size of the clone progressively increases as its cells divide, although the rate of growth is widely different from one cancer to another. In most cases, the primary growth causes no symptoms or signs until it consists of a total population of several hundred million cancer cells. The first indication of its presence is often from the mechanical pressure it exerts on surrounding tissues: a tumour of the brain, by taking up space within the confined volume of the skull, results in a rise of pressure there, giving headache, and the brain may have its blood squeezed out of it with the production of fits and coma; a tumour within the intestinal canal will eventually cause an intestinal obstruction requiring emergency surgery.

Unfortunately, the demarcation between a malignant growth and the surrounding normal tissues is rarely well defined; strands of cancer extending from the main mass *invade* and *infiltrate* the surrounding tissues in a way that obscures definition of the boundary. Hence, radical removal of cancer by surgery requires that the excised part includes a wide margin of apparently normal tissue: cancer of the larynx may require removal of the entire organ; cancer of the femur will nearly always require amputation of the leg. Nevertheless, if the cancer is totally confined to the site of origin, complete eradication of the disease can be achieved by adequate surgery.

The cure rate for surgical treatment of cancer would certainly be very much higher than it is, if cells of the cancer clone were confined to the primary tumour. But this is rarely the case. It has been estimated that, of the patients treated for cancer, two out of three die of their disease; and it has to be allowed that a fair proportion of those who do not, die of some other disease before failure of treatment becomes manifest. A large part of this death rate is due to extension of the disease to sites of the body remote from the primary tumour. The cells composing an organ such as the liver are very tightly adherent to one another, to the extent that even when a piece of liver is minced, teased, and treated with a variety of digestive enzymes, it is exceedingly difficult to separate off a single, intact, viable liver cell. It has long been appreciated that whereas many cancers may appear to have the same coherent solidity as liver, most will respond to disruptive manipulations by releasing single, viable, cancer cells. In accordance with this defective cohesiveness, it is known from studies of human and animal cancer that individual viable cancer cells very commonly separate off from primary tumours during their growth and are carried away in the blood or lymph vessels. A large majority of these disseminated cells are destroyed in the blood by mechanical trauma. A minority are arrested in the lymph or blood vessels, settle down there, and proliferate into fresh tumours. Thus, *secondary tumours*, or *metastases*, become established in parts of the body remote from the site of the primary tumour. At this stage of the disease, no local treatment of the primary tumour can cure the patient of the disease. The secondary tumours have the same structure and grow at approximately the same rate as the primary tumour, of which they can be regarded as 'seedlings'. Calculations from the time course of the disseminated disease in humans, and the results of animal experiments, show that in the majority of cases of disseminated cancer the process of dissemination begins before the primary is large enough to cause symptoms or can be detected by the most modern diagnostic techniques. Half the cases of cancer of the breast or of cancers arising in the head and neck region have evidence of secondary spread at the time the patient first reports with symptoms of the primary tumour. The sites at which secondary tumours arise show some variation with the type of cancer, but common sites are: the lymph nodes (commonly called 'glands') draining the site of the primary tumour; the lungs; and the bone marrow. Infiltration of the bone marrow by secondary disease commonly becomes so extensive that the production of normal blood cells is grossly depressed and the victim develops anaemia and infection; involvement of bone may also cause pain.

Leukaemia is different from other forms of cancer in that a primary tumour does not form at the site where the first cancer cell arises: the cells of the leukaemia clone become disseminated as soon as proliferation begins, and the first signs of the disease are of a constitutional nature—principally anaemia, multiple haemorrhages, and a high susceptibility to various local and septicaemic infections. Leukaemia accounts for one-third of the cancers arising in childhood.

The purpose of the above outline of the nature of cancer is to give some understanding of the size and difficulty of the problem confronting cancer researchers. It is being tackled at all levels. In some fields of enquiry there is little place for animal experiments; in others, it is difficult to see how research can continue without them. It will be appreciated from subsequent sections that the use of animals to supplement observations on man is not exclusively determined by ethical differentiation in relation to the species. A very large part of contemporary cancer research would not be *technically* feasible using members of our own species. Indeed, if the investigator were free of all ethical restrictions and were presented with an indefinite number of normal or cancerous human volunteers for his experiments, he would have considerable difficulty in devising a programme of research which could take advantage of these exceptional facilities.

THE LIMITATIONS OF CLINICAL OBSERVATIONS CONDUCING TO ANIMAL EXPERIMENTATION

Very great restrictions to understanding the biology of cancer are imposed on clinical doctors and on veterinarians, both of whom encounter miscellaneous cases of cancer who come, or are brought, to them as patients. Their role is to do their best for the individual cases in the light of existing knowledge and facilities for treatment. It would be contrary to the ethics of both professions to employ their patients merely to gain information of possible benefit to others, although the practical experience they gain in their dedication to individual welfare will certainly accrue to general advantage.

To understand how great this restriction is, especially in human cancer, some knowledge is required of the time scale of the disease. As already described in the preceding section, cancers result from the proliferation of a single transformed cell. It is now known that the rate of increase in the number of cells in a cancer clone follows a simple pattern. For any given cancer, the number of cells in it are doubled in successive equal intervals of time; the rate of growth can be defined by the doubling time (DT). In any species there is quite a wide range in DT values between different cancers. An average DT value for human cancers is 2 months. Thus, only two cells will be present 2 months after initiation of the cancer, four cells after 4 months, eight cells after 6 months, 16 cells after 8 months, and so on. Having in mind the very small size of one cell, it would seem that most of a lifetime would elapse before a sufficient number of cells has accumulated to give a visible tumour. We can, of course, calculate the tumour size to be expected with time elapsed since initiation of the tumour: after $3\frac{1}{2}$ years it would be only 1 mm^3 in size—quite symptomless and undetectable by any method of diagnosis available today; only after 5 years would it reach a mass 1 cm in diameter, and whilst it would then be detectable if it were in some directly observable site such as skin or tongue, it is likely still to be undetectable and symptomless in most internal sites. If the tumour were not treated, it would be expected to weigh 1 kg in a further 18 months, and would then certainly be threatening life even if it were so sited as not to interfere with a vital structure. Such is the time scale of the disease in an average case (DT = 2 months). The fastest growing tumours (DT = 1 week) would reach 1 cm in diameter after only $7\frac{1}{2}$ months. On the other hand, a very slowly growing tumour (DT = 1 year) would not be expected to reach this size until *30 years* after initiation. If the growth rate of a patient's tumour could be easily determined by some short investigation to be made about the time of diagnosis, that information would be of considerable assistance to the clinician in his consideration of what causative influences might have been involved, what effects might be expected from treatment, and what prognosis might be advised. As it is, since the disease is progressive, the therapist feels an obligation to initiate treatment as early as possible; there is usually no opportunity to determine growth rate; and once treatment has been given, the effects of treatment obscure any measurements of it that might be made. It will thus be appreciated that the therapist who is presented with a new case of cancer is unavoidably ignorant of a great many features of the disease which are most relevant to its history and potential: he does not know when the disease was initiated and cannot review the appropriate period of the patient's life in which environmental factors may have conduced to the disease; he can seldom predict the response to treatment or foretell the expected time of recurrence if treatment fails to cure—both of which features are strongly correlated with growth rate; if evidence of secondary spread of the disease is not already clearly manifest, he cannot tell whether it has occurred; above all, he cannot select the treatment most appropriate to an *individual* patient and can refer only to statistical information concerning the overall response of a large number of miscellaneous cases of the disease. Optimum management of the individual case at the present time is beset by the difficulty that the results of treatment, specified in terms of the length of disease-free survival, is the resultant of two independent influences on prognosis—the intrinsic growth potential of the cancer and the proportion of cancer cells killed by the treatment administered (on the understanding that nearly all the contemporary forms of cancer therapy act by removal (surgery) or killing (radiotherapy or chemotherapy) of a fraction of the cells in the cancer). The very large but indeterminate influence of the intrinsic growth rate of a cancer can be appreciated from the widely different survival times of patients who have not received treatment but have remained under observation. Bloom (1968) reviewed the survival times of 250 cases of cancer of the breast which received no specific treatment: no patients recovered spontaneously; half survived for over 2 years; over 10 per cent lived for 6 years; and two patients lived for over 15 years. Attempts to assess or compare the effectiveness of treatments superimposed on such an astonishingly wide range of disease potentials cannot possibly yield discriminating information about the relative value of different therapeutic measures.

It follows from the above account of the influence of case individuality on treatment effectiveness that clinical trials provide very limited prospects for deciding the relative merits of two rival treatments. Bross and Blumenson (1971) have calculated that over 10 000 cases of cancer of the breast would be required to discriminate the relative values of radical surgical procedures and much more limited surgery carrying a smaller risk of complications. It should be added that the conduct of a formal clinical trial is an expensive undertaking which makes heavy demands of medical, nursing, technical, and administrative staff. It is unlikely to yield definitive information within 5 years after the entry of patients into it, and a period of some years may be required for its planning. Recent publication of the results of a clinical trial of a modification of radiotherapy technique emerged *15 years* after planning was begun (Windeyer, 1978). Considerable hazards to patients may be encountered from clinical trials of a new drug: it may prove to be less effective than an existing treatment with which it is being compared, and this will not be apparent until many patients have been inferiorly treated; serious toxic effects, even carcinogenicity, may not appear until some time after treatment has been given.

This extended account can leave no doubt that development and trial of therapeutic innovations for cancer present formidable problems and hazards when the response of patients is the measurement used. This is so even when preliminary pre-clinical studies have been exhaustive and have included experiments in animals. It can well be imagined that if animal experiments were excluded, few clinical therapists would be prepared to depart from existing forms of treatment; their limitations would be accepted for the advantage of knowing the hazards and complications they entail; the prospects of progress would be dimmed.

It is in the sphere of therapy that the most convincing case can be made for resort to research using animals. And I would repeat here that this resort is not just by way of subjecting animals to the risk of injury so that humans can avoid it.

The fact is that suitably designed experiments using animal cancers can entirely eliminate the technical limitations to useful osbservation which are presented by a series of human patients or, for that matter, by a series of animal patients presented to the veterinarian in general practice. As will be explained in more detail in the following section, suitably bred colonies of small rodents enable the investigator to assemble a series of genetically identical animals bearing transplants of the same tumour in the same site of the body; in this way is avoided the very wide range of tumour growth potentials and time factors which characterize a series of patients. It becomes possible to compare the effectiveness of two treatment schedules applied to separate but identical groups of tumour-bearing animals, and a statistically significant advantage of one treatment over another can often be established using a relatively small number of animals. Another considerable advantage in the use of small rodent experimental cancer systems accrues from the much more rapid growth of cancers in the smaller mammals. Different cancers in any species of animal exhibit a range of growth rates or doubling times; but the *average* tumour in a mouse, for example, grows 25 times faster than an average human cancer. What this means is that an effect of treatment which it takes 5 years to measure by clinical trial can be measured in

2–3 months in the experimental system. It is also possible to include in a therapy study a group of cancer-bearing animals (controls) which receive no treatment, thus permitting an absolute assessment of the effects of treatment. Although such an untreated control group would not be ethically allowable in a clinical trial, it must not be inferred that animal controls need to be allowed to suffer the extremities of terminal disease. Whilst it is true that far too many experimenters still employ death of animals as an end-point in cancer therapy studies, such a callous and crude practice is neither a sound nor a necessary part of the scientific requirements and in my view should be discouraged. What must be acknowledged, however, is that the use of cancer-bearing animals for explorative comparisons of different therapeutic techniques does indeed enable accurate and repeatable information to be obtained within a matter of months, and that to dispense with this valuable facility would undoubtedly delay clinical application of scientific innovations for many decades.

What is commonly questioned is whether the likeness between different species (even of the same class, such as mammals) is sufficient to give us confidence in applying to any species information gained from observations made in another. No general answer can be given; but it is fair to say that *comparative* anatomy, physiology, pathology, etc., have been subjects of intensive study for some centuries by physicians and biologists and that those with some awareness of the accumulated knowledge are mostly alert to the *kind* of differences between species which are to be given significance when questions of relevance to man are posed by observations in animals. The wide difference of time scale in the progression of cancer in man and mouse has been fully accounted in preceding paragraphs and has indeed been referred to as conferring a particular value on therapy experiments using mice. What animal experiments have enabled us to do is to determine the *kind* of relationship which exists between the dose of a therapeutic agent and the proportion of cancer cells whose reproductive capacity is destroyed. It is now certain that this is not subject to wide species differences; indeed, it has been established, for example, that the dose of X-rays required to destroy 99 per cent of the cells in a cancer is the same for human as for mouse cancers. On the other hand, there is wide awareness that the toxicity of certain drugs, and their rate of destruction in the body, may show large differences between species. The fact is that certain *principles* of cancer therapy can be established in animals which cannot be ascertained from clinical observation, and it is a matter for clinical consideration how species differences would be expected to modulate expression of these principles. These remarks are made necessary by the common broad assertion that unspecified species differences must always disqualify animal experiments as guides to the clinical situation. This is not the case. It cannot be denied, however, that mistakes have been made in some assertions of the clinical relevance of data from animal experiments. The example I shall discuss in the following section betrays not a defect in our knowledge of comparative pathology but a neglect of that knowledge.

The value of animal experiments in therapy research has been discussed first because it is in this area of research that they have the greatest capacity to escape the tight limitations of clinical research. Their place in research directed to improvements in methods of diagnosis or to the recognition and control of environmental causes of cancer is less significant and often more difficult to justify, except in so far as very general advances in biological knowledge may have some impact on *any* biologically orientated interest.

Despite the publication of many claims to the discovery of a 'blood test' for the general detection of 'cancer' which have appeared over the years, no such simple facility has proved itself in practice. The fact is that the common constitutional changes which cancers may bring about arise only at a very late stage of the disease and are not specific indications of cancer but the manifestations of any debilitating disease. The techniques used to diagnose cancer in humans are all of an *ad hoc* character in the sense that they relate to the site in which a cancer is suspected. Cancer on a directly observable surface (the skin, the mouth, the lower intestine, vagina, etc.) can often be detected by direct inspection; for cancer of the lung, upper gastrointestinal tract, kidney, liver, and brain, radiological methods are largely used; for suspected cancer of the thyroid gland, radioisotopes may be used; in many sites (lung, stomach, womb) preliminary diagnosis may be made by examination of stained smears of cells extruded from these organs. It will be appreciated that, because most clinical methods of diagnosis do not jeopardize the health of patients, relate to the particular sites of cancer which affect man, and are specially adapted to man's body size, their development has been almost exclusively by progress in clinical research. Indeed, this is one of the several areas in which veterinarians have been able to bring to the care of animals facilities which have been perfected by clinical research in humans.

It is in the sphere of cancer research directed to *prevention* of cancer that the usage of animals has been most forcibly questioned and in which protests against their use can be most convincingly sustained. This is so because man has failed to control dominant cancer-causing influences known to arise from his correctable habits, and because the continual influx of potentially carcinogenic agencies into the social environment is the consequence of his own technological aspirations and commercial expediency. The primary means of identification of environmental causes of cancer in man is epidemiological studies, whereby significant associations can be demonstrated between the liability to cancer at a specified site and the pursuit of certain occupations, the following of certain customs or diets, or indulgence in some personal habits. Many of these associations have become obvious to physicians from their local experience. (For example, Sir Percival Pott observed in 1775 that chimney sweeps were unusually liable to develop cancer of the scrotum.) But where any doubt clouds impression, the reality of an association can be tested by formal survey and statistical analysis. It was by this means that the strong association between cigarette smoking and lung cancer was proved, to the entire satisfaction of all but those with a vested interest in the industry. Another proven association is that between adolescent sexual promiscuity and cancer of the womb. Since most causative environmental factors are peculiar to man, and have been identified by studies confined to man, and since they are usually quite within the control of individuals informed of them, the use of animals for their more detailed study is both scientifically and morally questionable. In the case of smoking, recent usage of animals for the testing of alternative smoking materials, motivated as it is by a relentless commercial endeavour to retain the massive financial gain to be had from exploiting a widespread habituation, has outraged animal defenders and cast a shadow of disapprobation on all experimental studies employing animals. The example of 'smoking' experiments refers to a large general question concerning the moral orientation of our own species in relation to our treatment of others. Is it defensible conduct to advance into an indefinitely broadening future of chemical innovation holding animals before us to defend ourselves against the possible consequences of miscalculation or irresponsibility? A case can be made for animal studies in the search for remedies for human or animal disease: it can hardly be made for studies, often sanctified by legal requirement of them, which do no more than monitor the trivial diversification of products that makes up competitive appeal to the vanity, indulgence or convenience of healthy persons.

It should be realized that among a range of randomly selected substances subjected to testing for carcinogenicity only a very tiny proportion would be expected to display this potential; and where cancer *is* induced, there is not the slightest need to allow the condition to progress beyond the stage at which diagnosis can be made by microscopic examination of the lesion. As in all cancer experiments in animals, the avoidance of distress from advanced malignant disease requires regular and skilled surveillance of the animals put at risk; no scientific information of value is to be obtained from allowing the disease to progress to a terminal stage, and after death the prospect of obtaining useful information diminishes by the hour.

Readers may detect in the context of this section a suggestion that I have graded the justification of animal experimentation in accordance with the objective served by it. I do this knowingly, in spite of my personal view that there is *no* human objective which can justify experiments that are intended to cause distress or may prove to do so without the intention. There is no anomaly in my discrimination of objectives. I believe that the mental hygiene of the experimenter, representing that of the society to which he belongs, requires him to forego the implication of innocent beings in approaching objectives which are of questionable moral status.

SPECIES USED IN EXPERIMENTAL CANCER RESEARCH

Since cancer may arise in all classes of the larger multicellular animals (fish, amphibia, reptiles, birds, and mammals) it might appear that the researcher has a wide and free choice of species. Certainly, scientific studies of cancer have from time to time been made in selected species of all these classes—and without disclosure of any fundamental difference in the nature of the condition from one species to another. However, the choice of species is in practice restricted by the logistics of research facilities and by a need to ensure the relevance to man of the observations made in animals. Use of a mammal is generally regarded as procuring sufficient relevance; but it has to be accepted that there is no mammal which closely resembles man in respect of the relative incidence of spontaneous cancer in the different organs.

Quantitative studies of cancer may be expected to require the use of moderate numbers of animals and, in Britain at least, it is usually necessary for animals to be bred by the institution in which the research is done or in a special breeding facility under its control. Economy demands that the animal to be used is small enough to allow some hundreds or thousands to be accommodated within limited floor space. To ensure a regular flow of young adult animals, the species must be one which breeds well in captivity and has high fecundity and a short generation time. It is convenient if the dietetic requirements can be met by a commercially available pelletted formula. Since the animals are to be maintained in fairly large numbers in confined premises, they should not be excessively susceptible to infectious diseases when reasonable standards of hygiene are maintained; if the animals require isolation or continual treatment with antibiotics to keep them in health, difficulties and complications are encountered in relation to experimental procedures, especially those which reduce their resistance. To the extent that any of these desirable features are diminished, the cost per animal will rise. In practice, it has been found that economical and reliable logistics are best obtained by the use of rodents, the common laboratory species being rabbits, guinea-pigs, rats, hamsters, and mice.

Whilst economy must always have had a dominant influence on the choice of animal for cancer research, the experimenter must also consider whether a species exhibits the biological characteristics which will provide for the experiments he proposes to undertake. If he is to study the behaviour of cancer arising in the species used it must clearly be one having a significant incidence of spontaneous tumours, with the possible alternative that it be one in which cancers can be induced to form by the administration of carcinogenic agents. Since the growth of cancers may require to be monitored by serial examinations and measurements the animal should not be aggressive and should readily adapt to repeated handling without the need for anaesthesia.

It is clear that no protracted study of a particular cancer can be undertaken if observations are confined to the animal in which it arises; within a few weeks of its appearance it will have grown to a size which demands killing of the animal bearing it. To continue study of the tumour and of its response to various experimental procedures, it is necessary to keep it in being by serial transplantation from one animal to another. But transplantation can be undertaken only when the colony of animals used for the experiments has been so highly inbred that all individuals in it are genetically identical; that is, *isogenic* animals are required for sustained research using particular tumours. Where animals are not isogenic, transplantation from one individual to another will fail; or, where isogenicity is imperfect, the transplants will succeed but will be subject to immunological influences generated in the recipient. These influences, affecting both growth of the tumour and its response to therapeutic agents, will constitute *artefacts* of the experimental system imposed by the imperfect isogenicity among individuals of the animal colony. The cancer literature of this century, up to the present time, is littered with misconceptions and misinterpretations concerning cancer biology which are entirely the product of unrealized usage of incompletely isogenic animal colonies for cancer studies requiring transplantation of the tumours.

The production of a fully isogenic animal strain of any rodent species by intensive inbreeding takes at least 5 years and may be frustrated by progressive weakening of the animals and the appearance of actual defects; commonly, fertility may be grossly reduced. It is understandable that very few cancer researchers undertake this lengthy and hazardous task; the overwhelming majority use strains which were originated by distinguished animal geneticists many decades ago and have since been maintained by continual inbreeding. The number of isogenic strains of rabbit, hamster, or guinea-pig available to workers is very small; less than 5 per cent of the animals used for experiments employing tumour transplants are of these species, the remaining 95 per cent being represented by mice and rats in the ratio 2:1 (Hewitt, 1978). The well-known predominance of the mouse as an experimental animal has prevailed throughout the history of cancer research, and the preference is justified on all counts: being one of the smallest of the readily available rodents it is cheapest to breed and maintain: it has very high fecundity—one female may rear 50 or more young in her breeding life; a very wide variety of different isogenic strains are now available which display a remarkable range of characteristics including high or low incidences of spontaneous cancer in particular sites, certain congenital diseases and deficiencies, and some physiological peculiarities; the adult is of such a size that it can be comfortably held and restrained in one hand, allowing a single operator to carry out minor procedures, such as injection, without assistance; most strains remain active and healthy under proper standards of laboratory care and the individuals adapt very well to repeated handling, so that the stress associated with it is progressively reduced; following minor operations under short-acting anaesthetics, they quickly return to normal activity, resume grooming, and do not appear to carry a memory of the event which makes them resistant to subsequent handling. The experimenter who carries out modest procedures with skill and kindness, and who does not allow animals to survive which develop any untoward complication of experimental interference is, in my experience, not given any sense of having been inhumane or having subjected the mice to indignity. Having kept two colonies of mice free of epidemic infectious disease for 25 years with no more than moderate hygiene precautions, I am not convinced that the expensive and elaborate regimes of hygiene and surveillance currently recommended and adopted for mice are at all necessary for general purposes, and I suspect that they are sometimes harmful in that they may deprive the animals of natural stimuli to the development of resistance.

It will be appreciated that the predominant use of mice as experimental animals during all this century has had a cumulative effect in recommending their use. There is now no mammal, with the possible exception of man, of which we have a greater stock of experience and knowledge concerning its breeding, genetics, physiology, or immunology, about the laws governing transplantation, or about its predilection for cancer.

So great are the advantages now presented by the mouse for the experimental study of cancer that some special requirement, not met by the mouse, has to be envisaged to justify use of any other species. The second commonest in use is the rat, which is required for its larger body size when certain operations or procedures are to be undertaken which would be too difficult using mice. Although several isogenic strains of rat are available for studies involving transplantation, the choice of strains is very much more restricted. Having in mind the greater cost of buying, accommodating, and feeding rats, their widespread usage in cancer experiments that do not require the larger animal seems to indicate a waste of financial resources; the preferential usage of rats commonly does not meet any special scientific requirement for this species and represents no more than the following of a local tradition due for review. It is said that rats tend to be more sensitive and intelligent than mice; if this is so, their unnecessary usage implies that there is room for some reduction of stress by a more appropriate selection of species.

Chickens are distinguished in the history of cancer research because this was the first species to provide evidence that viruses can cause cancer. The discovery was made in the very early part of the century and opened up a large field of research which continues to this day. An infectious leukaemia-like disease (Marek's disease) presents as a commercially significant problem demanding research for its solution; at a more fundamental level the condition provides opportunities for study of the complex interactions between viruses and their host cells. Apart from this limited area of interest, chickens are little used in contemporary cancer research.

It is questionable whether completely isogenic strains of rabbit have yet been obtained, and certainly very few transplantable tumours of rabbit origin are available for research which do not behave as 'foreign' transplants. Several viruses which give rise to *benign* tumours in the rabbit were isolated in the 1930s, and these discoveries led to a peak in the usage of rabbits in the following few decades. It is probable that the commonest current use of rabbits is for the production of *antibodies* against various fractions of tissue cells—representing an important part of immunological research concerning the structure of cells. The procedures employed involve only the injection of bland biological materials and subsequent (repeated) bleeding from an ear vein. Not only are rabbits considered to be 'good' producers of antibodies, but their relatively larger size enables adequate volumes of antisera to be collected at each bleeding.

In accordance with the universal liability of mammals to develop cancer, the disease arises commonly in many of the species which man has domesticated. Owen (1978) refers to the finding that, where a survey was made, the incidence of cancer in dogs exceeded the incidence of cancer of all types in human residents of the same locality; a high incidence is also found in cats; and a large proportion of grey horses develop malignant pigmented tumours (*melanomas*). Cancer is also found in some animals kept in zoos; 50 per cent of Tazmanian devils dying in a zoo in California were found to have malignant disease (Griner, 1979). From time to time, it is suggested that domestic pets which have developed cancer might be specially suitable for certain investigations. It is claimed, quite rightly, that because the disease has arisen naturally, such cancers represent reliable models of the human disease. However, the suggestion deserves very careful consideration of its technical, ethical, and humanitarian implications. It will be appreciated from the preceding section that the individual case of animal cancer presented to a veterinary practitioner is subject to the same limitations of observation as obtain for a human case: the disease is fully established when it is first encountered; there is no knowledge of the length of its biological history and a very limited opportunity for ascertaining the rate of growth or determining the prognosis; it has the same limitations of individuality and unrepeatability as any human case of cancer; because the animal is not of an isogenic strain, the cancer cannot be perpetuated by reliable transplantation. I was at pains to describe in the preceding section how the case for using animals for cancer research rested on the *technical* advantages over clinical observation which could be achieved by the use of multiple transplants of tumours whose history and potential are known. There is, indeed, no reason to believe that a spontaneous tumour which has been transplanted under conditions which take full account of present knowledge is not precisely equivalent in its behaviour to a naturally occurring cancer progressing in the animal in which it arose. Moreover, a considerable concession to humanity is achieved by the fact that the experimenter can place a transplant where it will be least likely to inconvenience or pain the recipient animal or give rise to uncontrolled complications; his previous knowledge of the transplanted tumour enables him to predict the course of events and to forestall painful eventualities before they occur by suitable euthanasia of the experimental animal. In a properly designed and conducted animal experiment, the experimenter remains in control of the situation, respecting both the technical and the humanitarian aspects of the experiment.

It is difficult to escape the conclusion that recommendation of the *use* of veterinary cases of cancer presented as 'patients' does not proceed from the prospect of advantageous technical control but from a decision to set aside the ethical restrictions accepted for human patients. Veterinarians play a large part in biological research and much of their work contributes to advances in medical science. But a clear distinction must be expected to be drawn between the ethics proper to a scientific role, and those appropriate to general practice. It must be assumed that the owner of a pet found to have developed cancer desires to take advantage of any known treatments which can relieve symptoms or prolong happy life, but to avoid subjecting the animal to any treatment which would cause sustained distress, especially if it offers little chance of cure; 'putting down' would be reluctantly accepted to avoid distress either from progress of the disease or from the application of drastic or dangerous treatment. It is questionable whether an owner would be happy to hand over a pet for experimental study: either it is well enough to return to its home, or sick enough to warrant euthanasia. The veterinary practitioner is expected to provide his expert knowledge in these implicit decisions and even to advise against the owner's wishes where these conflict with the animal's welfare. Thus, the contemplation of experimental usage of veterinary 'patients' requires to be examined with great circumspection. The challenging question must always be: 'Why cannot this investigation be made on human patients?' There are practically no answers to this question which can refer to purely technical considerations.

THE SELECTION OF ANIMAL CANCERS

It was made clear in previous sections that the experimenter who confines his attention to the behaviour of an individual tumour examined and manipulated in the animal in which the tumour has arisen is in no better position to make significant observations than the clinician with his patients. If the biology of a given tumour is to be studied in depth over a period or if it is to be employed to assay the effectiveness of therapeutic agents, his experimental system must provide for sequential studies of the tumour over a period which will extend far beyond the life of the animal in which the tumour originated. In practice, this can only be achieved if the system allows for serial transplantation of the cancer from one animal to another. If artefactual phenomena concerning the relationships between the cancer and the hosts are to be avoided, as is required if the experimental system is intended to be a *model* of the human disease, then the animals used must be of a highly inbred, or *isogenic*, strain. Under this strict condition, each mouse of the colony is the equivalent of an identical twin of every other, and the behaviour of the tumour in any graft recipient will be identical with that in the animal of origin. It was explained in the preceding section that these critical conditions are most readily obtained using mice, which species will, indeed, be used to exemplify the principles to be discussed here.

Given a productive breeding colony of an isogenic strain, the researcher's next requirement is one or more cancers which have arisen in that strain. He has two alternatives: either to wait for a tumour to turn up spontaneously in one of his mice; or to treat some of the mice with carcinogenic agents or certain cancer-causing viruses and so hasten the appearance of a tumour. His choice is of a *spontaneous* or of an *induced* cancer. Now it has been established that artificially induced cancers in mice, and most probably in other species, display certain biological peculiarities associated with the induction which are sufficiently influential on the behaviour of the tumours to distinguish them from spontaneous tumours. This distinction has been repeatedly demonstrated during the last few decades. Since the animal tumour system is commonly conceived and intended to be a model of human cancer from which it is hoped to extract information that is relevant to the clinical situation, it is a matter of great importance to decide whether spontaneous or induced animal tumours represent the most appropriate models of human cancer. Since human cancers are *not* artificially induced by the agents used in laboratories to induce animal tumours we can hardly escape the conclusion that spontaneous origin animal tumours are to be used when a clinically relevant model is intended.

The frequency with which normal, untreated mice develop cancer spontaneously varies considerably between different inbred strains. There are some strains which exhibit a very high incidence of either leukaemia or breast cancer, and the prospective researcher may expect to encounter several 'cases' soon after his inbred strain has been established. What is now well known is that these 'high cancer strains' owe their peculiarity to contamination of the animal strains by specific cancer-causing viruses which are continually transmitted from the mothers to their young, either through the placenta *in utero* (leukaemia) or via the milk during suckling (breast cancer): the agents maintain an endemic infection passing from one generation to the next. Cancers acquired in this way are certainly closer to spontaneous tumours than are those induced by viruses or chemicals administered by the experimenter. But there is no evidence that human tumours are acquired in this way, and there is now general agreement that these 'high cancer strains' are laboratory 'freaks' highly dependent on the condition of isogenicity: and human communities are not isogenic. The cancer researcher who is fastidious in excluding all laboratory artefacts that are unnecessary ingredients of a workable transplanted tumour system may be inclined to reject 'high cancer strains' on the grounds that the tumours or leukaemia arising in them are, in a sense, induced.

The researcher who elects to confine his attention to tumours of spontaneous origin arising in his own strain of animal has to face the disadvantage that he has to await the fortuitous appearance of a cancer in one of his animals and must adopt for his general usage whatever type of tumour turns up. The incidence of cancer in what are called 'low cancer strains' of mice is usually no higher than in a human community, and a long time may elapse before a tumour is encountered. However, since the liability to cancer, as in humans, increases with age, the chances of isolating a tumour is greatly increased if part of the animal colony is set aside to allow animals to live out their full life span of 2–3 years; if the individual animals of the ageing population are kept under regular and skilled surveillance, the presence of cancer can be detected at a relatively early stage; but if surveillance is inadequate, animals will die of their cancers before these can be transplanted. In our laboratory, we have attached supreme importance to the exclusive use of spontaneous origin tumours for clinically relevant experimental research; we have isolated and maintained over 40 different mouse cancers during the last 20 years (Hewitt, *et al.*, 1976); these cover a wide range of types overlapping the common types of clinical tumour. However, it is evident that our practice is rather rarely followed: many large and important centres of cancer research over the world do not have available a single spontaneous origin mouse tumour isolated from, and maintained in, their own animal colonies; we have been asked to supply transplanted tumour systems to 20 cancer centres in 10 countries; and many centres employ tumours, albeit of spontaneous origin, which were imported from other laboratories and are now over 25 years from their time of origin, with the consequent risk of incorporating a wide variety of artefactual properties affecting the results of experiments.

The danger of poor quality control in the selection and maintenance of transplanted animal tumour systems is that the results of experiments done with them may reflect very strongly the influence of laboratory artefacts arising out of faulty selection of tumours or their transplantation to animals which are not, in fact, isogenic with the animal in which the tumour arose. In these circumstances, experimental findings may be given spurious clinical significance; physicians and surgeons may be encouraged to employ methods of treating the human disease (which is not subject to such artefacts) that have little or no prospect of success where the relevant artefacts are not exerting an influence. It is my view, which has not been seriously refuted and which is gaining increasing support at the present time, that the basis of a widely practised form of treatment of human cancer known as 'immunotherapy' draws very heavily on evidence adduced from experiments done with animal tumours that have very questionable status as valid models of human cancer. I have reviewed and discussed this topic in detail elsewhere (Hewitt, 1978, 1979).

It remains to be seen whether the criticisms I have raised against the assertions of clinical relevance from this very large area of animal cancer research will be generally conceded. Certainly, they are supported by a large volume of evidence from our own laboratory, have commanded a large measure of agreement from scientists and doctors, and have not yet been refuted by conflicting data. If they are sustained, the matter carries an important implication for the justification of animal experimentation which is now demanded. Legislation proposed for reform of the *Cruelty to Animals Act 1876* would place an obligation on licensees to ensure that 'alternative' techniques are not available for whatever experiments are proposed. It would be reasonable to insist also that, in respect of cancer research, some form of quality control of the experimental animal systems be exercised to ensure that they meet standards of relevance to the clinical situation where that is what the licensee proposes for his research. The approach to such an ideal can be most cogently made by the scientific committees of organizations which award grants for research on the basis of projects represented to them. But it is a responsibility which, by long neglect, they have not yet the experience to assume.

SOME COMMENTS ON TECHNICAL PROCEDURES

My intention in this section is to discuss a selected number of procedures that are commonly employed in cancer research using animal tumours. My approach will be to consider the design of the experiments as the art of obtaining scientific information without jeopardizing a high standard of humanity. And this is not a matter of achieving a compromise but of retaining control over the investigation. The aim of the experimenter must be to interpret the effect of an experimental interference in as small a context as possible. If the interference is productive of stress attributable to pain or trauma the interpretation of its effects requires reference to an exceedingly complex array of constitutional changes which usually lie outside the experimenter's own brief and tend to be unjustifiably neglected as part of the interpretation. It follows that a high quality of science goes hand in hand with a high standard of humanity. It has to be remembered that even the necessity to employ anaesthesia for a procedure introduces constitutional changes which may complicate the issue under investigation; one of the most commonly used anaesthetics produces a quite profound fall of body temperature in small mammals. It is therefore always an advantage to be able to refine a procedure to an innocuous level permitting omission of anaethesia.

Induction of Cancers

The screening of compounds or materials for carcinogenic activity was touched on in an earlier section, and it was stated there that such tests constitute one aspect of chronic toxicity testing. Provided that a substance is not associated with painful or disturbing acute toxic effects, such as ulceration, its possible later effect in inducing formation of a tumour is most unlikely to be associated with the supervention of distress unless there is failure to terminate life before the tumour progresses to an advanced stage; in any case, only a small proportion of random screening tests are likely to reveal carcinogenicity. It was described in the last section that known carcinogens or virus suspensions are very commonly used to induce tumours in animals which are to provide the basis of transplanted tumour systems for extended research. The efficiency of carcinogenesis by this means is very high, but the commonest materials and routes of application used are not productive of pain or distress at any stage of induction.

Transplantation of Tumours

Once a tumour has arisen in an animal spontaneously or has been induced by carcinogenic application it can be maintained as an ongoing research facility only by serial transplantation from one animal to another (although it may be stored in liquid nitrogen for indefinitely long periods when its usage is to be temporarily suspended). Quantitative studies of a tumour or of its response to therapeutic agents will usually require the simultaneous production of a quite large number of animals bearing transplants of the tumour, to enable the assembly of several identical groups of tumour-bearing animals to receive the different treatments to be compared. Thus, transplantation of tumours constitutes the commonest procedure undertaken in cancer research using animals, and therefore deserves careful consideration of the humanity of the techniques to be used. Refinement is possible in respect of both the method of introducing the cancer cells and the site in which they are placed. The most common tissue layer to receive the cells is the loose tissue immediately beneath the skin (subcutaneous); this is so because it is the easiest to enter and because the tumours arising from these can be seen, palpated, and measured. The method of introducing the inoculum depends on how it is prepared from the tumour to be transplanted. Methods are available for reducing most tumours to a very fine cell suspension, enabling the inoculum to be injected through the smallest available hypodermic needles; such an inoculum has the further advantage that the number of viable cancer cells can be accurately determined and the innoculum standardized. A uniform inoculum can also be prepared from tumours reduced to a fine mince injectable through a rather larger hypodermic needle; the use of sharply pointed needles permits injection without the need to grasp and countertract the overlying skin with forceps, which is certain to bruise the grasped tissue. It is our experience with a large range of animal tumours that a texture of tumour that resists breakdown into an injectable preparation is very rare. Nevertheless, one of the commoner techniques of transplantation currently in use consists in cutting the solid tumour into small cubes and introducing these singly into the subcutaneous sites through a hollow instrument called a *trochar*. It is not, in my view, a favourable technique: the trochar is considerably wider and blunter than a hypodermic needle and its introduction therefore causes more trauma. Moreover, different cubes of tissue from a tumour may contain widely different numbers of viable cancer cells and cannot be regarded as equivalent samples as can equal volumes of mince or cell suspension. Solid tumours are not commonly transplanted into muscles or into the abdominal cavity or brain. One type of tumour found only in rats and mice grows as a fluid suspension of cancer cells in the abdominal cavity and is normally maintained by serial intra-abdominal transplantation. Animal leukaemias also are transplanted to this site. Transplantation into the skin (*intradermal*), as distinct from under it, is of particular value for experiments in which the transplanted tumour is required to be surgically excised.

An exceedingly important consideration for humane practice is selection of the region of the body to which a subcutaneous transplant is made. What is required is that the tumour which subsequently grows from the graft does not give rise to pain or interfere by its bulk with any normal function of the animal; to avoid pain from pressure, the graft site should not overlie a bony surface such as the spinal column, pelvis, or breast-bone. The most favourable site in small mammals is the posterior part of the loin, but it is by no means generally adopted. A common site of transplantation is the hind leg or foot, which is often employed where some local treatment of a tumour is to be given such as radiotherapy or *hyperthermia* (heat treatment). The advantage of the leg as a tumour site is that it can be retracted away from the rest of the animal, which can in some way be shielded from radiation or from the heat of a hot water-bath. The disadvantage of this site is considerable *to the animal* unless the tumour is kept to a very moderate size. If not, interference will ensue with mobility, grooming, access to food or water, or self defence against aggressive cage mates. The tails of mice or rats are not uncommonly used as sites for tumour transplantation, the technical advantage being that the tumour can be removed, when this is required, by amputation of the tail. The operation is simple and minor, but the tail is less of a disposable appendage than might be supposed: it is used by the mouse to assist balance and to provide stability when the animal is in a semi-erect posture; it also plays a part in temperature control; and handling of the animal for removal from the cage or for 'clinical' examination is surprisingly more difficult using tail-less mice.

Limitation of Tumour Size

The normal course of any spontaneous or transplanted tumour, once its growth is detectable, is progressive growth at a steady rate until, by producing one or more of a variety of complications distressing to the animal, it causes death. This is a universal termination of the natural disease. Except in the case of some artefactual tumour systems of the kind described in the earlier section on cancer selection, spontaneous arrest or regression of the growth is almost unknown. In 25 years' experience of a wide range of naturally arising mouse cancers kept under close observation, I have never encountered a single example of spontaneous regression of a tumour, although the number of animals observed with established transplanted tumours exceeds 30 000. It follows that a primary humane consideration in the design of any animal tumour experiment must be to make a rule concerning the maximum tumour size to be allowed; and if the experimenter has rare occasion to allow tumours to grow beyond a generally agreed size suitable for most experiments, the decision about termination of the experiment should be surrendered irrevocably to animal-house staff.

The principle of establishing a maximum tumour size as part of the experimental design serves not only to maintain a good standard of humanity, but also to restrain an experiment within the bounds required for precise investigation and valid interpretation of results. A growing tumour makes an increasing demand on an animal's constitutional resources; at a small size, the extra demand is met by the animal's reserves without the need for stimulation of bone marrow and other organs to a higher level of cell production; at a later stage, health can only be maintained by abnormal stimulation of these organs; eventually the animal manifests rapidly progressive constitutional deficiencies—anaemia, a fall in blood protein, loss of body weight, and listlessness leading to reduced intake of food and water. The experimenter who makes observations on tumour growth or behaviour after an animal has reached this moribund cachectic state without regard to its influence on what he is observing or measuring, cannot possibly make valid interpretations of his data; and several preposterous conclusions about the biology of cancer have depended on such unscientific and inhumane negligence. The tumour size at which constitutional delapidation begins is rather variable with tumour type and growth rate; in the mouse, I have seen some tumours bring about loss of body weight when they are under 1 g in weight, and others which have attained 4 g without inducing detectable deterioration.

The experimenter who sees his animal cancer as a 'model' of human cancer needs to take account of the size of a tumour in relation to the body weight of the animal bearing the tumour, and to appreciate that, for the same ratio of tumour weight to body weight representing a moderate sized clinical tumour, a mouse tumour would be only 2–3 mm in diameter. However, this is not a realistic size for most animal experiments. The maximum weight of tumour in a mouse which meets most scientific requirements and conforms to an acceptable standard of humanity is about 1 g, which would form a spherical mass of about 13 mm diameter. Nevertheless, a clinical tumour representing the same proportion of body weight would approach a weight of *3 kg*. Thus, it will be appreciated that a maximum size of 1.3 cm diameter for a spherical tumour in the mouse, when placed in a suitable site, meets a good standard of humanity, avoids the constitutional complications which confuse scientific deductions from the data, and approaches the extreme size for acceptance as a model of human cancer. For species of larger body size, proportionately larger tumours are allowable.

My own observations in laboratories and my reading of the literature reveal that reasonable restraint of tumour size is commonly not ensured. In one reported experiment, mouse tumours *in the leg* were allowed to grow to almost 10 grams (DeWys, 1972); I have been shown mice which had had tumours transplanted over the breast bone and were barely able to walk or to take food from the hopper on account of the large size of the tumours; a very senior researcher habitually allowed mice to continue with expanding abdominal tumours until they died of their condition; and to my plea for more humane practice replied that they 'just hadn't the time' to adopt the more refined endpoint that was possible by careful daily surveillance of the animals!

Studies of Metastasis

Dissemination of cancer to parts of the body remote from the primary tumour is very largely responsible for the high death rate from many forms of human cancer, and only too frequently cancels the satisfaction of patient and therapist with the success of eradicating the primary tumour by surgery or radiotherapy. It is reasonable that continuing attempts should be made by research to improve our understanding of the mechanisms underlying dissemination and our prospects of controlling secondary disease. Many important biophysical studies relevant to metastasis are made using isolated cells, whereby the factors controlling adherence, release, and seeding of cancer cells can be investigated. However, it is clear that natural metastasis involves the whole body and is affected by various physiological effects exerted by the whole constitution. Moreover, therapeutic control of disseminated disease can only be advanced by resort to animal models of the human disease which exhibit metastasis. Clinical endeavours to improve the treatment of disseminated disease are currently directed to cases in which the secondary disease is said to be 'occult'; that is, where the secondary tumours are *assumed* to be present but are microscopic in size and undetectable by any currently available method of diagnosis. Animal experiments can provide guides to these developments because the timing of dissemination and the size of secondary deposits at the time of experimental therapy is known, whereas in the clinical situation they can only be conjectured.

As in most lines of animal cancer research, the experimental system most often employed for metastasis research is provided from a transplantable tumour in the mouse. It may here be noted that a high proportion of the artificially induced cancers referred to in the section on cancer selection include among their artefactual features a failure to exhibit metastasis (Alexander, 1977). In our experience of a great many mouse tumours of spontaneous origin, metastasis almost always ensues in a proportion of the animals with transplants.

A technical problem confronting the prospective researcher into metastasis may be described as follows: the transplanted primary tumour is usually initiated by the injection of a rather large number of cancer cells and may be expected to grow to a visible tumour before cells are disseminated from it; thus, the secondary tumours, starting mostly from single cells and growing at approximately the same rate as the primary tumour, remain much smaller than the primary tumour; it follows that the period of observation has often to be extended beyond the time at which the primary tumour attains a maximum tolerable size if the secondary tumours are to grow sufficiently large to form visible, countable tumours. A humanely designed metastasis study therefore requires that the primary tumour be surgically excised to allow observation of the progress of secondary disease. The commonest way to procure this condition is to place the primary transplant in the leg of the animal and excise the tumour by amputating the tumour-bearing leg. It is true that an operator who is trained in the principles of surgery may complete the operation under anaesthesia with minimum trauma or risk of complications, and that the operated animal quickly adapts to the loss without serious impediment of its mobility within the cage. However, it certainly cannot be recommended for general use: I estimate that at least half of those who are prepared to undertake the operation have no training in the principles of surgery, and may not appreciate, for example, that exposure of a fractured bone end may be a potent source of pain; it may also be urged that to encourage familiarity with, and acceptance of, obviously mutilated animals in animal house staff must obviously lower the standard of humane practice they look for. Such a mutilating procedure is not at all necessary for the scientific requirement described. Our own procedure for experiments requiring surgical removal of tumours is to transplant the primary tumour *to the skin* of one flank and allow it to grow to about 0.5 g in size. It can then be simply and radically removed by excising a small ellipse of skin containing the tumour; the small wound is closed with two or three clips and heals almost without trace in a few days, when the clips are removed; there is no mutilation and no compromise of the animal's full and normal movements; blood loss is minimal.

An important humane consideration arises concerning the fate of animals whose primary tumours have been removed and which are kept under observation for the manifestations of secondary disease. The attainable ideal is to gain the information sought from an experiment before secondary disease has advanced to a stage at which it causes pain or discomfort to the animal. Beyond that stage, complications and systemic disorders of the animal, progressing to death, begin to confuse the very issue which the experimenter seeks to elucidate. The need for deciding upon some *end-point* which ensures humanity to the animal and serves the experimenter's purpose has to be incorporated in the experimental design. There are several courses open to the observer. Perhaps the most humane is to kill all the animals at some suitable pre-determined time after removal of the primary tumour—the optimum time being determined by preliminary observations and being well within the time at which animals develop sickness; after death, the animals are dissected to determine the extent of metastasis by direct inspection. This approach is exemplified by the humane practice of Sadler and Castro (1976). Our own practice has been to kill animals as soon as external evidence of secondary disease can be detected, which, in the mice whose intradermal tumours had been excised, was manifested in one of three ways: there was regrowth of tumour near the operation scar, which was palpable or visible at a very early stage; or there was enlargement of the lymph node draining the tumour site, also easily palpable through the skin; or the animals showed very early signs of slight sickness due to internal secondary tumours, when they were immediately killed. No animal was allowed to survive after its well-being had come into question, and they were kept under regular and frequent surveillance. This practice was found to yield very satisfactory data in a variety of studies (Hewitt, 1976).

Many animal experiments are done to assay the effectiveness of various therapeutic agents in controlling secondary disease. The usual design is to allow a primary transplant to grow to a moderate size and then to remove it surgically or eradicate it by a high dose of radiation. A series of animals identically treated in this way is divided into two or more groups; one group receives no further treatment; others receive the treatment(s) under test for their potential to inhibit or delay the progress of secondary cancer. Regrettably, the end-point most commonly used to measure any advantage of the treatments over one another, or over the omission of treatment, is comparison of the mean *survival times* of the animals in the different groups. This is a scientifically crude measure of effect and must be deprecated for its inhumanity; the animals are compelled to endure terminal secondary disease until it kills them; although not all such deaths are heralded by pain or sustained distress, some are, and it is the experimenter's duty not to abandon control of either the scientific or humanitarian implications of his manipulations. The experiments of Sadler and Castro (1976), referred to above, provided a perfectly satisfactory measure of the restraint of secondary disease by the agent they were testing. What must be observed is that the death end-point conveys a spurious sense of accuracy just by its finality, and that it is the end-point that is established with least trouble to the researcher. Death is the culmination of progressive pathological changes leading to fatal complications, and it is almost always possible to detect their earlier stages by sacrifice and inspection of the animals at a pre-determined time.

As in all animal cancer experiments, adjuvant procedures may be practised as a routine which are not an intrinsic part of a particular experimental design. These require consideration when reviewing the humanitarian status of an experiment. One particularly horrific practice that has appeared in the literature was encouraged by the observation that metastasis can be increased to a more scientifically useful level by physical trauma to the tumour-bearing animal. This was forcibly to crush *both* thighs of rats under anaesthesia using engineer's pliers (Ivarsson and Rudenstam, 1976). This practice, apparently confined to one European laboratory, has been followed for many years. And although the authors give reassurance that the measure does not appear to result in sustained distress, this is surely one of the occasions on which the experience of man can be used as a model for inferring the effect on the rat. It needs little imagination to conceive the consequence of severe bruising of the tissues and smashing of both thigh bones to the comfort and mobility of any animal.

Experimental Studies of Therapeutic Agents

Apart from the trauma and shock, from which recovery can be expected, surgery does not damage the constitution, and it is the treatment of choice for cancer when the required operation is feasible. Hormones can restrain the growth of some tumours and even bring about temporary regression of disseminated disease; their side-effects, such as feminization of men treated by hormones for cancer of the prostate, are neither painful nor lethal. Chemotherapy and radiotherapy have the limitation that they are as damaging to the normal tissues as to the cancer. The doses employed in therapy represent a compromise between inadequate treatment of the tumour and the infliction of severe damage on the normal tissues which may be associated with intolerable sequelae or be fatal. Animal experiments designed to evaluate and compare prospective therapeutic agents and techniques therefore imply separate measurement of the restraint of the tumour and the level of toxicity as determined by local or systemic damage; and the value of a treatment is expressed in terms of a therapeutic index signifying the relative values of these two effects. It is to be understood that exploratory experiments of this kind will often involve examination of a range of doses, from those which barely affect tumour growth to those which inflict a high mortality or unacceptable local damage. It is in this area of research that the cogency of clinical need is likely to be urged as a justification for inferior standards of humanity in animal experiments.

Potential new drugs are likely to be tested for toxicity in the animals to be used, by determination of the LD_{50}—the dose which gives 50 per cent mortality among animals receiving it. The technique is to administer a wide range of doses to groups of 5–10 animals, then to plot the mortality against dose and read off the interpolated dose for 50 per cent mortality. It is very simple to perform and gives an accurate and repeatable answer. But it is very far from being a humanely conceived method, especially where the toxicity may produce pain or distress; far more animals are injured than is required for the amount of information needed. In testing the acute toxicity of unknown compounds presented to us as potential radiosensitizers, we have found it sufficient to prepare a range of twofold dilutions of the substance and to inject *single* animals with each dose. The next dose is not injected until the animal receiving the lower proximate dose has been carefully watched for signs of an effect for a period of at least several minutes. The first of the ascending doses to have an effect gives an opportunity for the nature of the toxic effect to be observed and alerts the observer to the kind of signs to be looked for after the next dose up. The limiting dose is taken as that which either produces unacceptable signs of distress demanding that the animal be put to death, or which itself rapidly kills. If necessary, a finer range of doses is examined in the same painstaking way. The dose to be used for study of the potential of the drug to restrain tumour growth or enhance radiation effects can be decided by reference to the toxicity data obtained in such a limited assay, which has subjected a minimum number of animals to distress. Using highly uniform animals of specified body weight a wide range of animal susceptibility to a toxin is not to be expected. Indeed it is probable that the pattern of LD_{50} test commonly followed is a heritage from the use of non-uniform animals.

Chronic toxicity tests obviously require prolonged retention and surveillance of treated animals which have survived acute effects of the substances tested. In the case of chemotherapeutic agents, later effects on animals are very likely to be manifestations of damage to the gut or bone marrow, giving rise respectively to diarrhoea or to anaemia and increased susceptibility to infection. Such effects follow whole-body exposure of animals to radiation, rather higher doses being required to induce gut death than marrow death. LD_{50} tests are commonly used to compare the biological effectiveness of different forms of radiation. Judging by the effects of radiation on human victims of the atom bombs in Japan it is probable that death from gut damage is the more distressing. Death from marrow damage can, in any case, in my experience, be predicted from the results of a simple blood test done on the 10th day after exposure to radiation, thereby eliminating the need to establish actual mortalities.

A variety of techniques are available for measuring the dose-related effects of therapeutic agents on animal tumour growth, and it is convenient to use radiotherapy to exemplify their application. Humane considerations must take account of the fact that the very large doses of radiation required to cure quite small animal tumours locally exposed to the beam are sufficient to burn the adjacent skin severely, although these effects are not fully manifested until many days after exposure. Equivalent degrees of radiation burn in human cases are known to be painful and to demand the administration of analgesics (which are very rarely, if ever, given to small experimental animals). One method of measuring the effect of radiation on a tumour is available which avoids any need to retain the animal until skin damage becomes manifest: the animal is put to death immediately after irradiation; a suspension of cells is prepared from the tumour and these are assayed to determine accurately the proportion of cells which have survived the irradiation; the method is capable of detecting a cell surviving fraction of one in several million (Hewitt and Wilson, 1959). Another humane technique was perfected by Thomlinson (1960). Tumours significantly damaged by radiation undergo regression to a smaller size; but if they are not actually eradicated, growth is resumed and they return to their size at the time of irradiation. This sequence of events can be plotted graphically from accurate serial measurements to determine the time taken for the tumours to recover their original size; and this time is found to increase in proportion to the dose of radiation delivered to the tumour. A further method, perfected by Suit *et al.* (1965), involves the assembly of a large series of equivalent tumour-bearing mice; groups of mice then receive a range of doses of radiation to their tumours; the mice are retained under observation for some months to determine the proportion of tumours in each group which have been cured. Again, a graphical arrangement or computerized analysis of the data allows calculation of the dose of radiation required for cure of 50 per cent of the tumours. Usually, but not always, animals are anaesthetized during local irradiation merely to restrain them from movement whilst under the beam. None of these techniques impose great stress or discomfort on the animals used, provided only that they are carried out with skill and that the treated animals are kept under close and regular surveillance. The Hewitt and Wilson (1960) technique obviates any possibility of distress to the animal from irradiation: its exposure to the rays is such that minimal restraint of the animal is required; and it is put to death long before any symptoms of radiation damage appear.

Experiments Involving Induction of Stress

The technical understanding of 'stress' in the context of physiology refers to a particular state of the constitution of an animal which is induced by the threat, or actual application, of stimuli which are interpreted by the animal as potentially harmful. The translation from a state of rest to one of stress implicates an extremely wide range of physiological mechanisms, as may be appreciated from a consideration of the 'alarm reaction', extensively investigated by Hans Selye and reported on in a voluminous series of publications over several decades. An animal in a state of alarm exhibits: protrusion of the eyes, giving a wider field of vision; erection of the hairs, which increases apparent body size, and so may discourage a potential aggressor; and a posture of tense alertness. Underlying these obvious signs of alarm are profound alterations of physiology, including: increase of blood pressure; redistribution of blood from the digestive organs to the brain and muscles; massive release of certain hormones; and alteration of the coagulability of the blood. The condition is not a state of disease but represents an orderly redeployment of the resources of the animal to meet circumstances in the environment which must be given absolute priority in securing survival. The state of alarm does not necessarily result from a conscious assessment of the portent of signals in the environment but is perhaps most often triggered by stimuli registered subconsciously. However, the *sensation* of the bodily changes constitutes the mental state of *anxiety*. The critical issue which sets off the alarm reaction is normally resolved, one way or another, in a relatively short time. But the bodily changes are so profound and so exclusively directed to an emergency situation that their prolonged stimulation *will* lead to disease; in a state of alarm, for example, both appetite and digestion are almost totally inhibited—leading eventually to exhaustion and undernutrition. Perhaps the simplest understanding of 'distress' is that it constitutes abnormal prolongation of the state of stress. Unfortunately, in the description of many animal experiments, the word *stress* is most often a euphemism for *distress*.

Peters and Mason (1980) have recently reviewed the experimental studies which have been made in search of an effect of stress on the initiation or growth of cancer. They reached the general conclusion that the yield of information, subject as it is to the influence of many technical artefacts of the kind discussed in the earlier section on cancer selection, hardly justifies some of the gross cruelties it has entailed. The following are among the manipulations which these authors refer to as having been used to 'stress' animals: exposure of rats for 6 hours to a temperature of $-6\,°C$, amputation of the leg, forced walking for 16 hours daily, electrical shocking, 15 days confinement in copper meshes, production of neuroses, exposure to flashing lights for 8 hours daily, induction of convulsions, and forced swimming.

THE MAINTENANCE OF STANDARDS OF HUMANITY

Since cancer researchers who conduct animal experiments are financially supported by grants from public funds or from charities, they carry an implicit responsibility to respect public opinion about the means as well as the objectives of their professional activity. Many who make charitable donations to cancer research would withdraw their support if it came to their notice that animals are subjected to distress in its prosecution; and I have encountered patients with cancer who would rather forego certain treatments than take the benefit of advances that had depended on 'cruel' experiments. But even the researcher who is concerned to meet the expectations of public opinion about his standards of humanity may be unable to distinguish a consensus within the wide range of opinion that prevails. The very contemplation of animal experiments implies a rejection of the extreme view, formally represented by anti-vivisection societies, that all animal experiments are morally unjustifiable and should be proscribed. On the other hand, it is to be hoped that very few researchers would allow themselves *unlimited* licence to inflict distress on animals under cover of a submission that their experiments may serve to save or extend human life. It is true that experimenters are controlled, in Great Britain, by restrictions imposed on them by the provisions of the *Cruelty to Animals Act 1876*. But it is increasingly felt, even by some experimenters, that this Act fails, under present day conditions, to ensure the standard of animal welfare which it was intended to maintain, that many of the legal demands it makes are fulfilled only ritualistically, and that the resources for its effective administration are insufficient. What is evident is that widely different standards of humanity are observed by different licensed researchers, ostensibly without any breach of the regulations: the standard is evidently set by the sensibility, temperament, ingenuity, dexterity, and self-restraint of the individual, with some diffusion by personal example. Inevitably, the discussion of standards of humanity is a presentation of my personal orientation—the product of my personal experience, of my observation of the practices of other licensees within the institutions in which I have worked, and of my reading of detailed reports of experiments done in this and other countries.

Anti-vivisection societies have always emphasized the *number* of animal experiments per year as a measure of the enormity which they would proscribe. This is inevitable in the absence of available information which permits classification of experiments according to the degree of pain or distress they cause; and if there is no significant change in the proportion of 'cruel' experiments over the years, then the *number* of animals subjected to interference is, indeed, a measure of the amount of suffering caused. However, I must confess that my own concern is really not with the *number* of animals, in the sense that I should be more upset by my having caused *one* animal to suffer by my neglect or ineptitude than I should be by administering euthanasia to 50 animals at the termination of an experiment in which none had been caused suffering. The question the prospective animal experimenter has to ask himself is whether he considers that the painless taking of animal life is itself an immoral act. For me, it is not; and it is not because sudden termination of a purpose-bred animal's life does not entail suffering or constitute any interference with the ecology of wild life. Furthermore, if we grant perfect euthanasia in all situations, I do not see the moral distinction between the taking of animal life in the laboratory, in the slaughter house, or in the course of pest control.

Humane killing of animals is one matter: *death* of animals under experiment is quite another. In my view, there are very few exceptions to a general rule that any experiment which carries a mortality due to the interference must entail distress, one of the more obvious exceptions being when the animal dies under anaesthesia during an initial procedure. I hold this view because death resulting from interference must be the culmination of pathological processes and their complications, which are unlikely to be devoid of symptoms of illness. It should be added that the surviving animals in a uniformly treated group exhibiting a mortality must be expected to endure similar symptoms before their recovery. Thus, *all* the animals in such an experiment will be exposed to distress, varying in severity with the experimental procedure, the duration of illness, and the mode of death. What may be observed about experiments carrying a mortality is that there would in most cases be a measurable alteration of some physiological characteristic which portends the moribund state; it may have to be sought by separate research; but, once found, it can obviate any necessity to allow animals to endure the distress of terminal disease. Although the death end-point is scientifically unsound and inconsistent with a high standard of humanity it continues to be very widely used to determine and compare *survival rates* and *survival times* following experimental interference. In the case of many cancer experiments, 'non-survival' means that the animal has been allowed to pass through all the progressively deteriorating phases of terminal cancer. Mortality and survival time have also been very widely used to measure the effects of radiation on normal tissues such as the lung, kidney, or intestine. What may certainly be said is that a colony of experimental animals which regularly displays an incidence of sick, dead, and dying animals is the mark of a careless and inhumane experimenter whose practice must erode the morale of good quality animal-house staff. I shall discuss some general causes of such negligence later in this section.

The attainment of a high standard of humanity demands that the animal experimenter consistently observes two conditions: firstly, he must deny himself knowledge, however valuable it may appear, which cannot conceivably be obtained without the infliction of suffering; secondly, he must be prepared to devote a large part of his diligence and ingenuity to the design of procedures and end-points which enable the information he seeks to be gained before the onset of distress resulting from interference. To refuse the first condition is arrogant; to fail in meeting the second may often betray a want of research ability.

It has been proposed that any new legislation to control animal experimentation should include a clause which prohibits the infliction of pain or distress, or prolonging the life of an animal that unexpectedly displays such reaction: current legislation makes such prohibition subject to fulfilment of the experimenter's requirements. It has been argued against this proposal that pain and distress are indefinable and that observers could not agree about their recognition in animal species. Granted that such difficulties exist, it may nevertheless be noted that similar difficulties in assessing the degree of distress in our fellows are not permitted to impede the estimation of damages in cases in which a plaintiff claims to have suffered physical or mental distress. In such cases, the issue is decided essentially by *sympathy*—by the exercise of vicarious imagination; judicial decision takes account of the capacity to pursue usual enjoyment, and tends to favour the plaintiff where doubt remains. There is no reason why a similar approach should not be made to the assessment of distress in animals. It is reasonable to assume that procedures or conditions which are painful to man would be so in higher animals; and that an animal is not seriously distressed if its movements, and its desire and capacity to eat, drink, and groom itself show no detectable impairment; attendants familiar with the normal animals of a species can hardly fail to distinguish one that is sick, especially if the disease is sustained or worsening. What must be acknowledged is that animals cannot possibly suffer those forms of psychic stress or fear which are projected from the kind of knowledge which they cannot possess. Patients commonly fear that they have been misled about the true diagnosis of their complaint; they may be afraid because they have knowledge that their disease has a high mortality; or they may be preoccupied with a regret that they may die or be crippled while still in their prime of life. We can envy animals their ignorance of the calendar of life, of which our knowledge encourages in us some speculation about our present age in relation to expected life span. Animals are aware of neither. The charge of 'sentimentality' sometimes made against abolitionists is quite valid when fear is attributed to animals which could only be generated from knowledge gained by verbal or written instruction in a highly developed language. Animals can certainly enjoy the bliss of ignorance more often than we can.

Among avoidable causes of distress to animals under experiment, the following have come to my notice: unsuitable basic qualification or poor practical training of licensees; failure to select the most suitable animal system for an investigation; failure of the experimenter to give top priority among his responsibilities to ensuring the welfare of his animals; commitment to an experiment of more animals than can be kept under close and frequent surveillance; ineptitude and insufficient foresight in the design of experiments;

disregard of informed advice concerning the suitability of a technique or the condition of an animal; failure to give complete authority to animal attendants to administer immediate euthanasia to an animal found in distress; failure to appoint a suitable deputy for surveillance of animals during absence; and grim determination to stick to an experimental protocol after it has become apparent that it is inconsistent with humane practice.

All of the above derelictions of humane practice are being made more likely by a progressive alteration over recent decades of the sociological climate of bioresearch in general and of cancer research in particular. Some 30 years ago such work was rather poorly remunerated, gave little security, and tended to attract only those whose aptitude and interest distracted them from consideration of their personal status and career prospects. These prospects are now so much improved that 'research' now attracts a wider variety of aspirants: the research 'careerist' has become prominent on the scene, familiar to critics as the 'jet-set scientist'. By nature and resolution, he is not content with the remote chance of *distinction* as a bonus from patient, dedicated application to practical research, but actively seeks *prominence* by self promotion. This requires him to devote an extravagant part of his time to the writing of many scientific papers and attendance at an excessive number of conferences and congresses, many abroad; a reputation has to be *made* rather than be left to chance. There is no doubt that these tactics often succeed in their questionable intention; but they are not conducive to the display of originality or to the maintenance of a high standard of humanity in animal experimentation. Moderately senior animal experimenters of this demeanour are likely to hurry experiments through merely to gain some marginal competitive advantage in priority; their frequent absences from practical laboratory work lead to neglect of the supervision and teaching of younger scientists, for whom they are likely to be responsible. It is a common complaint of younger laboratory workers that 'the Chief is never there'.

It may not be generally realized that quite a high proportion of those licensed by the Home Office to perform experiments on living animals have had no basic training, and have no qualification, in biology, anatomy, surgery, physiology, or pathology. Moreover, several cancer research laboratories with extensive animal facilities are scientifically directed by appointees with no appropriate biological qualification; in the principal laboratory of a well-known cancer research organization, in which animal experiments are the predominant activity, the Director is a physicist by training, the Deputy Director is an electronics engineer, and even the *Animal Curator* is a physicist, with no separate authority being delegated to a senior member of staff with qualification or experience in biological science. This situation, which goes against former recommendations of the Research Defence Society, is hardly conducive to maintenance of a high standard of humanity. Whereas it is unthinkable that a geneticist or veterinarian should be allowed to tinker with the circuits of an expensive and complex piece of electronic apparatus, it is evidently considered that no basic knowledge or expertise is required for experimentation on animals or for the supervision of those who do such experiments.

It is true that a standard primary procedure employed in animal research may be taught by apprenticeship to anyone of average intelligence and dexterity; but the control and responsibility of an experimenter has to extend over the entire period of surveillance of the animals; and during that period complications from the experimental interference may arise which require medical or veterinary experience for their anticipation, recognition, investigation, and control. The neglect of such complications may lead not only to animal distress but to misinterpretation of experimental results. The influx of biologically unqualified animal experimenters is largely from those whose basic training and qualification are in one of the exact sciences, such as chemistry or physics, and it is reasonable to infer that, by their free choice of career, such persons have already expressed a temperamental leaning to study of the inanimate. The causes of the migration of graduates from the exact to the experimental biological sciences are complex. Perhaps the most potent influence has been that the distribution of university places between faculties has not reflected the opportunities for employment in the respective professions to which they lead.

REFERENCES

Alexander, P. (1977). 'Back to the drawing board—the need for more realistic model systems for immunotherapy', *Cancer*, **40**, 467–470.

Ames, B. N. (1972). 'A bacterial system for detecting mutagens and carcinogens', in *Mutagenic effects of environmental contaminants* (H. E. Sutton and M. I. Harris, eds), Academic Press, New York, pp. 57–66.

Bloom, H. J. G. (1968). 'Survival of women with untreated breast cancer—past and present', in *Prognostic factors in breast cancer* (A. R. M. Forrest and P. B. Kunkler, eds), Livingstone, Edinburgh, pp. 3–19.

Bross, I. D. J. and Blumenson, L. E. (1971). 'Predictive design of experiments using deep mathematical models', *Cancer*, **28**, 1637–1645.

DeWys, W. D. (1972). 'Studies correlating the growth rate of a tumor and its metastases and providing evidence for tumor-related systemic growth-retarding factors', *Cancer Research*, **32**, 374–379.

Everson, T. C. and Cole, W. H. (1966). *Spontaneous regression of cancer*, Saunders, Philadelphia.

Griner, L. A. (1979). 'Neoplasms in Tasmanian devils (*Sarcophilus harissii*)', *Journal of the National Cancer Institute*, **62**, 589–592.

Hewitt, H. B. (1976). 'Projecting from animal experiments to clinical cancer', in *Fundamental aspects of metastasis* (L. Weiss, ed.), North Holland, Amsterdam, pp. 343–357.

Hewitt, H. B. (1978). 'The choice of animal tumours for experimental studies of cancer therapy', *Advances in Cancer Research*, **27**, 149–200.

Hewitt, H. B. (1979). 'A critical examination of the foundations of immunotherapy for cancer', *Clinical Radiology*, **30**, 361–369.

Hewitt, H. B. and Wilson, C. W. (1959). 'A survival curve for mammalian leukaemia cells irradiated *in vivo* (implications for the treatment of mouse leukaemia by whole-body irradiation)', *British Journal of Cancer*, **13**, 69–75.

Hewitt, H. B., Blake, E. R., and Walder, A. S. (1976). 'A critique of the evidence for host defence against cancer based on personal studies of 27 murine tumours of spontaneous origin', *British Journal of Cancer*, **33**, 241–259.

Ivarsson, L. and Rudenstam, C. M. (1976). 'Trauma, Dextran 40 and metastases formation after intravenous tumour cell injection in rats', *Acta chirurgica Scandinavica*, **142**, 226–230.

Owen, L. N. (1978). 'Therapy of metastatic disease: canine and feline models', in *Secondary spread of cancer* (R. W. Baldwin, ed.), Academic Press, London, pp. 131–161.

Peters, L. J. and Mason, K. A. (1980). 'Influence of stress on experimental cancer', in *Mind and cancer prognosis* (B. A. Stoll, ed.), John Wiley, Chichester, in press.

Sadler, T. E. and Castro, J. E. (1976). 'The effects of *C. parvum* and surgery on the Lewis lung carcinoma and its metastases', *British Journal of Surgery.*, **63**, 292–296.

Suit, H. D., Shalek, R. J., and Wette, R. (1965). 'Radiation response of C3H mouse mammary carcinoma evaluated in terms of cellular radiation sensitivity', in *Cellular radiation biology* (M. D. Anderson Hospital, University of Texas, ed.), Williams and Wilkins, Baltimore, pp. 514–530.

Thomlinson, R. H. (1960). 'An experimental method for comparing treatments of intact malignant tumours in animals and its application to the use of oxygen in radiotherapy', *British Journal of Cancer*, **14**, 555–576.

Windeyer, B. (1978). 'Hyperbaric oxygen and radiotherapy. The Medical Research Council's Working Party', *British Journal of Radiology*, **51**, 875.

Animals in Research
Edited by David Sperlinger
© 1981 John Wiley & Sons Ltd.

Chapter 8

Animal Experimentation in the Behavioural Sciences

ROBERT DREWETT AND WALIA KANI

INTRODUCTION

The behaviour of animals, like their internal workings, raises two types of question: one recent author has called them *Cartesian* and *Darwinian* questions (Dobzhansky, 1968). Cartesian questions concern the mechanism of a phenomenon; in the case of behaviour, the mechanism that controls the behaviour. It was Descartes who, much impressed by Harvey's work on the circulation of blood, heralded modern physiological research by proposing the view that animals were naturally occurring machines:

> 'This will not appear in any way strange to those who, knowing how many different automata or moving machines the industry of man can devise, using only a very few pieces, by comparison with the great multitude of bones, muscles, nerves, arteries, veins and all the other parts which are in the body of every animal, will consider this body as a machine, which, having been made by the hands of God, is incomparably better ordered, and has in it more admirable movements than any of those which can be invented by man' (*Discourse on Method*, translated by F. E. Sutcliffe, 1968)

One task of biology, then, is to seek understanding of the way the machinery of the body works, and a part of this task is to investigate how behaviour is controlled by the 'neuronal machine' that we call the brain. A second task is to understand the *function* of behaviour. Function, in this context, is a Darwinian term: understanding the function of a behaviour pattern means understanding its adaptive usefulness and the way it has evolved. These two are different types of understanding. Each is important, and one cannot substitute for the other.

It is generally true that investigations of the Cartesian type are more likely to be morally objectionable than investigations of the Darwinian type. The Darwinian tradition in the study of behaviour has tended to use field observations, or field experiments that are minimally disruptive to the normal life

175

of the animal (for those who are not familiar with this type of research, Tinbergen (1953, 1972) would provide an illuminating introduction). There are, of course, exceptions. An instance of an unpleasant and unacceptable field experiment is reported by Berkson (1970), who trapped baby monkeys on an island in the Gulf of Siam, blinded them, returned them to the wild and observed their progress until they 'disappeared'. In spite of some examples like this, we have concentrated on experiments of the Cartesian type in this chapter because they generally do give rise to more concern. A second reason is that they are probably the type that is most likely to lead to advances in medical practice. It is a quite widely held principle among the general public that if experiments that injure animals are to be conducted, they should at least be restricted to those that might reasonably be expected to improve medical care. We are not going to evaluate the principle, but we have made some effort to decide on the extent to which certain types of experiment might have benefits of this kind.

Experiments on the behaviour of animals are the province of three academic disciplines—zoology, psychology, and physiology. It sometimes surprises people to discover that psychologists carry out experiments on animals, and there have indeed always been psychologists who have thought that this was not among the tasks proper to the discipline. Ironically, the reason psychologists carry out Cartesian experiments on animals is precisely because they are not, in another sense, Cartesians; for Descartes believed not only that the body of men and other animals should be seen as a machine, but also that other animals comprised *nothing but* this machine, while men comprised an inexplicable association between their body, a machine, and their mind, which whatever else it was was certainly not a machine; a view which has since been stigmatized as that of 'The Ghost in the Machine' (Ryle, 1949).

This idea of a radical discontinuity between man and other animals would imply, were it accepted, that there would be little profit in approaching human psychology via the study of other animals. The widespread use of animals in psychology departments rests on a rejection of this view, in favour of the more Darwinian view that although human beings may have some unique characteristics, they nevertheless have many others that they share with animals of other species. What these species are, of course, will depend on the characteristic we are considering. The mechanism controlling thirst, for example, is one that we share with reptiles, birds, and mammals—i.e. with all vertebrate classes primarily adapted to life out of water. Suckling the young is a habit we share with all and only mammals. Colour vision appears to be very similar in humans and those apes and Old World monkeys that have been studied; but there is evidence that the perceptual world of some colour-blind humans (the protanomolous trichromats) approximates more closely to that of the New World monkeys. Faced with examples like these, it makes no sense to talk about the similarity or dissimilarity of 'man and animals'; worthwhile laws simply cannot be found at that level of generality. Some human psychological functions can be studied in some other animal species; others cannot. The fact that some can is sufficient justification, scientifically, for psychologists mainly interested in human beings to study psychology comparatively, in other animal species. It is also here, of course, that the central moral question raises itself most acutely; for the same evolutionary insights that lead us to believe that there can be no radical discontinuity between man and other animals in psychological functions ought also to make us wonder whether it is proper to continue to treat them as radically different in morals.

As regards the number of animals used in psychology departments, we have reasonably accurate figures for Great Britain from a useful recent survey carried out by the British Psychological Society (1979). This results are reproduced as Table 8.1. The total number of animals used in the year of the survey was 43 196. This is less than 1 per cent of the $5\frac{1}{2}$ million experiments carried out annually in this country. Behavioural experiments are very time consuming, and it is not surprising that the number of animals used in psychology departments is small in relation to the total number of animals used. The total number of animals used is not likely to be reduced by the development of alternatives of a technical kind, in the way that radio-immunoassay has reduced the number used in endocrinology, and tissue cultures may reduce the numbers used in some areas of cancer research. If one is to study the behaviour of an animal, one has to have an animal. Computer simulation is used in the study of behaviour, along with other forms of formal modelling; but these are types of theorizing rather than types of empirical investigation, and cannot substitute for empirical work.

We will first consider briefly the 'non-experimental' use of animals in column (a); the survey indicates that 1174 animals a year are being used for purposes of this kind, including 136 monkeys.

It is quite legal in Great Britain to use animals of any species in experiments that are not 'calculated to cause pain' without a licence from the Home Office, and hence, of course, without including them in annual returns to the Home Office. It seems unlikely that this happens on any substantial scale in medical departments, but it certainly happens during the course of behavioural work in departments of psychology and zoology. Most behavioural experiments do not themselves involve the death of the animal; but animals of species commonly used in behavioural experiments, rats and chicks for example, are generally killed after a single experiment. It seems to be widely believed that keeping animals in captivity, and killing them, are, in themselves, of no moral significance; this belief presumably underlies the exclusion of experiments of this kind from the *Cruelty to Animals Act*. This view is deeply rooted in the Christian moral traditions of our culture, but it is worth at least considering other views.

As regards death, the problem was raised in a characteristically clear form by J. B. S. Haldane, in his very interesting introduction to *The biology of mental defect* by Lionel Penrose (1949), a great work of human biology and one that every psychologist ought to have read:

Table 8.1 Numbers of animals per year subjected to various types of procedure

Animal	(a) Non-experimental	(b) Experimental other than (c)–(h)	(c) Drugs	(d) Surgery	(e) Shock	(f) Food/water deprivation as incentive	(g) More severe food/water deprivation	(h) Other forms of deprivation
Monkey	136 (4)	35 (3)	15 (2)	84 (6)	9 (1)	85 (5)		
Cat	12 (1)			20 (1)				
Rabbit	90 (2)	42 (1)	40 (1)	128 (2)	230 (1)	70 (1)		
Rat	616 (13)	8478 (19)	5837 (18)	3750 (20)	3550 (20)	6779 (29)	634 (8)	5916 (4)
Guinea-pig		20 (1)	20 (1)			60 (1)		
Gerbil	20 (2)	40 (2)	50 (2)		10 (1)	10 (1)	10 (1)	
Mouse	150 (2)	1210 (5)	639 (5)	105 (2)	70 (2)	400 (2)	40 (2)	
Other mammals	54 (2)	160 (2)				7 (1)		
Pigeon			170 (3)	70 (3)	60 (3)	738 (15)	24 (2)	
Chick		1000 (1)		600 (1)				
Other birds	92 (3)	600 (1)				18 (1)		
Reptiles	4 (1)							
Fish			80 (1)	4 (1)		5 (1)	100 (1)	
Totals	1174	11 585	6851	4761	3929	8172	808	5916

Figures in parentheses indicate number of departments involved in each total.
Reproduced by permission of the Scientific Affairs Board of the British Psychological Society. (We have corrected the total under (e), which appears as 3939 in the original.)

'Apart from its practical aspect, I believe that the study of mental defect is of considerable philosophical importance. The question of why people are different or what determines their individuality is of the greatest interest, and to my mind is one of the questions which shows up the weakness of idealistic philosophies. It can be answered in a few cases. And the answers may be very surprising. "John Smith is a complete fool because he cannot oxidize phenylalanine" discloses a relation between mind and matter as surprising as transubstantiation, and a good deal better established. On the ethical side it raises the problem of human rights in a rather sharp form. Has a hopeless idiot the right to life and care, though he or she is not a rational being nor likely to become one? If so, has a chimpanzee with considerably greater intelligence similar rights; and if not, why not?'

It would be interesting to see an animal psychologist who works on primates take up this challenge, because psychology has a double interest here; it is both a discipline that uses animals in the framework of the conventional Christian moral tradition, and a discipline that is professionally concerned with comparisons of the mental life of man and other animals. Haldane himself, we can be sure, would have answered, 'Yes', to both questions.

Captivity can injure animals in two ways. It can cause suffering directly, if the animals are not properly fed, housed, and so on. In our experience the standard of animal husbandry, in this restricted sense, is extremely high in Great Britain; there is not much ground for complaint here. We might consider, as well, though, that keeping animals in captivity can injure them in a second way by depriving them of their 'proper pleasures'; and that this is also of moral significance (an issue that is discussed by Clark, 1977). Consider the following observation, by a psychologist who has worked a good deal with monkeys:

'Some years ago I made a discovery which brought home to me dramatically the fact that, even for an experimental psychologist, *a cage* is a bad place in which to keep a monkey. I was studying the recovery of vision in a rhesus monkey, Helen, from whom the visual cortex had been surgically removed (Humphrey, 1974). In the first four years I'd worked with her Helen had regained a considerable amount of visually guided behaviour, but she still showed no sign whatever of three-dimensional spatial vision. During all this time she had, however, been kept within the confines of a small laboratory cage. When, at length, five years after the operation, she was released from her cage and taken for walks in the open field . . . her sight suddenly burgeoned and within weeks she had recovered almost perfect spatial vision. The limits on her recovery had been imposed directly by the limited environment in which she had been living. Since that time, in working with laboratory monkeys I have been mindful of the possible damage that may have been done to them by their impoverished living conditions.' (Humphrey, 1976)

What this shows rather clearly is that keeping animals of at least some species in laboratory cages may be against their interests, even if normally accepted standards of animal husbandry are assured, because the animal is deprived of the opportunity to exercise its capacities and engage in its normal instinctual life. And if this is so, one ought not to consider even those experiments that involve keeping animals in cages and using them in ways that do not cause pain as of no moral significance, even if they need no licence from the Home Office. We just mention this here, since it is mostly in the behavioural sciences that such experiments are likely to be conducted.

We turn now to consider three areas of psychological research in which the injuries done to the animal are rather more obvious. The first is sexual behaviour; the second, pain; and the third, vision, and in particular that subset of the many experiments on vision that is concerned with the development of the visual system of mammals. We have chosen the first because we have some familiarity with the kind of work involved at first hand; the second because some of the work done in this area is a disgrace, and we take this opportunity to say so; and the third because it has been the source of considerable public concern in Great Britain. From our limited acquaintance with published work we are inclined to think that this concern may have some justification.

SEXUAL BEHAVIOUR

Recent comprehensive reviews of research on sexual behaviour in vertebrates can be found in Hutchison (1978). The behavioural method basic to the study of sexual behaviour is the mating test; and the evidence, if evidence is needed, is that mating is as rewarding to other mammals as to human beings. There is little to object to in the mating test itself. The physiological methods are another question. Castration and ovariectomy are operations which must cause a certain amount of discomfort after recovery from the anaesthetic. A guess based on observations of operated animals would be that they involve moderate discomfort for 1 or 2 days after the operation. Replacing the hormones by injection is painful only to the extent that the injection is painful, which depends on the skill of the experimenter. (It is worth noting at this point that the vast majority of experiments in this country are done under an A certificate, i.e. without anaesthetic: for example, in 1977, 4 319 321 out of 5 385 575; and in 1978, 4 276 160 out of 5 195 409 (Home Office, 1978, 1979). It is probably safe to assume that many of these would involve injecting the animal, and that the injection procedure would not itself be more than trivially painful. But it would be quite wrong to assume that this large body of experiments therefore involves no suffering. How much the animal suffers depends on what is injected; some substances cause suffering: others do not. And we have no way of knowing what proportion of the experiments involves the injection of substances that do cause suffering It is not even always necessary to specify on a Certificate A what one is going to inject.)

Other methods that have been used in the study of sexual behaviour are the production of lesions in the brain, and direct stimulation of the brain of conscious animals with hormones or electrically. These are techniques also used extensively in the study of hunger and thirst, and in other kinds of behavioural work. Brain lesions are produced under anaesthetic, by a variety of means. The discomfort produced by the operation itself is hard to assess, but would probably last for several days. More permanent and traumatic effects may occasionally result from the neurological effect of the lesion. It is impossible to generalize here, but it is worth noting that one lesion, the lateral hypothalamic lesion, which has been used widely in the study of hunger and also, to some extent, in the study of sexual behaviour, does seem to have a traumatic general effect:

> '[There is]. . . not only loss of the specific alimentary drive, loss of hunger, but at the same time a lack of some kind of general drive, absence of energy, loss of the "joy of life". The dogs were sad, indifferent, did not jump, did not play, did not greet their experimenters. This general look of deep sadness was very striking and reminiscent of patients during the depressive state or schizophrenia. Nothing is worth an effort.'
> (Fonberg, 1969)

Stimulation of the conscious brain depends upon the use of implants in the skull, either of cannulae containing minute amounts of hormones (as in Harris and Michael, 1964), or of electrodes (as in Delgado, 1961; Miller *et al.*, 1961). The use of chronic techniques of this kind is indispensable in behavioural work; for it is only possible to show that stimulation of a discrete brain site by a hormone, or electrically, results in a particular pattern of behaviour if the animal is conscious and free to move. There is no reason to think that such implants cause distress to the animal after recovery from the operation as long as they remain secure; but unfortunately these implants often come loose from the skull after they have been in use for some time, as a result of local infection or mechanical failure of some kind. This is not an isolated happening; it has been a problem in most of the laboratories we have known that use implants of this kind in rats. It is obviously traumatic, and leaves the animal with an open wound to its brain. The animals are generally killed when they are so found, but this might be several days after the implant comes loose, unless inspection procedures are very strict.

Work in this area has led to several major advances in our understanding of the control of sexual behaviour of animals. We know that sexual receptivity in most mammalian species depends absolutely in the female on one or both of the steroid hormones of the ovary; and that sexual activity in the male depends on the androgens of the testis. (Those who think that these facts are too obvious to need demonstration might care to re-read what Kinsey *et al.* (1953) had to say on these relationships.) Detailed studies are available of the patterning of sexual interaction in a few species, and studies of hormone uptake and of the effects of hormone implants in the brain have gone some way to localizing the sites at which steroid hormones act. Developmental studies have shown that the brain of some mammalian species is *sexually differentiated* at about the time of birth as a result of the action of testicular androgens, which produce permanent changes in its anatomical structure and physiological function, and so permanent changes in behaviour.

This is a rich harvest of biological facts, and we do not think that their scientific value is open to any doubt. In addition to their obvious interest as regards the control of behaviour, and the mechanisms by which common hormonal mechanisms integrate sexual behaviour with ovulation and spermatogenesis, studies of this kind had a considerable historic importance in the development of what is now known as 'neuroendocrinology', i.e. the study of the relationships between the nervous system and the endocrine system, the two major control systems of the body. Earlier treatments of these two systems as separate gave way during the 1930s to the idea of closely integrated systems in which steroid hormones controlled behaviour, and their own release, via the central nervous system, and the central nervous system controlled the release of hormones via its control of the pituitary gland. The isolation, characterization, and synthesis of the hormones that the brain elaborates to control the pituitary gland has been a major gain to medicine, which opens up new possibilities for the treatment of infertility and for the control of conception.

The extent to which studies of the control of sexual behaviour in animals has contributed to the solution of clinical problems of sexual behaviour in humans, however, has been much less. There are two reasons for this. One is that the relationships between sex hormones and behaviour in humans seem to be essentially unlike those of most other mammals, with the possible exception of some other primates. If ovarian oestrogens affect human sexual interactions at all, they do so to an extent that is so slight and variable as to be almost undetectable (James, 1971; Bancroft, 1978). It must follow, then that the very large body of work dealing with the dramatic all-or-none effect of oestrogens on sexual behaviour in other species cannot be directly relevant to the study of human sexual behaviour. Similarly, the sexual differentiation of the brain (or at any rate of those mechanisms that control ovulation) seems to be something that does not occur either in rhesus monkeys or in humans (Herbert, 1978), so the large body of work directed to its elucidation is not directly relevant to humans. This last could hardly have been predicted; but there can never have been a time at which it would have been reasonable to suppose that change in ovarian hormones during the menstrual cycle would have a major effect on sexual behaviour. Recent work on monkeys (Herbert, 1978) has indicated that adrenal androgens may stimulate sexual behaviour in the female, and this may also be true of women. While this notion may certainly be of importance in clinical practice, it was itself first derived from clinical practice (see, for instance, Money, 1961); and the work on monkeys has not given it the unequivocal confirmation that one might have hoped for (Michael et al., 1978). The second reason why studies of sexual behaviour in animals do not seem to have made a great impact on human clinical practice is that the commonly presenting problems of human sexual psychology are not generally problems of hormone–behaviour interaction. Burnap and Golden (1967) provide a list of sexual problems presenting to general practitioners, with their frequencies; a more general discussion is to be found in Bancroft (1972). It is clear from these sources that the vast majority of sexual problems in psychiatric practice are of a quite different kind. A very occasional case of impotence in men may be due to low testosterone levels (e.g. Beumont *et al.*, 1972), but most are not; indeed, one could argue that our understanding of impotence might have advanced more quickly had less importance been attached to the possibility that it might have an hormonal basis. Premature ejaculation, the other common problem of men, does not involve any hormonal abnormality, nor do the main problems of women: lack of orgasm, painful intercourse, and the group of difficulties unpleasantly and unhelpfully referred to as 'frigidity'. What we think is remarkable here is the extent to which the very large body of behavioural work on animals has not had any major clinical pay-off in the treatment of human sexual problems. The research workers who have made major contributions here have been those pioneers who have worked directly on human sexual behaviour, particularly the zoologist, Kinsey (Kinsey *et al.*, 1948, 1953) and the gynaecologist and social psychologist Masters and Johnson (1966, 1970).

It is certainly true that research on the sexual behaviour of mammals contributed to the early development of the field now known as neuroendocrinology—the study of the relationships between the brain and the endocrine glands. But we are inclined to think that psychologists and others who are now genuinely interested in advancing medical progress in this area ought to be working directly on the human case, and not on animals. The reason that 'our understanding of the relationships between hormones and human sexual behaviour is at a primitive stage' (Bancroft, 1978) is not that it is impossible or unethical to carry out the relevant studies in humans. We now have good assays available for all the relevant hormones, and it is not difficult to arrange the level of medical supervision needed to take blood samples. Methods of behavioural analysis are also available which are adequate, if not perfect (see, for example, Udry and Morris, 1967; James, 1971; Beumont *et al.*, 1972). What has been lacking has simply been a sufficient investment of time and effort. To some extent this may have been a result of too great a concentration on work on animals.

PAIN

To psychologists must go the credit, if such it be, for the initiation and development of research concerned with the behavioural effects of painful stimuli, and particularly of electric shock. (We refer here to shock applied to the skin, not the electrical stimulation of brain tissue, which can be, but it is not necessarily, painful.) The first study using electric shock as a punishing stimulus was by Elmer Gates (1895); Yerkes (1903) contributed another early study. A useful source of information on research on punishment is Boe (1969), a bibliography of papers on punishment from 1895 until 1967 which gives an indication of the growth of research using aversive stimulation for its first 65 years. In six leading American journals devoted to research in experimental psychology, the number of papers on punishment published was less than five each year until about 1958, and then suddenly increased sevenfold to about 35 a year in 1966. The total number of papers published increased about three times over the same period. In interpreting these numbers it is important to note the constraints on Boe's bibliography. It is concerned only with studies in which punishment was itself the focus of attention and not with those, for instance, in which it was used as a tool in the investigation of memory, or of drug effects. Nor does it include studies of active-avoidance or escape learning, in which the animal learns to *do* something to escape or avoid an aversive stimulus: it deals only with those in which it must learn *not* to do something in order to avoid an aversive stimulus. It also deals with all studies of punishment, and so includes studies on humans, though most of the studies are on animals, coming as they do from the salad days of behaviourism. More recent data on the extent of the use of electric shock in departments of psychology in Great Britain can be found in the British Psychological Society report (British Psychological Society, 1979), which shows that nearly 4000 animals a year are used in experiments involving shock; about 9 per cent of all animals used. In the country as a whole, the number of animals subjected to aversive stimuli (electrical or other) was 54 363 in 1977 and 57 790 in 1978 (Home Office, 1978, 1979).

All intensities of electric shock above the detection threshold are aversive (Campbell and Masterton, 1969). The detection threshold sets the lower limit to the range of useful currents. The upper limit is set by the death of the animal, and psychologists have not shirked the stern duty of using electric shocks throughout this range. Consider, for instance, Ulrich and Azrin (1962). This study involved an examination of reflexive fighting in response to aversive stimulation. Briefly, rats paired together will fight if they are given an electric shock. This has been known for many years (O'Kelly and Steckle, 1939). The special contribution of Ulrich and Azrin was to investigate 'several possible determinants of this fighting reaction'. To this end comparisons were made of the effects of different rates of presentation of shock (0.1–38.0 per min) and different intensities of shock (0.50–5.00 mA). The authors' comment on the effect of the 5 mA shock on Sprague-Dawley rats is as follows: 'The slight decrease in fighting behaviour at the highest intensity (5 mA) appeared to be partly a consequence of the debilitating effects of the shock. Prolonged exposure to this intensity often resulted in a complete loss of fighting because of the paralysis of one or both of the subjects.' Two out of four rats of another strain (Wistars) died after exposure to 2 mA shocks. A further aspect of this study was an investigation of 'reflex fatigue'. The fighting reflex proved extremely resistant to fatigue; this point was confirmed by subjecting a pair of rats, over a period of $7\frac{1}{2}$ hours, to a total of 18 000 shocks. At the end of this time the rats were 'damp with perspiration and appeared to be weakened physically.'

Dealing with the scientific importance of these results would be easier if the authors dealt with it themselves; but following the ascetic tradition of the *Journal of the Experimental Analysis of Behavior* they simply state their results with no theoretical elaboration. Clearly the authors have answered a series of questions; what is not at all clear is why they have asked them, and whether they are important. Fighting between animals of the same species that is elicited by a naturally occurring stimulus could be of scientific interest, but fighting elicited by a wholly artificial stimulus is not necessarily of any interest; still less, of course, are *parametric* studies of the effects of an artificial stimulus necessarily of any special interest.

It is true that studies of this series are cited in a recent widely used textbook of psychology (Brown and Herrnstein, 1975; Figure 5.1) but the citation does little to reassure us of their value. The figure shows a picture, based on Azrin *et al.* (1964), of a squirrel monkey seated in a restraining chair; in front of it is a covered ball hanging from a chain. The monkey is given electric shocks to its tail. The legend tells us that

'The constant stimulus object is the ball. In the absence of shock the monkey shows no interest in the ball. When a brief shock is delivered to the monkey's tail he attacks the ball and sinks his teeth into it. So we have variable behaviour in the presence of a constant stimulus. To explain this variability we plan to invoke an internal drive of aggression, a regulator thrown off by shock.'

As a piece of psychological theorizing, this seems quite vacuous. The facts are clear enough: the monkey attacks the ball when it is shocked and does not when it is not. To describe this as showing that there is 'variable behaviour in the presence of a constant stimulus' seems rather strange; one stimulus—the ball—is constant; the other—the shock—is not. In any case, we do not need to give electric shocks to monkeys in order to convince ourselves that animals sometimes respond in different ways to the same stimulus. It is a common enough observation. As to the suggestion that there is 'an internal drive of aggression, a regulator thrown off by shock' this seems to add nothing at all to the facts as already stated; what 'a regulator thrown off by shock' is intended to imply is hard to imagine. The initial experiment is not, of course, the responsibility of the writers of textbooks. What we think they should consider is the effect on the student of using examples of this kind. Whatever else they may learn of psychology, it seems clear that one thing that they will learn is that giving electric shocks to restrained monkeys is a perfectly conventional and acceptable thing to do. Some authors may believe this; those that do not would help a good cause if they would make their position plain, following the example of Archer (1979) who describes these procedures as 'barbarous': an opinion from which we would not dissent.

It may sound surprising that these studies should be cited in relation to clinical practice, but so they are, in Rachman and Teasdale's (1969) careful and critical work on aversion therapy. This is a method used by clinical psychologists to alter behaviour patterns (in, for example, homosexuals and alcoholics) by associating them with punishing stimuli, sometimes chemical, but more recently generally electrical. The use Rachman and Teasdale make of Ulrich and Azrin's data is as follows. They note that they establish that pain can lead to aggression. They draw attention to the critical parameters; aggression is a function of shock frequency, duration, and consistency. And they suggest on the basis of the animal work that a possible improvement in the design of aversion therapy may be to prevent the occurrence of pain-elicited aggression by administering the noxious stimulus remotely. We intend no criticism of Rachman and Teasdale, but there are three points that we would like to make about this use of Ulrich and Azrin's data. Firstly, the extent to which it can serve in a defence of the animal work must depend partly on one's evaluation of aversion therapy as a method of psychiatric treatment; like the authors of the book, we have mixed feelings about it and 'recognize that it is an unpleasant form of therapy and one which is open to abuse.' Those who think its use can never be justified ought, nonetheless, at least to consider the way in which a humane psychiatrist might use it in his patients interests (see, for instance, Bancroft, 1974, Chapter 9). Secondly, we need to consider whether these suggestions depend *critically* on the experimental work carried out by Ulrich and Azrin, and we would take some convincing on this point. That people might respond aggressively if they are deliberately given electric shocks, that the extent of the aggression might depend on the number and strength of the shocks, and that the aggression might be managed by ensuring that there was nobody else in the room at the time are sensible enough suggestions; we find it hard to believe that they could only occur to a therapist who had read Ulrich and Azrin. And thirdly, and, as far as we are concerned, centrally, even if experimental work were needed on these points it is difficult to see on what basis animals could properly be used as subjects. Those levels of shock that can be used in clinical practice are those that are safe. Since they are safe, they can be used for experimental work on humans, and work on human volunteers would obviously be more directly relevant. Of course, the human volunteers may decide that the general good to be gained from the results would not be enough to persuade them to accept the pain involved. But if this is so, it is hard to see why we should believe that it does justify inflicting it on animals of other mammalian species; the shocks are the same shocks, and the general good the same general good.

Ulrich and Azrin's enthusiasm for unnecessary parametric studies is not unique; another example can be found in Campbell and Masterton (1969). This paper reports investigations into what the authors like to call the 'psychophysics of punishment', in which the detection, aversion, tetanization, and death thresholds of rats are plotted systematically, using shock generators of different source impedance.

Now one can understand that there might be some rationale for determining systematically the detection and aversion thresholds (i.e. the lowest currents that the rat detects, and avoids, respectively). And obviously, of those experiments that use shock, the ones that in the light of this information confine themselves to shock at or close to the aversive threshold will be less traumatic than those that use higher shocks. The tetanization threshold is much higher; for example, using the two constant current shock sources the tetanization thresholds are 134 and 406 times greater than aversion thresholds. Obviously currents of this level will be among the most painful that it is possible to use; and one might expect to find a correspondingly greater justification given for their use. The authors seem to be aware of the need, since they themselves say that 'the investigation of these processes is one that should be undertaken only after serious consideration of the scientific value of the findings that may be obtained'. The fruits of this serious consideration are hard to detect. Whatever else may be open to argument, it seems obvious that currents above the *tetanization* threshold would be of no value in behavioural work; for the animal that is tetanized is 'completely immobilized and unable to release the stainless steel grid'. To continue, as Campbell and Masterton do, to then raise the shock to determine the 'death threshold', which in the case of a constant current d.c. source is $2\frac{1}{2}$ times higher again than the tetanization threshold, is an example of work of exceptional cruelty which as far as we can see is of no scientific worth at all. Even their tetanization threshold seems to have been measured in the wrong way; for if the importance of the threshold is that it marks the 'upper limit of usefulness of electric shock as a psychological stimulus' then clearly it is the current that produces tetanization in any animal, not half the animals, that represents this relevant upper limit. The use of the 50 per cent mark seems to be an ill considered and quite inappropriate mimicking of its use in the LD_{50} test.

We have been dealing here with experiments by psychologists that involve severe pain, deliberately inflicted. We want to consider now what is undoubtedly an important practical objective, namely the understanding of the mechanisms of pain. One might think that research in this area must necessarily involve causing pain to animals. We do not think that this is necessarily so; or at least, that anything approaching the levels of pain found in the experiments cited above would be necessary.

In the first place, the neuronal systems involved in pain can be studied in anaesthetized animals just as those involved in vision can. The isolation of the skin nociceptors involved electrophysiological experiments of this kind (Iggo, 1959, 1977); and the so-called 'gate-control' theory of pain, the publication of which (Melzack and Wall, 1965) was undoubtedly the seminal event in recent research on pain mechanisms, was founded on work on neuronal interactions in the spinal cord which was not itself concerned with pain at all (Nathan, 1976).

In addition to work involving fully anaesthetized animals, there is much to be learned from human beings in this area. The development by Vallbo and Hagbarth (1968) of a method for recording the electrical activity of peripheral nerves in conscious humans has been of major importance, and has led to the discovery in humans of peripheral nociceptors of a kind comparable to those found in other mammals (Torebjörk and Hallin, 1974). Studies of this kind are, of course, of exceptional importance. In the first place, they dispense with the need for anaesthesia, which is always a major gain in studies of the nervous system; secondly, they allow neurophysiological and psychophysical work to proceed in parallel, another major gain; thirdly, they elucidate, not an animal model, but the actual pain system that is clinically relevant: a third major gain. Many different types of psychological investigation of pain are also, of course, possible in human volunteers; indeed, one of the most important facts of pain is the extent to which it can be controlled psychologically, for instance by relieving anxiety.

Even when conscious animals are used in experiments on pain, the pain does not necessarily have to be of an unacceptable level. Consider a recent experiment of Mitchell and Hellon (1977). In electrophysiological experiments these authors investigated the responses of neurones of the dorsal horn of the spinal cord and the thalamus to a stimulus that would be painful in conscious animals, namely the heating of the tail. They found a sudden response when the tail was heated to 42 °C. In one of their behavioural experiments, they allowed conscious rats to rest their tails in water, at 30 °C. They then heated the water. The rats removed their tails from the water (on average) when it reached the temperature of 43.7 °C. This result shows, therefore, that the rat removes its tail just above the temperature at which the neurones studied in the electrophysiological experiment increase their activity, and this is an important additional piece of evidence that they are indeed involved in pain responses. Yet the actual discomfort involved, if the rat is free to remove its tail at any time, would be no greater than we subject ourselves to in testing out a bath with one toe while running the hot tap.

Melzack (1976) argues that such threshold tests are not adequate. This is on the grounds that

'there is convincing evidence that procedures that effect pain at intense levels may have little or no effect on pain at threshold level. The tritration technique, for example, has missed finding the analgesic properties of midbrain stimulation. Stimulation appears to produce analgesia to intense, noxious stimulation, but seems to lower the threshold to shock'.

It is hard to see on what basis these claims are made. Melzack (1976, p. 139) cites two papers on the analgesic effect of midbrain stimulation. One, by Reynolds (1968), shows that midbrain stimulation produces analgesia not only to surgical laparotomy, but also to mechanical pressure from a haemostat applied to the paws and the tail. In the other (Mayer and Liebeskind, 1974) a series of tests was used, including the 'pinch' test, the 'jump' test—which measures the response threshold to electric shock—and the tail flick test. The analgesic properties of midbrain stimulation was illustrated in all cases—the author's summary reads 'stimulation of the central and periventricular gray matter produced analgesia equal to or greater than 10 mg/kg morphine *on all tests*' (our italics).

Melzack is distressed at the criticism to which he has been subjected. 'Having personally been the object of attacks by the anti-vivisection press', he writes, 'I know how much it "hurts" to be considered an evil, brutal sadist when my aims were, in my estimation, genuinely humanitarian'. We agree that Melzack's aims are genuinely humanitarian, and we have a very great respect for the variety and quality of his research. But it is, of course, means, as well as aims, that need to be genuinely humanitarian. For those interested in this problem, the carefully considered views of the Editor of *Pain* are well worth reading (Wall, 1975, 1976). A valuable principle put forward in these editorials is that 'we shall refuse to publish any reports where the animal was unable to indicate or arrest the onset of suffering.'

VISION

The study of vision is perhaps the most successful of all areas of brain research (Hubel and Wiesel (1979) provide an accessible recent review.) Although the ablation of parts of the visual system has made a contribution here, there is no doubt that the use of electrophysiological methods for analysis of responses of single neurones has been principally responsible for the advanced state of the area, and we are going to restrict ourselves to experiments of this type.

Virtually all electrophysiological experiments involve the death of the animal involved; the extent to which they also involve suffering depends critically on the type of anaesthetic employed. The principles here are general, and apply to areas other than vision research. At one extreme is the case in which an animal is anaesthetized with a general anaesthetic at the start of the experiment and killed before the effect of the anaesthetic wears off. The animal will suffer only if the anaesthetic level is not correct; and it is readily obvious to the experimenter from the animal's movements if the anaesthetic level is too light. This is not the case if the animal is treated with a paralysing agent such as curare or one of its analogues. These block the activity of muscles, without acting as anaesthetics. So an animal treated with a muscle paralysing agent cannot respond in a way that would indicate its pain. The use of these agents is now extremely common in electrophysiological work; for example, they were used in 11 of 27 studies on the hypothalamus listed by Cross and Silver (1966), and in almost every one of the many electrophysiological studies of the visual system published in the *Journal of Neurophysiology* over the last few years. If the animal is fully anaesthetized, there is no objection to using a paralysing agent in conjunction with the anaesthetic, for example to reduce eye movements. The problem is ensuring that this condition is fulfilled. So it is critically important to examine the anaesthetic conditions involved.

For an animal to be paralysed and not anaesthetized at all is unusual. An example which at least comes close to this can be found in Gur and Purple (1978). These authors used ground squirrels that were initially anaesthetized with sodium pentobarbital. No local anaesthetics were used. A tracheotomy was performed, and a vein and artery cannulated. The upper eyelid was removed and an area of sclera dissected away so that a microelectrode could be lowered into the eye. At the end of these procedures, the animal was paralysed with gallamine triethiodide. The barbiturate was then replaced by halothane, but in some experiments neurones were studied in ground squirrels 'emerging from pentobarbital anaesthesia before halothane was titrated in to stabilize anaesthetic state'; that is to say, in animals that were not anaesthetized.

Much more common is the use of a combination of a paralysing agent with local anaesthetics. One experiment published in *Nature* recently (Duffy *et al.*, 1976) used the following procedure. Five kittens were deprived of the vision in one eye by sewing up the eyelids in the fourth week of life. Seven months later they were paralysed with gallamine. They were then held with their heads in a clamp and the skull opened using a local anaesthetic. In addition, the paralysed cats were given drugs at different stages in the experiment which sent them into epileptic fits. Experiments of this kind are open to two objections. The first is that the paralysis may itself cause suffering. We are not told the length of the experiment, but experiments of this kind often continue for 48 hours or more (Barlow, 1975). It would be extraordinary if holding a cat immobile and unable to breathe spontaneously for as long as this did not in itself cause suffering. Secondly, the paralysis makes it quite impossible to know to what extent the local anaesthetic is effective in preventing pain resulting from the operative procedures. As far as we are aware, it is not possible to obtain a licence to carry out electrophysiological experiments under these conditions in Great Britain. This is as it should be, and it seems a pity that a journal of *Nature*'s standing should nonetheless be prepared to publish them.

In a third type of experiment, the paralysing agent is combined with a general anaesthetic. This is, of course, satisfactory if the general anaesthetic is itself adequate. An example in which this is open to question is the combination of nitrous oxide and oxygen used in conjunction with paralysing agents in a long series of experiments on the development of vision in cats (e.g. Blakemore and Cooper, 1970; Eggers and Blakemore, 1978).

It is clear that nitrous oxide is not itself a general anaesthetic in cats (Russell, 1973). Its use in combination with paralysing agents has been based on a claim (Blakemore *et al.*, 1974) that it maintains adequately anaesthesia first induced with barbiturates. An independent report shows that this claim is mistaken (Richards and Webb, 1975; and see too Hammond and Mackay, 1977). Richards and Webb found that nitrous oxide only maintained anaesthesia induced with a barbiturate for about 2 hours, and then only if the barbiturate dose was large. It appeared not even to have an analgesic effect. It therefore seems quite wrong that this combination should continue in use, since the balance of evidence is that the cats, although paralysed, would not be adequately anaesthetized. In one American study (Cynader and Berman, 1972) cats were used in electrophysiological experiments for 36 hours, and monkeys for 3–5 days. Adequate general anaesthesia was used during the surgical preparation, but thereafter the animals were paralysed and no local anaesthetics used. The general anaesthetic during the recording (and while the animals were perfused with saline and formalin at the end of the experiment) was 60 per cent nitrous oxide with 40 per cent oxygen. Even in the original paper of Blakemore *et al.* (1974), nitrous oxide in a higher concentration than this (65 per cent) was firmly described as 'definitely inadequate'.

What is the importance of these studies on the development of vision? There is little room for doubt as to their scientific value. By demonstrating that the physiological responses of the nervous system are modified by visual experience, they have made accessible in a new and sharpened form the old problems often characterized as the empiricist/nativist controversy. The comparison of different species is important, too, for it has in it the germ of a new integrated understanding of the relationships between ecology, developmental psychology, and brain physiology. And it is not wholly unreasonable to think that the kind of change found in these experiments might turn out to have a more general importance either for studies of the embryology of the nervous system, or for work on the mechanism of learning.

As to their importance in guiding clinical practice, however, there is perhaps more room for doubt. Amblyopia is abnormality of visual function, characterized by reduced visual acuity, in the absence of ophthalmoscopically visible abnormalities of the eye. It results from causes such as strabismus (misalignment of the two eyes), anisometropia (unequal optical power of the two eyes), and congenital cataract. It is clear from the work we have been considering that conditions that may be analogous to human amblyopia can be produced experimentally in animals, by monocular deprivation, for example, or by surgically produced strabismus; and there is a tendency now to claim considerable clinical importance for such findings. The Report of the Medical Research Council for 1977, for example, reads as follows:

> 'Work on the development of the mammalian visual system may lead to explanation of human visual disorders and to the identification of sensitive periods during which the visual system is capable of modification . . . The programme grant held by Professor H. B. Barlow and Dr C. B. Blakemore . . . has been renewed. Basic research in this field should facilitate the investigation of more applied problems such as the treatment of amblyopia . . .'.

The conclusion of one of the papers referred to above (Eggers and Blakemore, 1978) reads as follows: 'The development of an animal model for the common human disorder of amblyopia offers hope for the design of more effective methods of treatment or prevention.'

Such claims are now commonplace; yet it is difficult to see what new principle of clinical importance derives from these findings. The animal work originates with Wiesel and Hubel (1963) and Hubel and Wiesel (1965); but it was by then already well established in clinical practice that there was a critical period in the development of the human visual system. Worth (1903) believed that binocular vision had to develop within the first 6 years of childhood, or it would not develop at all, and the importance of correcting optical or muscular defects in the early years of life was noted by Peter (1931), Chavasse (1932), and Bock (1960). Juler (1921) and Broendstrup (1944) present data showing that cataracts acquired before 6 years of age left much greater visual loss after extraction and optical correction than those occurring in older children; both papers present clear documentation of the effect in the form of lists of tested cases. Indeed, Wiesel and Hubel (1963) themselves point out that the difference between the effects of deprivation in kittens and adult cats is 'a difference one might have expected from the profound visual defects observed after removal of congenital cataracts in man, as opposed to the absence of blindness on removal of cataracts acquired later in life'; and Hubel and Wiesel in (1965) that 'it is recognized that squint in a child must be corrected in the first few years of life if capability of using both eyes in binocular vision is to be retained.' So the fact that there is a critical period in the development of vision, and its clinical implication, that visual defects should be detected and corrected as early as possible, did not derive from this work on animals. It was already known.

TEACHING

We turn now to consider a rather different use of animals; their use in teaching. Two textbooks have recently been published to provide for laboratory courses in physiological psychology. One, an American work, *Experimental psychobiology: a laboratory manual*, edited by B. L. Hart, and written by a group of authors, is published by W. H. Freeman and Co.; the other, a European work, *Techniques and basic experiments for the study of brain and behavior* is written by J. Bureš, O. Burešová, and J. Huston, and published by Elsevier.

Both these books set out to provide experimental exercises for the use of students. Thus Hart, in his preface, states that,

'Up to the present time students in introductory psychobiology courses have not been able to do much experimental work because the procedures used by professional researchers are complex and their equipment is expensive. This is unfortunate because it is direct familiarity with laboratory results that makes many of the significant advances in the field most meaningful. In writing this manual, we have tried to make the intriguing laboratory experiments of psychobiology more accessible to students . . .'.

And Bureš *et al.*, in theirs, say that,

'The aim of the volume is to teach the student the basic skills of behavioural research by providing him with a wide repertoire of reproducible experiments. Most of the experiments can be completed within a few hours, which makes them suitable also for classroom demonstrations and laboratory courses for students.'

An example of an experiment from the Hart volume, contributed by R. L. Isaacson and M. L. Woodruff, runs as follows. Rats are deprived of water except for 1 hour a day. On the third day of water deprivation they are run down a runway, and rewarded with water when they reach the end. On the following day, they are trained in the same way except that in addition they are given 0.5 mA electric shocks when they drink the water ('passive avoidance training'). After this training the rats are anaesthetized, the skull is opened, and aspiration lesions made in different areas of the brain. They are run again in the passive avoidance apparatus, and then killed for histological analysis of the brain lesion. Of the 13 experiments suggested in this book, five involve brain surgery, and three other types of surgery; in all cases, with the animal recovering from the anaesthetic after operation.

The other work, by Bureš et al., contains such a catalogue of cruelty that it is hard to know what to quote. Chapter 2, for example, contains a series of experiments (p. 57) on 'enforced locomotor activity'. Rats are to be put to swim in a tank full of water. 'Rats', we learn, 'can swim for hours before fatigue is manifested by a slower rate of leg movements and occasional cessation of swimming, causing submersion of the head'. The rats are to be left to swim thus until they remain submerged for longer than 30 seconds, something that indicates 'considerable exhaustion'. Suggested experiments utilizing this measure as a dependent variable include varying the temperature of the water between 5 and 35 °C, treating the rat with low doses of anaesthetics or paralysing agents, and taping 5 and 10 g lead weights to the rats' tails (sic: p. 52, line 7; it would be hard to think how cruelty and triviality could be more successfully combined). Rats are to be kept without water for 2 days, or food for 3 days (p. 47), for no purpose other than to provide students with an opportunity to make behavioural observations using an event recorder. The experiments of Ulrich and Azrin on shock-induced aggression (see page 184 above) are suggested as exercises. Rats are to be given shocks of up to 2 mA, lasting up to 5 seconds at a rate of up to 50 shocks a minute. Shocks of 2 mA, the authors state 'should be sufficient to induce the mutual upright fighting posture'. They should indeed: they were sufficient to kill half the Wistar rats used in experiments of this type by Ulrich and Azrin (1962). A further suggestion is that the effects on this response to shock of 'various hypothalamic and amygdaloid lesions' might usefully be examined (p. 64, line 12). In Chapter 3 an experiment on autonomic conditioning is recommended. Rats are paralysed with gallamine triethiodide, and maintained under artificial respiration. In this paralysed state they are subjected to electric shocks given to the tail, in order to examine the effects of conditioning on heart rate. Another learning experiment, on conditioned taste aversion, involves poisoning groups of rats with lithium chloride, apomorphine, and physostigmine. In the chapter on 'Pathological states', experiments on epileptic seizures are recommended. In one experiment conscious rats are immobilized by being taped with adhesive tape to a restraining board; fixation of the extremities in this way is described as a 'highly stressful stimulus that is poorly tolerated by rats' (p. 225, line 14). While thus restrained, the rats are to be subjected to a combination of Metrazol, a convulsive drug, and two alarm bells generating a sound intensity of 90 dB, and thus driven into epileptic fits. Readers whose taste runs to further examples must indulge it with the original book; these are enough to exemplify what is proposed.

It is hard to know quite how to comment on books that recommend experiments as cruel as these for the routine stuff of student practical classes. Many aspects of our use of animals are open to debate; what is proposed here is so obviously wrong that one would have hoped that simply to draw attention to the contents of these books would be enough to ensure that they were condemned. Yet clearly it cannot be so. For publishers must know the content of the books they publish, send the drafts out to referees, have them evaluated and approved before proceeding with publication. Can it really be that no voices were raised against them? And if so, is this because we are so bemused by the jargon of techniques and procedures that we can no longer even see what is being proposed; or is it because a substantial body of psychologists really does believe that there is nothing to object to here? It is hard to understand how publishers of the standing of Elsevier could disgrace themselves by promoting a book like this.

A brief note on these books (Drewett, 1977) suggested that it would not be possible to carry out experiments of this kind in Great Britain; but it appears now that we cannot be entirely confident of that. Doubts here come from the Report of the Working Party on Animal Experimentation of the British Psychological Society (with which we began). What the report says about the use of animals in teaching reads as follows:

'. . . a proportion of studies in which undergraduates play an active role involve procedures which are licenceable. It appears that parts of an overall experiment which might include the administration of drugs or surgery are carried out by a licence holder in preparation for more observational studies by students. For example, a licence holder might inject a drug, but unlicensed students may then observe the behaviour of the animal. There does appear to be, however, some uncertainty among Heads of Departments as to the extent to which the 1876 Act relates to undergraduate teaching, and the guidance of the Home Office Inspectorate is frequently sought on such matters. At the time of the survey only 2 undergraduate students were licensed to carry out experiments on animals. Accordingly, the bulk of such licenceable procedures as undergraduates are involved in is carried out under the supervisor's licence.'

Whatever uncertainty may be felt by Heads of Departments, the status of these experiments is perfectly clear. They are illegal. The *Cruelty to Animals Act* authorizes only two types of experiment. The first is the type which is performed 'with a view to the advancement by new discovery of physiological knowledge or of knowledge which will be useful for saving or prolonging life or alleviating suffering.' This is not the intention behind the experiments referred to above; they are not performed to advance knowledge by new discovery, but for the teaching of students. They therefore come under the restriction of Certificate C, the only Certificate which can authorize experiments for teaching purposes. But Certificate C states that in any experiment carried out for this purpose the animal must during the whole of the experiment be under the influence of an anaesthetic. The experiments referred to are not of this type; for instance the report specifically talks of a licence holder injecting a drug, and 'students then observing the behaviour of the animal'. In other cases the animal is observed after surgery, or given electric shocks. The animals are not anaesthetized throughout the experiment; and the experiments are therefore illegal. Whether the Home Office approved these experiments we cannot tell; but this would not, of course, affect their legal status, for a Minister of the Crown cannot lawfully authorize what is illegal.

A second way in which the safeguards of the *Cruelty to Animals Act* can be circumvented in teaching is for British students to train in another country. The European Training Course in Brain and Behaviour Research runs an annual course in the Laboratory of Physiological Psychology at the University of Bergen, Norway, on 'Basic methods of brain and behaviour research in animals'. In this 10-day course students are given an opportunity to learn and practise such techniques as making brain lesions, implanting animals with electrodes, and avoidance training with electric shock. For reasons given above, it would be illegal to run a course of this kind in Great Britain. And if we believe that the provisions of our own *Cruelty to Animals Act* are proper, then clearly we ought not to evade them covertly by sending our students abroad to carry out exercises that would be illegal in this country. The *Cruelty to Animals Act* does not prevent the training of students; it requires only that, like surgeons, they should be trained by an apprenticeship system, working on experiments with an established researcher until they are sufficiently competent to work unsupervised. We believe that this is the correct system. It may, at times, be somewhat less than convenient. Its justification, however, is not that it is convenient, but that it avoids causing unnecessary suffering.

CONCLUDING REMARKS

It is clear that one can do a great deal of research on the behaviour of animals without taking the subjects out of their natural environment and without acting against their interests. This is the Darwinian tradition of ethological research. It has done a tremendous amount to increase our understanding of the natural world, and has also had a beneficial influence, of a methodological kind, on human psychology, including clinical psychology (see, for instance, Hutt and Hutt, 1970).

The chief value of the laboratory tradition of comparative psychology seems to us not that it provides a general method of understanding the animal species in question, nor that it provides a general means to understanding human behaviour. For as regards the former, it is clear enough by now that the behaviour of an animal is adapted to its natural environment, and so needs to be studied in its natural environment if it is to be fully displayed and understood; and as regards the latter, it is, we believe, generally by direct observational and experimental work on human beings that human behaviour is to be understood. What the laboratory tradition of comparative psychology does provide is methods for analysing psychological processes—learning, motivation, perception, and so on—which allow behavioural research and physiological research to be integrated. It is in getting to grips with Cartesian questions, therefore, that the strength of this tradition lies.

An example of the power of this kind of work can be found in the analysis of the reward systems of the brain. The idea that behaviour is controlled by rewards and punishments is of course an old one; the contribution of the early learning theorists was to provide a method and a technology for assessing whether an event was rewarding and for investigating the way different types and schedules of rewards control behaviour (Skinner, 1938, is the classic in this field). Progress towards the analysis of the neural mechanisms of reward came when Olds and Milner (1954) ingeniously combined these behavioural methods with the method of chronic electrical stimulation of the brain developed by Hess (see Delgado, 1961) so as to define *reward pathways* of the brain, thus introducing an entirely new concept of brain function. These pathways have since been explored in a brilliant series of electrophysiological experiments (Deutsch, 1964; Gallistel *et al.*, 1969; Rolls, 1975). We mention this example because it is so clearly a product of the combination of laboratory behavioural analysis and physiological methods. A more general view of the progress that has been made in brain research can be found in *Scientific American* for September 1979. There is, perhaps, no need to argue that a scientific understanding of the brain is an objective worth attaining: one could make a case that it is among the most important tasks of biology, and of science as a whole. And we do not think that there can be much doubt that progress towards this understanding would be very severely hampered if behavioural and neurophysiological experiments were not carried out on animals. (That said, perhaps we should also make it clear that we are not saying, and do not believe, that showing thus that our interests are served by experiments on animals is enough to show that they are right. I do not justify taking your purse just by explaining how useful the money will be to me. But we are not going to discuss these further problems here.)

Whether behavioural research on animals is responsible for clear substantial advances in clinical practice that could not be attained without it is something that is more difficult to assess. The rider is important. The critical question, for those who wish to make a serious attempt to reduce the number of animals used in experimental work, is not whether research on animals makes a contribution, but whether it makes an indispensable contribution. To take an example, the group of methods for treating phobias and other neurotic disorders that is known as behaviour therapy is one of the most important of the contributions that psychology has made to clinical practice. (An introduction to these methods can be found in Gelder (1972); aversion therapies comprise but a small part of them.) Learning theory, the branch of experimental psychology that underlies this therapeutic method, has developed partly by research on animals. But it would be curious to argue that it would have been *impossible* to carry out the relevant research on human subjects; for how could a therapeutic method for use with human beings be based on principles of learning which could be demonstrated and investigated in dogs and rats, but which could not be demonstrated and investigated in human volunteers? Indeed, one could argue that the development of behaviour therapy might have been more rapid if more of the relevant research had been carried out on human volunteers rather than on animals (for instance, the importance of imagery would probably have been defined earlier). It is here, of course, that we find the general alternative to the use of animals in psychology; the general alternative is to use ourselves.

Psychological research has recently come to be especially sharply attacked by groups opposed to the use of animals in experiments. We do not think that this singling out of psychology is wholly unjustifiable. For, firstly, psychology is a subject that necessarily uses conscious animals, and a substantial proportion of psychological experiments involves deliberately causing pain to animals; in some cases, severe pain. Secondly, the undeniable benefits that a humane and scientific psychology can bring to medical practice do not usually depend critically upon experiments on animals in the way that advances in, for example, endocrinology or pharmacology do. The roots of psychology are much more varied, and the possibilities for direct research on human beings much more common. It is these two facts taken together that make psychology especially vulnerable as a target. And it is these same facts that ought, we feel, to make psychologists more unwilling than they are to carry out experiments that cause suffering to their animal subjects; here, as so often elsewhere, the most creative response to criticism may be to listen to it.

REFERENCES

Archer, J. (1979). *Animals under stress*, Studies in Biology No. 108, Edward Arnold, London.

Azrin, R., Hutchison, R. R., and Sallery, R. D. (1964). 'Pain-aggression toward inanimate objects', *Journal of the Experimental Analysis of Behavior*, **7**, 223–228.

Bancroft, J. (1972). 'Problems of sexual inadequacy in medical practice,' in *Psychiatric aspects of medical practice* (B. M. Mandelbrote and M. G. Gelder, eds), Staples Press, London, pp. 243–259.

Bancroft, J. (1974). *Deviant sexual behaviour*, Clarendon Press, Oxford.

Bancroft, J. (1978). 'The relationship between hormones and sexual behaviour in humans', in *Biological determinants of sexual behaviour* (J. B. Hutchison, ed.), John Wiley, Chichester, pp. 493–519.

Barlow, H. B. (1975). 'Visual experience and cortical development,' *Nature, London*, **258**, 199–204.

Berkson, G. (1970). 'Defective infants in a feral monkey group', *Folia primatologia*, **12**, 284–289.

Beumont, P. J. V., Bancroft, J. H. J., Beardwood, C. J., and Russell, G. F. M. (1972). 'Behavioural changes after treatment with testosterone: a case report', *Psychological Medicine*, **2**, 70–72.

Blakemore, C. and Cooper, G. F. (1970). 'Development of the brain depends on the visual environment', *Nature, London*, **228**, 477–478.

Blakemore, C., Donaghy, M. J., Maffei, L., Movshon, J. A., Rose, D., and Van Slutyers, R. C. (1974). 'Evidence that nitrous oxide can maintain anaethesia after induction with barbiturates', *Journal of Physiology*, **237**, 39P–41P.

Bock, R. H. (1960). 'Amblyopia detection in the practice of pediatrics', *Archives of Pediatrics*, **77**, 335–339.

Boe, E. E. (1969). 'Bibliography on punishment', in *Punishment and aversive behavior* (B. A. Campbell and R. M. Church, eds), Appleton-Century-Crofts, New York, pp. 531–587.

British Psychological Society: Scientific Affairs Board (1979). 'Report of the Working Party on Animal Experimentation', *Bulletin of the British Psychological Society*, **32**, 44–52.

Broendstrup, P. (1944). 'Amblyopia ex anopsia in infantile cataract,' *Acta ophthalmologica*, **22**, 52–71.

Brown, R. and Herrnstein, R. J. (1975). *Psychology*, Methuen, London.

Bureš, J., Burešová, O., and Huston, J. (1976). *Techniques and basic experiments for the study of brain and behavior*, Elsevier Scientific Publishing Company, Amsterdam.

Burnap, D. W. and Golden, J. S. (1967). 'Sexual problems in medical practice', *Journal of Medical Education*, **42**, 673–680.

Campbell, B. A. and Masterton, F. A. (1969). 'Psychophysics of punishment', in *Punishment and aversive behavior* (B. A. Campbell and R. M. Church, eds), Appleton-Century-Crofts, New York, pp. 3–42.

Chavasse, B. (1932). In Symposium on non-paralytic squint, *Transactions of the Ophthalmological Society*, **52**, 348–352.

Clark, S. R. L. (1977). *The moral status of animals*, Clarendon Press, Oxford.

Cross, B. A. and Silver, I. A. (1966). 'Electrophysiological studies on the hypothalamus', *British Medical Bulletin*, **22**, 254–260.

Cynader, M. and Berman, N. (1972). 'Receptive-field organization of monkey superior colliculus', *Journal of Neurophysiology*, **35**, 187–201.

Delgado, J. M. R. (1961). 'Chronic implantation of intracerebral electrodes in animals', in *Electrical Stimulation of the Brain* (D. E. Sheer, ed.), University of Texas Press, Houston, pp. 25–36.

Descartes, R. (1968). *Discourse on Method and the Meditations* (translated by F. E. Sutcliffe), Penguin Books, Harmondsworth.

Deutsch, J. A. (1964). 'Behavioral measurement of the neural refractory period and its application to intracranial self-stimulation', *Journal of Comparative and Physiological Psychology*, **58**, 1–9.

Dobzhansky, T. (1968). 'On some fundamental concepts of Darwinian biology', in *Evolutionary biology*, Vol. 2 (T. Dobzhansky, M. K. Hecht, and W. H. Steere, eds), Plenum Press, New York, pp. 1–34.

Drewett, R. F. (1977). 'On the teaching of vivisection', *New Scientist*, **76**, 292.

Duffy, F. H., Snodgrass, S. R., Burchfiel, J. L., and Conway, J. L. (1976). 'Bicuculline reversal of deprivation amblyopia in the cat', *Nature, London*, **260**, 256–257.

Eggers, H. M. and Blakemore, C. (1978). 'Physiological basis of anisometropic amblyopia', *Science, New York*, **201**, 264–266.

Fonberg, E. (1969). 'The role of the hypothalamus and amygdala in food intake, alimentary motivation and emotional reactions', *Acta biologica experimentalis, Warsaw*, **29**, 335–358.

Gallistel, C. R., Rolls, E. T., and Green, D. (1969). 'Neuron function inferred from behavioral and electrophysiological estimates of refractory period', *Science, New York*, **166**, 1028–1030.

Gates, E. (1895). 'The science of mentation and some new general methods of psychological research', *The Monist*, **5**, 574–597.

Gelder, M. G. (1972). 'The behaviour therapies', in *Psychiatric aspects of medical practice* (B. M. Mandelbrote and M. G. Gelder, eds), Staples Press, London, pp. 88–100.

Gur, M. and Purple, R. L. (1978). 'Retinal ganglion cell activity in the ground squirrel under halothane anesthesia', *Vision Research*, **18**, 1–14.

Hammond, P. and Mackay, D. M. (1977). 'Responses to visual texture in striate cortex neurones', *Experimental Brain Research*, **30**, 275–296.

Harris, G. W. and Michael, R. P. (1964). 'The activation of sexual behaviour by hypothalamic implants of oestrogen'. *Journal of Physiology, London*, **171**, 275–301.

Hart, B. L. (ed.) (1976). *Experimental psychobiology: a laboratory manual*, W. H. Freeman, San Francisco.

Herbert, J. (1978). 'Neuro-hormonal integration of sexual behaviour in female primates', in *Biological determinants of sexual behaviour* (J. B. Hutchinson, ed.), John Wiley, Chichester, pp. 277–318.

Home Office (1978). *Experiments on living animals: statistics 1977, Cmnd 7333*, HMSO, London.

Home Office (1979). *Statistics of experiments on living animals: Great Britain 1978, Cmnd 7628*, HMSO, London.

Hubel, D. H., and Wiesel, T. N. (1965). 'Binocular interaction in striate cortex of kittens reared with artificial squint', *Journal of Neurophysiology*, **28**, 1041–1059.

Hubel, D. H., and Wiesel, T. N. (1979). 'Brain mechanisms of vision', *Scientific American*, **241**, 130–144.

Humphrey, N. K., (1974). 'Vision in a monkey without striate cortex: a case study', *Perception*, **3**, 241–255.

Humphrey, N. K. (1976). 'The social function of intellect', in *Growing points in ethology* (P. P. G. Bateson and R. A. Hinde, eds), Cambridge University Press, Cambridge, pp. 303–317.

Hutchison, J. B. (ed.) (1978). *Biological determinants of sexual behaviour* John Wiley, Chichester.

Hutt, S. J. and Hutt, C. (1970). *Behaviour studies in psychiatry,* Pergamon Press, Oxford.

Iggo, A. (1959). 'Cutaneous heat and cold receptors with slowly conducting (c) afferent fibres', *Quarterly Journal of Experimental Physiology*, **44**, 362–370.

Iggo, A. (1977). 'Cutaneous and subcutaneous sense organs', *British Medical Bulletin*, **33**, 97–102.

James, W. H. (1971). 'The distribution of coitus within the human intermenstruum'. *Journal of Biosocial Science*, **3**, 159–171.

Juler, F. (1921). 'Amblyopia from disuse. Visual acuity after traumatic cataract in children', *Transactions of the Ophthalmological Society*, **41**, 129–139.

Kinsey, A. C., Pomeroy, W. B., and Martin, C. E. (1948). *Sexual behavior in the human male*, W. B. Saunders, Philadelphia.

Kinsey, A. C., Pomeroy, W. B., Martin, C. E., and Gebhard, P. H. (1953). *Sexual behavior in the human female,* W. B. Saunders, Philadelphia.

Masters, W. H. and Johnson, V. E. (1966). *Human sexual response*, J. A. Churchill, London.

Masters, W. H. and Johnson, V. E. (1970). *Human sexual inadequacy*, J. A. Churchill, London.

Mayer, D. J. and Liebeskind, J. C. (1974). 'Pain reduction by focal electrical stimulation of the brain: an anatomical and behavioral analysis', *Brain Research*, **68**, 73–93.

Medical Research Council (1977). *Annual Report 1976–77*, HMSO, London.

Melzack, R. (1976). 'Pain: past, present and future', in *Pain: new perspectives in therapy and research* (M. Weisenberg and B. Tursky, eds), Plenum Press, New York, pp. 135–145.

Melzack, R. and Wall, P. D. (1965). 'Pain mechanisms: a new theory', *Science, New York*, **150**, 971–979.

Michael, R. P., Richter, M. C., Cain, J. A., Zumpe, D., and Bonsall, R. W. (1978). 'Artificial menstrual cycles, behaviour and the role of androgens in female rhesus monkeys', *Nature, London*, **275**, 439–440.

Miller, N. E., Coons, E. E., Lewis, M., and Jensen, D. D. (1961). 'Electrode holders in chronic preparations: B. A simple technique for use with the rat', in *Electrical stimulation of the brain* (D. E. Sheer, ed.), University of Texas Press, Houston, pp. 51–54.

Mitchell, D. and Hellon, R. F. (1977). 'Neuronal and behavioural responses in rats during noxious stimulation of the tail', *Proceedings of the Royal Society B*, **197**, 169–194.

Money, J. (1961). 'Components of eroticism in man: the hormones in relation to sexual morphology and sexual desire', *Journal of Nervous and Mental Diseases*, **132**, 234–248.

Nathan, P. W. (1976). 'The gate-control theory of pain', *Brain*, **99**, 123–158.

O'Kelly, L. E. and Steckle, L. C. (1939). 'A note on long enduring emotional responses in the rat', *Journal of Psychology*, **8**, 125–131.

Olds, J. and Milner, P. M. (1954). 'Positive reinforcement produced by electrical stimulation of septal area and other regions of rat brain', *Journal of Comparative and Physiological Psychology*, **47**, 419–427.

Penrose, L. S. (1949). *The biology of mental defect*, Sidgwick and Jackson, London.

Peter, L. C. (1931). 'Advancements and other shortening operations in concomitant squint', *Archives of Ophthalmology*, **6**, 380–388.

Rachman, S. and Teasdale, J. (1969). *Aversion therapy and behaviour disorders: an analysis*, Routledge and Kegan Paul, London.

Reynolds, D. V. (1968). 'Surgery in the rat during electrical analgesia induced by focal brain stimulation', *Science, New York*, **164**, 444–445.

Richards, C. D. and Webb, A. C. (1975). 'The effect of nitrous oxide on cats anaesthetized with Brietal', *Journal of Physiology, London*, **245**, 72p–73p.

Rolls, E. T. (1975). *The brain and reward*, Pergamon Press, Oxford.

Russell, W. J. (1973). 'Nitrous oxide—is it an adequate anaesthetic?', *Journal of Physiology, London*, **231**, 20p–21p.

Ryle, G. (1949). *The concept of mind*, Hutchison, London.

Skinner, B. F. (1938). *The behavior of organisms*, Appleton-Century-Crofts, New York.

Tinbergen, N. (1953). *The herring gull's world*, Collins, London.

Tinbergen, N. (1972). *The animal in its world*, Vol. 1, *Field Studies*, George, Allen and Unwin, London.

Torebjörk, H. F. and Hallin, R. G. (1974). 'Identification of different C units in intact human skin', *Brain Research*, **67**, 387–403.

Udry, J. R. and Morris, N. M. (1967). 'A method for validation of reported sexual data', *Journal of Marriage and the Family*, **29**, 442–446.

Ulrich, R. E. and Azrin, N. H. (1962). 'Reflexive fighting in response to aversive stimulation', *Journal of the Experimental Analysis of Behavior*, **5**, 511–520.

Vallbo, A. B. and Hagbarth, K. E. (1968). 'Activity from skin mechanoreceptors recorded percutaneously in awake human subjects', *Experimental Neurology*, **21**, 270–289.

Wall, P. D. (1975). Editorial, *Pain*, **1**, 1–2.

Wall, P. D. (1976). Editorial, *Pain*, **2**, 1.

Wiesel, T. N. and Hubel, D. H. (1963). 'Single cell responses in striate cortex of kittens deprived of vision in one eye', *Journal of Neurophysiology*, **26**, 1003–1017.

Worth, C. (1903). *Squint: its causes, pathology and treatment*, John Bale, Sons and Danielson Ltd., London.

Yerkes, R. M. (1903). 'The instincts, habits and reactions of the frog, 1. Associative processes of the green frog.' *Harvard Psychology Studies*, **1**, 579–638.

Animals in Research
Edited by David Sperlinger
© 1981 John Wiley & Sons Ltd.

Chapter 9

Ethology—The Science and the Tool

DAVID MACDONALD and MARIAN DAWKINS

A bullfrog croaks. A bat catches insects in the gathering dusk. A male stickleback builds a nest from water weeds cemented together with a glue produced by his own kidneys. These animals do not blunder about the world: their behaviour is remarkable in its appropriateness. The bullfrog croaks in response to other bullfrogs. The bat swoops on insects, but not on twigs or leaves. The stickleback switches from courting females to caring for the eggs in his nest at just the time when the eggs need the oxygen he can supply by fanning a stream of water through the nest with his fins.

Behind each of these examples lies a myriad of yet more striking adaptations. The pitch of the frog's croak influences his success at attracting females, and the bat which identifies its prey by echo-location, makes up to 200 calls in 1 second and listens to the returning echoes in the tiny intervals between each one. Adaptations such as these, and the mechanisms which produce them, are the material of ethology.

The science of ethology concerns the study of animal behaviour at a level of analysis lying between, but merging into, both physiology and population ecology. At one level ethologists are concerned with large units, individuals, and communities, but at another level they seek to understand evolution and the workings of genes. The subject matter may be invertebrates as diverse as sea anemones and stag beetles or it may be vertebrates such as birds, fish, or any of the 4137 species of mammal, including humans.

With such a broad scope, ethology embraces diverse studies and a spectrum of methodologies, so much so that it is sometimes hard to know exactly what ethology is! Perhaps an answer can best be seen in the roots of the science. Niko Tinbergen entitled one of his books *Curious naturalists* (Tinbergen, 1958) and that, in part, is what many ethologists are. But they are more than this, too. They seek to be rigorous and quantitative, to describe, hypothesize about, and test their theories and so better to understand animal behaviour and the evolutionary forces that fashioned it. In this chapter we want, first, to introduce a selection of ethological studies and so briefly to describe the science. Secondly, we will discuss how the findings of ethological studies have practical applications and can be used as tools in animal welfare and conservation. Lastly, we will focus on some of the ethical questions raised by ethological studies.

Several different sorts of question can be asked about the way animals behave. One sort of question concerns the evolutionary significance of the behaviour to the animal. Why, in other words, has natural selection favoured animals which behave in this way or that? For example, why do adult black-headed gulls fly away from their nests after their chicks have hatched, carrying with them the fragmented shells from which the chicks have just emerged? The adult gull drops the shell some distance from the nest, but in the meantime, the chick is exposed to the dangers of cannibalism by neighbouring gulls. Tinbergen *et al.* (1962) hypothesizd that the evolutionary significance of this apparently risky egg-shell removal behaviour was that, in the long run, it protected the chicks from being spotted by crows and other hunters. The white interior of the shell is so conspicuous that it disrupts the camouflage of the nest. This led to the suspicion that it endangered the chicks by attracting the attention of predators. The ethologists tested this idea by means of an experiment in which they put camouflaged eggs out on the sand dunes near the gull colony they were studying, with broken egg-shell at different distances from them. Sure enough, the nearer an egg was to a broken egg-shell, the more likely it was to be taken by a marauding crow.

So, it is clearly advantageous for the parent gulls to remove egg-shells from their nests, since their chicks are more likely to survive if they do. The reason why an animal behaves in a particular way cannot always be teased apart with such ingeniously clear-cut experiments especially because there are different sorts of answers to a single question (e.g. in the short or the long term). For example, black-backed jackals sometimes live in family groups comprising a mated pair and their offspring from the previous year. These yearling animals act as helpers, bringing food to their younger brothers and sisters (Moehlman, 1979). But why don't the helpers leave home, set up their own territories, and have their own pups? What is the advantage to them of staying to help their parents? In the short term, they seem to be missing out on the opportunity to rear young of their own, but, in the long term, this may be to their advantage. Firstly, as Moehlman has shown, more pups survive in families which have more helpers. So, by staying to help their parents, the young jackals contribute to the survival of more brothers and sisters and hence increase the number of their surviving relatives. (On average, an individual shares as many genes with a sib, as it would with its own offspring. So, by aiding relatives, the young jackals also help themselves, increasing their so-called inclusive fitness and perhaps also learning something of the art of pup-rearing.) Secondly, it is likely that local features of jackal ecology, such as a limited food supply or shortage of breeding dens, combine to lower the likelihood that young jackals which do emigrate during their first year will be able to reproduce successfully for themselves. So, in terms of passing their genes on to the next generation, they are better off in the long run staying at home in their first year and tending to their sibs.

This example clearly points to the importance of knowing not only about a particular behaviour (e.g. helping) but seeing it as part of a larger mosaic of behaviour (e.g. the social system). Most importantly, these have to be interpreted in the context of the species' ecological circumstances and the genetic relationships of the individuals involved. There has recently been a great upsurge of interest in these evolutionary aspects of animal behaviour. It is sometimes thought that this has heralded the birth of a new science variously called socio-ecology, sociobiology, or behavioural ecology. In truth it is just a new but compelling face for the much older science of ethology whose pioneers, such as Konrad Lorenz and Niko Tinbergen, constantly emphasized the evolutionary adaptations of behaviour.

Another powerful weapon in the ethologist's methodological armoury is that of comparison. Questions about the adaptive significance of behaviour can be answered by making comparisons between different species or between individuals of the same species behaving in different ways. For example, golden jackals sometimes hunt alone, and sometimes hunt in pairs. They are particularly inclined to hunt in pairs when searching for gazelle fawns, which are vigorously defended by their mothers. When two jackals hunt together, one can distract the mother gazelle while the other kills the fawn. Wyman (1967) found that by hunting co-operatively, a pair of jackals caught more than twice as many fawns as did single jackals. So, by observation alone, he was able to demonstrate the evolutionary advantages of hunting in pairs. In the study of social behaviour such comparisons take varied forms: why do jackals in different habitats appear to form contrasting societies? why between species such as foxes and hunting dogs are there different sex ratios amongst helpers? or why indeed, in other carnivorous species, is the whole social system quite different? (Macdonald, 1979; Frame and Frame, 1976; Rood, 1978).

A second sort of question that can be asked about animal behaviour concerns its mechanics—that is, the events going on inside the animal's body and in the outside world, that cause it to behave in particular ways. What is it about a male stickleback in breeding dress that causes another male to attack it, for example? How do digger wasps remember where their nesting hole is? Many questions such as these interest not only ethologists but also fall within the province of physiology. But the methods by which the two disciplines seek for answers are often very different.

The physiologist looks inside the animal's body and tries to observe directly how the various components work. He might, for instance, record nerve impulses travelling between the eye and the brain in order to study what the animal sees. The ethologist, on the other hand, seeking the same information, might keep the animal intact and watch how it responded to various visual stimuli, noting which ones it does and does not respond to. (This is not to say that ethologists do not also sometimes use more intrusive techniques—see below.) In some studies the two approaches may complement each other, but here we will mention cases where the ethological approach alone has been particularly fruitful. First, however, we would like to emphasize that the idea of deducing what is going on inside an animal's body from simply observing intact individuals has been an extremely fruitful one in the development of biology. One of its most striking successes was in the field of genetics. Long before DNA was known to be the carrier of genetic information, before molecular biology became fashionable or had even been thought of, much had been deduced about the genes themselves. It was known how genes behaved during the formation of gametes. It was known how they combined and separated again from one generation to the next. It was even known how they were arranged with respect to one another along the chromosomes. All this was discovered not by biochemical analysis but by the apparently unpromising method of breeding animals and counting the frequency with which progeny with various characteristics resulted. Mendel's deduction of the basic laws of heredity was a stroke of genius. He had nothing more to go on other than the numbers of various sorts of pea plant produced by crossing parents with certain physical characteristics. Similarly, with outstanding insight, Morgan and Sturtevant realized that genes could be accurately mapped on the chromosomes simply by looking at the offspring of carefully chosen parents.

It is of course a long way from genetics to deductions about the mechanisms of behaviour, but the point is that much can be learned about what is going on *inside* animals with a 'whole animal' approach. Niko Tinbergen pioneered the study of the mechanisms of behaviour without damaging the animal in any way—physiology without breaking the skin as he aptly called it. He used the method of watching the response of animals to dummies or model animals to probe their sensory worlds. One of his studies involved trying to find out how herring gull chicks recognize their parents (Tinbergen and Perdeck, 1950). To do this, he used cardboard models of gulls' heads, as he had found previously that the chicks would respond to them in much the same way as they did to a real gull's head. The gull chicks did not seem to notice whether the cardboard heads had bodies attached to them or not, nor whether the heads were painted red, white, or blue. What seemed to be most important was the appearance of the beak. Herring gull chicks have a very characteristic response to their parents' beaks when the parent returns to the nest with food in its crop. The chicks peck at a red spot on the lower mandible of the otherwise yellow beak. This pecking by the chick stimulates the parent to regurgitate the food in its crop.

The chicks pecked vigorously at many of the cardboard models that Tinbergen and Perdeck (1950) held out to them, but not at all of them. They pecked strongly at models which had long thin beaks, but much less at models which had short stubby ones. They pecked particularly at models where the beak was painted red, even though the natural background colour of the herring gull's beak is yellow.

Tinbergen could actually 'improve on nature' by presenting chicks with a long thin red needle with white bands on it. This headless model made the chicks peck even more than a real gull's head and led to the conclusion that the key features of 'parent' from the chicks' viewpoint were redness, thinness, and a strongly contrasting pattern to peck at. The fact that Hailman (1967) could subsequently refine, by more extensive tests, some of Tinbergen's original conclusions does not invalidate the general method of using models to find out about the animal's view of its world. It may take many trials with various controls to verify or refute a single hypothesis. The increasing use of computers and other electronic devices will give new scope in this context, by making it possible to present large numbers of stimuli relatively easily. For example, Herrnstein and Loveland (1964) used an automated system to present pigeons with a large selection of colour photographs. The pigeons were trained to peck when they saw a certain sort of picture, and in this way it was shown that these birds could abstract complex concepts from the photos, such as 'human being' or 'water' irrespective of the details of the picture, findings that would have been out of the reach of physiological methods altogether.

Ethological techniques have shed light on various other perceptual problems aside from vision. Capranica (1966) used electonically produced sounds to investigate the mechanism by which bullfrogs recognize the croaks of other members of their species. The croak of a bullfrog has two peaks of sound energy—one at 200 Hz and the other at 1400 Hz. Capranica found that bullfrogs croaked in response to his electronically produced croak only if the artificial sound also had sound energy at 200 and 1400 Hz. But if he *added* sound in the middle region, at about 400–700 Hz, the bullfrogs stopped croaking. To their ears, this mixed sound did not resemble a real croak. On the basis of such experiments, Capranica deduced that the hearing mechanism involved excitation of auditory nerves at the low (200 Hz) and high (1400 Hz) frequencies, but inhibition at frequencies in between. The neural basis of the excitation and inhibition was subsequently confirmed by physiological means (Frishkopf *et al.*, 1968).

Ethological methods have also been used to investigate more central mechanisms of animal behaviour. Without knowing the details of the physiological mechanisms involved, it is possible to learn about how the animal works from the interactions of various motivational systems such as feeding, drinking, and aggression. Van Iersel and Bol (1958), for example, studied the occurrence of various behaviour patterns in terns which were incubating their eggs. Their observations suggested a model to account for the occurrence of preening by these birds in terms of the interaction of various motivational states such as the tendency to incubate eggs and the tendency to take flight, which was derived and tested purely on the basis of the birds' behaviour.

McFarland (1977) and his coworkers are trying to measure the costs and benefits underlying the complex decisions which herring gulls make in each facet of their daily lives. Gulls which are incubating eggs have both to feed themselves and to ensure that the eggs are protected and kept warm. A decision to leave the nest in order to forage in nearby mussel beds will be influenced by such factors as the height of the tide, the behaviour of the bird's mate, the time since it last fed, the likely hatching date of the chicks, and so on. Many of these factors, such as food or water deficits, are known to have physiological effects, but even though these may not be known in detail, a great deal can still be learnt about the way the behaviour is organized and the interaction of many different factors to produce the final overt behaviour.

Another major field concerns the development of behaviour. What factors, for example, change the behaviour of a kitten into that of a cat? Bateson (1980) reports a study which has shown that the development of play by male and female kittens differs: female kittens play with objects less than males, but this sex difference is seen only in litters where all the kittens are of the same sex. When female kittens grow up with their brothers, both sexes show equal amounts of play with objects. The effect of having brothers is long lasting: female kittens reared with males continue to play like males even when their brothers are removed.

Studies such as these are important not only because they shed light on the behaviour of a few individual kittens or on one species of seabird in a colony in the north of England, but because they can be used to establish general principles of how animal behaviour may be organized. This is a point which is often misunderstood by incredulous laymen watching the antics of ethologists. How trivial it may seem to watch a great tit dismembering a meal worm or a butterfly defending a patch of sunlight from its fellows; how trivial the topics and how futile to expend time and energy in studying them. But these are just specific examples of much more general phenomena, and while the individual instances may be of ephemeral importance, the principles they illustrate are not trivial (in this case foraging and territorial behaviour) and affect our entire view of the animal kingdom, including that of ourselves.

In addition to being a challenging science, worthwhile in itself in the same way in which other explorations of Nature have been, ethology is also increasingly assuming a practical role. In almost every case where man takes decisions concerning other animals—in zoos, wildlife refuges, laboratories—a knowledge of animal behaviour can aid his understanding of the problems. Whether the aim is exploitative (persuading poultry to lay the greatest number of eggs), concerned with welfare (assessing possible adverse effects of intensive indoor husbandry), 'pest' control (limiting the damage caused by the bearers of disease or by pests on food crops), or conservation (reserve management and design), the key to success often lies in a proper understanding of animal behaviour. Many ethological experiments are undertaken with the aim of solving a particular practical problem in fields like these, and we will briefly outline the sort of work that is done and that could be done more in the future.

Domestic pigs sometimes develop a distressing habit of biting each other's tails and ears. When the victim is unable to escape, an initial wound will be followed by more serious ones, as the other pigs in the group join in, and major injury and even death can result. Somewhat similarly, chickens, stimulated by the sight of blood on the feathers of another bird, will peck at it until the wound becomes bigger, and the pecked bird may eventually be killed. Such apparently pathological behaviour is undesirable for the farmer, at least because of the financial loss involved. It is also disheartening for the stockman and, it seems fairly safe to assume, distressing to the victims. Experimentally investigating the factors which contribute to these behavioural disorders, seeking to understand their development, and clarifying whether they are anomalies of captivity or have parallels in the behaviour of wild animals, can help to provide solutions. In the case of tail-biting in pigs, some respite was achieved by the simple device of providing tyres, chains, and other diversions so that the pigs had something else to bite at besides each other (Colyer, 1970). Investigations into the basis of feather pecking amongst chickens and pheasants have focussed on the effects of different diets, different group sizes and stocking densities, food of different textures, and alternative objects to peck at (e.g. Hoffmeyer, 1969; Hughes and Duncan, 1972).

It is not, however, only in cases where behaviour leads to physical damage or ill-health that people have become concerned about the welfare of farm animals. Practices such as keeping chickens in battery cages, veal calves in small confined pens, and pigs in group cages have provoked violent criticism because of the possible *mental* suffering that they might cause through restriction of movement and disruption of social interaction. We can never know for certain what the mental experiences of other animals are (but then neither can we know for certain in other people). But despite this we can still accumulate circumstantial evidence of what their mental experiences might be through careful observation and manipulation of behaviour (Thorpe, 1965; Griffin, 1976). As a pragmatic rule-of-thumb, it seems desirable to err in the direction of assuming too great a mental sensitivity rather than too little. Indeed the Brambell Committee (1965) which was set up by the UK government specifically to look into the problems caused by intensive husbandry methods in agriculture, strongly emphasized the importance of behavioural information about the animals concerned, as an essential guide to their mental state.

It may turn out that abnormalities of behaviour, such as an increase in aggressiveness, pacing up and down, and other stereotypes in the cage might be used as indicators of pathological states to come (Ewbank, 1973). Intensive husbandry systems or badly designed cages in zoos may prevent animals from carrying out much of their normal behaviour, and as a result they may show evidence of conflict and frustration; this seems to apply to battery-caged hens prior to egg laying (Duncan and Wood-Gush, 1972). Information about the way in which different species express fear, frustration, and other emotions, and the conditions which provoke these, is essential to the evaluation of their well-being.

Another technique, and one which has auspicious beginnings for assessing stock housing from the inmates' point of view, is to give an animal a choice of a variety of living conditions and to see which it selects. This approach has the merit that the animal's behaviour can be used to rank its choices on a relative scale, for example on the basis of how hard it 'works' to avoid certain types of cage, or to gain entry to other types. Given this sort of choice, poultry have shown how different their selection may be from that predicted by human goodwill and guesswork. The Brambell Committee recommended that the floors of battery cages should not be made out of fine hexagonal wire as they believed that this was uncomfortable for the hens' feet. When given the opportunity to stand on either this type of wire or a coarser type which the Committee did recommend, the hens chose the finer mesh (Hughes and Black, 1973) and so cast doubt on the original, well-intentioned supposition. Of course, any interpretation of data derived from preference tests in terms of higher mental processes, for example that hens 'like' a given type of cage or 'dislike' (and even suffer in) another type is a very much more complex matter (Duncan, 1977; Dawkins, 1980), since animals do not necessarily like what is good for them. Clearly, there are cases familiar to us all (a dog snarling at a vet, or a child refusing to go to the dentist) where what is preferred and what is best for long-term welfare do not match; this is true for wild animals too. In general, however, wild animals are quite remarkable in the appropriateness of the decisions they make.

In an attempt to show how the question, 'Do hens suffer in battery cages?', might be answered, and by extension to advocate a methodology for analysis of stock welfare and husbandry practices in general, Dawkins (1977, 1978) has offered hens the choice between battery cages and outdoor runs, between various cage floors, and between the presence or absence of companions. The hens certainly do exhibit preferences, although these are, as might be expected, influenced by an array of complicating factors which must be carefully controlled if this type of experimental approach is to be valid. For instance, hens accustomed to life in a battery cage may, at least initially, continue to choose such confined housing even when offered more spacious accommodation.

One experiment illustrates the method. Individuals from two groups of hens (one accustomed to battery life, and the other to an outdoor run) were given the choice on emergence from a starting box, of walking along a passage and into a battery cage, or walking in the other direction and into an outdoor run, both of which they could see from their starting place (see Figure 9.1). The results showed that in spite of considerable individual variations in their behaviour, the hens accustomed to outdoor runs consistently chose the outdoor pen, while the battery hens initially selected the cage. However, as the experiment progressed, with repeated trials during which the battery hens had the opportunity to sample the outdoor pen, they began to change their 'minds', or at least their behaviour, and selected the outdoor run too. These and related trials do not yet constitute a

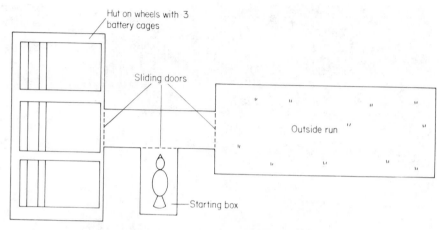

Figure 9.1 Diagram of a choice offered to hens. Hens were free to move into either a battery cage or an outside run. (From Dawkins, 1977)

firm basis for criticizing or endorsing any aspect of poultry husbandry, but they do constitute a test of a tool which may be used to this end in the future. For the time being one can conclude that at least this strain of hen (Sykes Tinted) does prefer outside runs (and from other trials, larger cages, grass underfoot, and the company of other hens), but that the preference is significantly influenced by past experience. A similar phenomenon may underlie the case of a fox farm where Arctic foxes were housed in cages which were slightly shorter than the foxes' nose-to-tail length (Figure 9.2); following an extensive newspaper report (Harrison, 1976) on the upsurge of this industry the fur farm was raided by someone who opened all the cages, driving the foxes to liberty. Within a week all but a handful had returned to their individual cubicles and all that the farmer had to do was close the latches behind them.

Dawkins (1980) emphasizes that one method cannot provide all the answers to these questions of animal welfare. Duncan and Filshie (1980) have reported on the use of another technique, biotelemetry, to investigate the welfare of battery hens. The critical assumption is that measures of physiological parameters, such as heart rate, can be taken to mirror the animal's internal experience (e.g. fearfulness). Certainly, the results provide some intriguing conundrums concerning attempts to interpret internal state from outward appearance. They tested two strains of hen in response to a presumably unnerving, if not outright frightening, stimulus, namely a loud noise. One strain reacted frantically, the other was more sedate. However, the calm exterior of the second strain was illusory, since its heart rate elevated and remained high, while the flighty bird's heart rate increased fast but then soon returned to normal. Which bird was the more disturbed? The solution to such questions could influence the selection of breed stocks and significantly contribute to the welfare of farm animals.

Figure 9.2 Three thousand arctic foxes live in pens 3 ft × 2 ft × 2 ft in this one fur farm. Clearly the conditions are unnatural, yet the animals are in good condition physically, as evidenced by the quality of their pelts, and when released many of them returned to their cages of their own accord. (Photograph by courtesy of the *Sunday People*)

In fact, the whole technique of biotelemetry, often involving implanted telemeters which transmit information from within the animal's body, offers a completely new opportunity to explore the physiological correlates of outward behaviour, perhaps to the considerable benefit of animal welfare (as suggested by Amlaner, 1977), but violating Tinbergen's ideas of ethology as physiology without breaking the skin. Techniques of this type which involve even a minor operation (but, on the face of it, much less cruelty than the 'old fashioned' monitoring techniques of experimental psychology, which necessitated the experimental animal to be bedecked in trailing wires linked to permanent electrodes within the confines of a small cage) inescapably highlight the question of whether a possible improvement in the lot of battery hens in general compensates for the possible (perhaps minimal, perhaps not) distress caused to the few who are studied. Indeed, the same question in slightly different garb can be asked of the merits of much research which purports to better the lot of any animals, including humans.

In fact the use of biotelemetry highlights a significant ethological contribution to scientific thought, namely the importance of ensuring that questions to which any sort of evolutionary answer are sought should be tackled under natural circumstances. The penalty for neglecting this has been elegantly shown by Butler (1980), who studied the respiratory and cardiac physiology of diving ducks as they submerged. Using radio telemetry (a small radio implanted in free-diving ducks and transmitting information on their heart and respiratory rates) he discovered that the real responses of undisturbed ducks were almost exactly the opposite of those described in the text books of a decade ago, based on laboratory bound, traditional, physiological techniques. Similarly, it is interesting to consider much of the work done on laboratory rats in the context of the behaviour of wild rats as studied by Taylor and Hardy (1980).

Studies of animal behaviour can, and should be, an important tool for the improvement of our behaviour towards wild as much as domestic animals. If a problem involving wildlife necessitates some sort of human intervention, as in conservation or control, then it is important that any action taken should be based on knowledge of the species' behaviour. Several examples were given by Macdonald and Boitani (1979), including the case of the wolf, whose conservation has been aided both directly and indirectly by intensive ethological studies. Such studies have contributed directly in the sense that knowledge of the species' behaviour has allowed assessment of the desirable (or at least the minimal) requirements for its conservation, the magnitude of the danger to its survival, and the plausibility of the accusations levelled against it. They have also contributed indirectly, by increasing public awareness of the animal and replacing outworn myths with far more fascinating accounts of its real behaviour, so that by getting a better 'public image' the species has a much greater chance of escaping persecution.

The claim that sensible management or conservation of a species has to be based on a knowledge of its biology would sound trite were it not so frequently abused. Think of the short-sightedness and inhumanity in our treatment of whales. Many of the cases where knowledge of animal behaviour is applied in practice involve some sort of conflict between animals and man, for example, between the wolves and the shepherds who want to protect their sheep, or the more efficient exploitation of a species, for example in fur harvests, but perhaps this merely indicates that most of our interactions with wildlife are tinged with either conflict or exploitation. It is surely better to have management than mismanagement, if only because the former may afford an opportunity to think again, while the latter only an opportunity to lament another extinction. Often ethological studies do have important applications in conservation.

There are many problems facing those who have to decide upon the optimal dimensions for National Parks. How much space is needed to maintain a viable population of a dwindling species? How many animals comprise a viable population? Without knowledge of the movement patterns, feeding habits, social structure, and reproductive biology, there can be no realistic answers to such questions. In Copo Park of the Chaco region of Argentina, the puma is being studied with a view to finding out how much wilderness should be set aside to ensure that it can survive there. In the Appenine region of Italy, both wolves and deer are being studied with the long-term aim of reintroducing red and roe deer into an area from which they have been hunted to extinction by man. It is hoped to create a sufficiently large park that a population of wolves can maintain itself by predating the deer. Even more ambitiously, a study of the behaviour of many species ranging from termites and civets in the Gunung Mulu National Park of Sarawak and of the interactions between nomadic tribesmen and their prey (wild pig and monkey), have all been integrated to develop the park's management plan (Jermy and Hanbury-Tenison, 1979).

Ethological and associated research can contribute to the welfare of both wildlife and men where problems of disease transmission arise. Two examples are the spread of rabies amongst fox populations and the possible role of European badgers in the transmission of bovine tuberculosis. In both cases, investigations of the behaviour of the species shows more promise of providing solutions which are both logistically and ethically satisfactory than any alternative approach. Cheeseman and Mallinson (1980) are studying the social behaviour of badgers in Gloucestershire, in a region where many of the badgers are infected with bovine TB. By finding out about the badger's family life and how the animals interact with other family groups, they hope to understand how the disease spreads from one badger to another. By then relating the incidence of the disease in cattle to its incidence in the local badger populations (in which there are many anomalies), they hope eventually to understand and then to combat the problem of infected cattle in a way which is effective, humane, and ecologically acceptable.

In the other example, rabies is transmitted by foxes throughout much of Europe. It is of great importance to understand how the disease spreads through fox communities and how it can be stopped. Most practical attempts to control rabies have been straightforward—simply to kill foxes. This has been costly in time, money, lives, and, we presume, suffering of the foxes; moreover, it has been noticeably unsuccessful in halting the spread of rabies. One of the reasons for this has been an ignorance of the habits of the foxes themselves (Macdonald, 1980). Work on foxes in various habitats, from farmland to suburbs and city centres, has shown that they are capable of adapting their behaviour considerably to the constraints and opportunities of their local environment. An average fox territory may be a few tens of hectares or well over a thousand, depending on the habitat. Clearly, the size of a fox's territory and the number of other foxes it meets is going crucially to affect how rapidly rabies will be spread. Alternative systems of controlling rabies should be explored and should take into account these features of fox behaviour. One possibility is to persuade foxes to eat baits, loaded with vaccine against the disease, distributed over a wide area from an aircraft (D. Johnson, personal communication). The diverse facets of Johnson's study included aspects of fox behaviour as varied as their food preferences (and hence the choice of bait) and the magnitude of the distances over which the juvenile animals disperse (and hence the area which needs to be baited).

An important feature of this scheme (in which some problems still remain), is that it does not involve killing the resident foxes and thus leaving a vacuum which might be rapidly filled by a transient animal moving into the area. It is known from work on other species, such as great tits (Krebs, 1977) that if a territory owner is removed, its territory does not remain empty for long before another occupant moves in. This may be one of the reasons why conventional rabies control has been unsuccessful. Empty territories would not arise, however, if vaccinated foxes were left *in situ*, and hence the movement of animals (with its increased risk of spreading disease) would be reduced, If Johnson's plans are successful, the result should be both cheaper and more effective rabies control.

Much of wildlife management, particularly in the USA, involves cropping the animals for meat, fur, or sport. Whether or not these activities are desirable at all is a separate issue, but where they do occur they can be conducted in ways which minimize the threat to the species and its habitat. A basic feature of many populations is an annual increase in the number of young, which boosts the population above the numbers which the environment can support. There is then an annual loss caused by decimating factors, such as starvation. A basic principle of 'cropping' is that there need be no danger to the population if human beings kill only the proportion of the population that would have died 'natural deaths' anyway. Deciding on the ecologically permissible cropping rate requires a thorough knowledge of the species' behaviour; where this is lacking the results can be disastrous. Macdonald and Boitani (1979) discuss the example of the Pribilof seal, and similar considerations may also apply to the grey seal, whose effect on fisheries has received a great deal of publicity. One estimate is that grey seals take about 65 000 tonnes of fish which would otherwise have been commercially exploitable (5–10 per cent of the total UK usable fish catch) and as seal numbers are evidently increasing at 7 per cent per annum (thus doubling every 11 years) this has caused alarm amongst fishermen. To make matters worse, the seals are a host for the codworm, *Phocanema decipiens*, which is also on the increase. In an effort to halt this trend, about 1000 seal pups have been killed annually in the Orkneys for some years. However, the policy of taking a fixed number of pups each year may not, in the long run, be the best one. There is a 6 year time lag before a generation of baby seals matures, so that the effect of any other factors acting on the population in the meantime would not be apparent until the culled age-group began to breed, and it might then be too late to avoid a disastrous drop in seal numbers. Further knowledge of seal behaviour (which is urgently needed) might permit an alternative policy based on a small annual reduction in the number of breeding females. But at the moment there is still insufficient information about the seal's behaviour and their effects on fish stocks.

Knowledge of the behaviour of various ungulates is paving the way from simply cropping animals to actually farming them. It is often more productive and less damaging to the land to farm local species, with a proven ability to survive in the area, than to import cows from abroad. Studies of eland and hartebeest at the Golana Game Farm suggest that they do better in the arid areas of East Africa than do introduced cattle, since they have better physiological tolerance and behavioural adaptations to water deprivation (King and Heath, 1975). Reindeer behaviour is similarly being studied with a view to better management in Finland. Plans are afoot in Venezuela to farm the large South American rodent, the capybara. These sheep-sized rodents, like gargantuan guinea-pigs in appearance, can live in marshes and esteros which are too wet for cattle (Ojasti, 1973).

In describing the ethological approach to the study of animal behaviour, we have made little reference to the feelings or emotions of the animals involved. Does this mean that ethologists generally do not believe that animals have such feelings? Do they see the animals which they study as intriguing but insentient automata? The answers to these questions are complicated and vary very much with an individual's personal philosophy, but in our experience, it is hard to find an ethologist who does not believe to some extent in the continuity of mental experiences between humans and other animals, particularly mammals. For one who has had privileged glimpses into the world of frolicking fox-cubs, squabbling monkeys, or sun-bathing lions, it would be hard to believe otherwise. That the layman may have gained the opposite impression of the ethologist's view of his subjects partly arises from the confusion between a methodology and an ideology: in answering questions about the function and causation of behaviour, the ethologist's unemotional methodology has been a powerful approach. To understand that the number of individuals participating in a given hunting party of spotted hyaenas is adapted to the particular type of prey that they are seeking and to their chances of success is enthralling. Such insights can be most clearly and elegantly gained without consideration of the hyaenas' feelings about each other's company, of whether they think about hunting with elation or as daily drudgery, or indeed, of whether they think at all. These are quite distinct questions, the subject of different investigations. Because the hyaenas' moods are not mentioned in, nor germane to, the discussion of their hunting success is not to deny them emotions. We can never know what another animal (human or otherwise) is experiencing. We can quantify and describe their behaviour, we can watch gestures and facial expressions, but the possible mental experiences that they might be having will remain hidden.

There are, however, signs of changing attitudes amongst scientists (e.g. Griffin, 1976) and a number of studies have been undertaken with the express aim of exploring the mental capacities of other animals. Griffin (1976) has given the term 'cognitive ethology' to such studies, but although the term is new, the subject itself is not. Experiments such as those of Wolfgang Köhler (1925) on chimpanzees and of Otto Koehler (1951) on birds had already pioneered the way.

Griffin (1976, 1978) argues that some of the recent findings of ethology, particularly those that have revealed some of the complexities of animal communication, should lead ethologists to reconsider their traditional reluctance to even try to learn about the mental experiences of animals. An example of one such study is by Savage-Rumbaugh et al. (1978), who studied the exchange of goods and information between two chimpanzees both trained to use the same system of symbols. The chimpanzees were housed in adjoining rooms with a view of each other. They had been trained to use a keyboard on which the depression of keys caused particular symbols to be projected on a screen which both animals could see. By pressing a key, one chimpanzee could show a chosen symbol to the other and by so doing request that the item represented by the symbol could be passed through a hole in the partition between their rooms. The item might be a tool which the first chimpanzee needed to gain access to some food, e.g. a key for a lock. A quotation from Savage-Rumbaugh et al.'s account of this experiment is particularly striking:

> On one trial Sherman requested key erroneously when he needed a wrench. He then watched carefully as Austin searched the tool kit. When Austin started to pick up the key, Sherman looked over his shoulder towards the keyboard, and when he noticed the word "key", which he had left displayed on the projectors, he rushed back to the keyboard, depressed "wrench", and tapped the projectors to draw Austin's attention to the new symbols he had just transmitted. Austin looked up, dropped the key, picked up the wrench, and handed it to Sherman.'

We cannot refrain from following this with another quotation from philosopher Anthony Kenny: 'To have mind is to have the capacity to acquire the ability to operate with symbols in such a way that it is one's own activity that makes them symbols and confers meaning upon them.' (Kenny, 1973)

One of the major contributions of experiments on animal behaviour may in the future be to give us a better understanding of the animals' intellectual abilities (Griffin, 1976) and of their capacity for emotional responses. Such ethological work may help us to make our treatment of them fair and humane.

We have outlined some of the many things that comprise ethology, mentioned some of its applications, and, we hope, given some indication of the overwhelming interest to be found in its subject matter. Nevertheless, the pursuit of ethology, like that of every other science, raises various ethical questions. Some published results, particularly when taken out of context, appear to be trivial (and some probably are) and some are acquired by seemingly barbarous techniques; mothers have been parted from their infants, senses tampered with, deliberately unpleasant stimuli foisted upon unwilling subjects, and encounters engineered between rivals, enemies, predators, and prey. Whatever the manipulation, be it great or small, or be the stress physical or mental, the problem is essentially the same: study involves interference and intrusion. The difficulty is to define what types of interference are legitimate, to formulate guidelines for delimiting the desirable from the deplorable, and somehow to bisect the grey zone of the merely acceptable. At the level of watching animals in the wild, or of doing 'physiology without breaking the skin' ethology seems to provide an ideal alternative to some other more intrusive or damaging procedures, but to tackle many questions does necessitate greater interference.

To emphasize to our scientific colleagues that we are not in any way attempting to adopt a 'holier than thou' attitude, we will begin by looking at some of the ethical problems that arise out of our own work.

One of us studies the behaviour of foxes, partly in the belief that knowledge about this species will be of benefit to human beings (e.g. in helping to devise more effective means of rabies control), partly in the belief that it will benefit foxes in general (e.g. in changing public attitudes towards an animal that has traditionally been seen as a 'pest'), and partly out of the sheer interest of the subject.

Now, studying the behaviour of wild animals generally involves trying not to interfere with their natural lives, so that the observations are minimally distorted by the presence of the observer. However, since foxes are active at night, range over a wide area, and tend to be secretive, many important topics simply cannot be studied by stealth and a pair of field glasses. Many of the same topics are assailable with the aid of radio tracking. Once fitted with a radio collar, the fox can go freely about its business, far away from (and yet monitored by) the ethologist. Yet, to be equipped with the collar, the fox must be caught. This involves some risk, for trapping wild animals is never foolproof and is undeniably frightening, however much effort and concern is put into streamlining the operation, selecting the best anaesthetic, minimizing the time in the trap, and so on.

It is clearly a very difficult matter to balance the transient, although possibly intense fear experienced by a fox on being caught with the many human and fox lives that *might* be saved in a new rabies control scheme. There is no guarantee that studying fox behaviour will save any lives at all, for the outcome of any scientific investigation is always uncertain. There is no simple equation from which we can calculate the units of knowledge gained or number of lives that must be saved in order to justify frightening one fox, or a dozen foxes, or a hundred foxes.

A similar dilemma arises in the study of the behaviour of chickens in battery cages. If ethology can help to provide methods for assessing the welfare of these animals, does this justify a study in which hens are kept in small cages? If hens are to demonstrate that they prefer larger cages, a number of hens would have to be confined in small cages in order to demonstrate this. How is one to balance the possible benefit to hens in general (through, perhaps, developing alternative systems of housing) against the possible stress imposed on the animals used in the study?

How convenient it would be if we could give clear answers to such questions. But we do not want to pretend that there are hard and fast rules when we believe that any rules that might be followed would have to be arbitrary. For example, we can see no obvious discontinuities among animals that would enable us to say that the welfare of one is important and that of another is unimportant. Similarity to human beings or large size seem particularly unconvincing criteria, even though they are widely used.

To illustrate the difficulties of making ethical judgements about ethological studies, consider one particularly controversial area, that of predation. An ethologist may observe a fox killing a rat, hunting dogs disembowelling a wildebeest, or a kingfisher falling like a jewel of death on the stickleback diligently fanning his eggs. These are naturally occurring events which in no sense constitute 'an experiment'. But consider the problems of drawing ethical distinctions within the following continuum:

(1) *The predatory behaviour of wild animals in studies with minimal human interference*   Much can be learnt about the predatory behaviour of wolves by following their footprints in the snow, looking at the remains of animals they have killed, and analysing the bones of the dead animals to see how they compare with healthy ones. From this kind of analysis, biologists discovered that wolves tend to pick out the old and sick animals from a herd (Mech, 1970). Similarly, by looking at the mutilated corpses which foxes leave behind after an occurrence of 'surplus killing' (when they kill far more than they can eat), Kruuk (1972) was able to begin to understand a phenomenon which has infuriated and intrigued generations of poultrymen and game keepers, by relating it to the phase of the moon and turbulence of the weather.

(2) *The predatory behaviour of a trained animal is studied in the wild*   By flying a trained goshawk at pigeon flocks (which at the time were destroying a farmer's brassica crop), Kenward (1978) showed that pigeons in larger flocks were safer, since they seemed to see the hawk and take to the air sooner, than single birds or birds in small flocks. This experimental study of the evolutionary significance of flocking and of the hawk's hunting strategy was done under natural conditions, except that the hawk was transported around the countryside to the best pigeon areas.

(3) *The naturally occurring predator—prey interaction is transferred to the laboratory*   Cats freqeuently kill mice and rats (and many a countryman would kill the cat that didn't). Biben (1979) introduced cats and mice to each other in the laboratory. Her experiments were aimed at a better understanding of predatory behaviour and shed light on, amongst other things, the same phenomenon of 'surplus killing' that Kruuk had studied.

It would be absurd to even consider, as some people do, whether it is undesirable or even reprehensible for predators to kill prey. But we have described a progression of studies of predatory behaviour, from those which merely involve watching to those that involve 'staging' in the laboratory events similar to those that occur naturally. We can think of no logically robust criterion for drawing lines between what is permissible and what is not. All these studies are interesting and significant, and some people would argue that all three approaches are quite acceptable. Others would see the laboratory studies as 'cruel', but would find analagous events in the wild quite acceptable.

The evolution of animals did not start with protozoa and 'end up' with human beings. It took place as a great ramification of millions of different life styles, more like a bush than a ladder. Some animals, such as chimpanzees, are very like us, both structurally and biochemically. Others, such as birds, have evolved along different lines and have rather different brain structures to mammals, but at the same time have many similarities of behaviour (Stettner and Matyniak, 1968). There are yet others, such as octopuses, which have had no evolutionary connection with us for at least 600 million years and are more closely related to snails and slugs. Yet they are highly intelligent animals (Wells, 1978), and it is arguable that they qualify for the same consideration given to mammals, and should be given some legal protection. If they were, could people be prosecuted for swatting flies, or for leaving them to die slowly on fly papers? If not, what would be the basis of the distinction?

In everyday speech one often hears a distinction drawn between intent and accident, but the distinction is probably less forceful to the victim; compare the interference of the ethologist trapping a field vole, albeit for speedy release, with the horrible realization that the nature-loving rambler risks crushing a blameless shrew beneath his unsuspecting foot, or at the least desecrating a network of carefully woven paths and scent marks.

In practice, the formulation of laws or an individual's own decision of whether or not to perform a particular experiment has ultimately to rest on some sort of moral cost—benefit analysis, where neither the costs nor the benefits can be precisely defined. In this context, no reader should be tempted to condemn or applaud any of the studies we have mentioned, solely on the basis of our summary—they should go to the original texts and thereafter consider them in the context of the general literature on that topic; the out-of-context quotational evaluation is an abominable device too often employed in discussions of animal experimentation. People who regard all 'experiments' on living animals as equally deplorable ignore the range of both the experiments and the motives behind them.

It is by no means obvious that an experimental operation on a living animal performed under anaesthesia imposes more suffering than destroying the nest of a wild herring gull whose behaviour is all tuned to avoiding just such an event (which is regarded as a humane method of gull control), or indeed removing a newly-born calf from its mother (which is standard agricultural practice). Nor is it obvious that killing animals for food, or in the name of medical research for the ultimate benefit of human health, is either any more or any less justified than is research aimed at increasing knowledge for its own sake. This is not to argue that two wrongs make a right, or that because a greater evil exists a lesser one should be tolerated, but rather to point to the difficulties of calibration which beset this topic.

An ethological approach to experimentation holds great promise for understanding certain aspects of animal behaviour, with minimal mutilation, or even without any at all. Inevitably, though, any study involves some intrusion or interference with an animal's life, even if it is only that the animal is disturbed by the breeze that carries the odour of a skulking ethologist.

Anyone who has read this chapter in the hope that they will be told what is right and what is wrong with ethological experiments will be disappointed. The subject is too complex for dogmatic answers. We feel it would have been both premature and arrogant to arbitrate on ethological experiments in this chapter. But we do feel that there is a need for ethologists to think about the ethical problems of their science and to work towards a set of guidelines for their research. We have tried to highlight both the potential of ethology and its ethical difficulties. In so doing, we have explained that any guidelines which are laid down in the future will have to contain many arbitrary distinctions. We hope that, nevertheless, these guidelines will be based on respect for animals and applied as consistently as possible. We trust that ethologists will be able to live up to the distinction given to them by the editor of this book (Sperlinger, 1979, p. 198) as scientists with a concern to learn *from* and *about* the animals they study.

ACKNOWLEDGEMENTS

We would like to thank P. P. G. Bateson and W. A. F. Macdonald for commenting on an earlier draft.

REFERENCES

Amlaner, C. J. (1977). 'Biotelemetry from free-ranging animals', in *Animal marking: recognition marking of animals in research* (B. Stonehouse, ed.), Macmillan, London, pp. 205–228.

Bateson, P. P. G. (1980). 'Discontinuities in developmental changes in the organization of play in cats', in *Behavioural development* (K. Immelmann, G. W. Barlow, M. Main, and L. Petrinovich, eds), Cambridge University Press, Cambridge, in press.

Biben, M. (1979). 'Predation and predatory play behaviour of domestic cats', *Animal Behaviour*, **27**, 81–94.

Brambell, F. R. (Chairman) (1965). *Report of the Technical Committee to Enquire into the Welfare of Animals Kept under Intensive Livestock Systems, Cmnd 2896*, HMSO, London.

Butler, P. J. (1980). 'The use of telemetry in the studies of diving and flying birds', in *A handbook on biotelemetry and radio tracking*, (C. J. Amlaner and D. W. Macdonald, eds), Pergamon Press, Oxford, pp. 569–577.

Capranica, R. R. (1966). 'Vocal responses of the bullfrog to natural and synthetic mating calls', *Journal of the Acoustical Society of America*, **40**, 1131–1139.

Cheeseman, C. L. and Mallinson, P. (1980). 'Radio tracking badgers', in *A handbook on biotelemetry and radio tracking* (C. J. Amlaner and D. W. Macdonald, eds), Pergamon Press, Oxford, pp. 649–656.

Colyer, R. J. (1970). 'Tail biting in pigs', *Agriculture*, **77**, 215–218.

Dawkins, M. (1977). 'Do hens suffer in battery cages? Environmental preferences and welfare', *Animal Behaviour*, **25**, 1034–1046.

Dawkins, M. (1978). 'Welfare and the structure of a battery cage: size and cage floor preference in domestic hens', *British Veterinary Journal*, **134**, 469–475.

Dawkins, M. (1980). *Animal suffering: the science of animal welfare*, Chapman and Hall, London.

Duncan, I. J. H. (1977). 'Behavioural wisdom lost', *Applied Animal Ethology*, **3**, 193–194.

Duncan, I. J. H. and Filshie, S. H. (1980). 'Using telemetry devices to measure temperature and heart rate in domestic fowl', in *A handbook on biotelemetry and radio tracking* (C. J. Amlaner and D. W. Macdonald, eds), Pergamon Press, Oxford, pp. 579–588.

Duncan, I. J. H. and Wood-Gush, D. G. M. (1972). 'Thwarting of feeding behaviour in the domestic fowl', *Animal Behaviour*, **20**, 444–451.

Ewbank, R. (1973). 'The trouble with being a farm animal', *New Scientist*, **60**, 172–173.

Frame, L. H. and Frame, G. W. (1976). 'Female African wild dogs emigrate', *Nature, London*, **263**, 227–229.

Frishkopf, L. S., Capranica, R. R., and Goldstein, M. H. (1968). 'Neural coding in the bullfrog's auditory system—a teleological approach', *Proceedings of the Institution of Electrical and Electronic Engineers*, **56**, 969–980.

Griffin, D. R. (1976). *The question of animal awareness*, The Rockefeller University Press, New York.

Griffin, D. R. (1978). 'Prospects for a cognitive ethology', *The Brain and Behavioral Sciences*, **1**, 527–538 (see also pp. 555–629).

Hailman, J. P. (1967). 'The ontogeny of an instinct', *Behaviour*, Supplement, **15**, 1–159.

Harrison, F. (1976). 'Born to die for fashion', *Sunday People*, 19 September 1976.

Herrnstein, R. J. and Loveland, D. H. (1964). 'Complex visual concept in the pigeon', *Science, New York*, **146**, 549–551.

Hoffmeyer, I. (1969). 'Feather pecking in pheasants—an ethological approach to the problem', *Danish Review of Game Biology*, **6**, 3–35.

Hughes, B. O. and Black, A. J. (1973). 'The preference of domestic hens for different types of battery cage floor', *British Poultry Science*, **14**, 615–619.

Hughes, B. O. and Duncan, I. J. H. (1972). 'The influence of strain and environmental factors upon feather pecking and cannibalism in fowls', *British Poultry Science*, **13**, 525–547.

Jermy, C. and Hanbury-Tenison, R. (1979). 'The Gunung Mulu expedition', *Geographical Journal*, **145**, 175–191.

Kenny, A. J. P. (1973). *The development of mind*, Edinburgh University Press, Edinburgh.

Kenward, R. (1978). 'Hawks and doves: factors affecting success and selection in goshawk attacks on woodpigeon', *Journal of Animal Ecology*, **47**, 449–460.

King, J. M. and Heath, B. R. (1975). 'Game domestication for animal production in Africa', *World Animal Review*, 23–30.

Koehler, O. (1951). 'The ability of birds to "count"', *Bulletin of Animal Behavior*, **9**, 41–45.

Köhler, W. (1925). *The mentality of apes*, Routledge and Kegan Paul, London.
Krebs, J. R. (1977). 'Song and territory in the great tit, *Parsus major*', in *Evolutionary ecology* (C. M. Perrins and B. Stonehouse, eds), Macmillan, London, pp. 47–62.
Kruuk, H. (1972). 'Surplus killing by carnivores', *Journal of Zoology, London*, **166**, 233–244.
Macdonald, D. W. (1979). 'The flexible social system of the golden jackal, *Canis aureus*', *Behavioral Ecology and Sociobiology*, **5**, 17–38.
Macdonald, D. W. (1980). *Rabies and wildlife: a biologist's perspective*, Oxford University Press, Oxford.
Macdonald, D. W. and Boitani, L. (1979). 'The management and conservation of carnivores: a plea for an ecological ethic', in *Animals' rights—a symposium* (D. Paterson and R. D. Ryder, eds), Centaur Press, Fontwell, Sussex, pp. 165–177.
McFarland, D. J. (1977). 'Decision making in animals', *Nature, London*, **269**, 15–21.
Mech, L. D. (1970). *The wolf*, Natural History Press, New York.
Moehlman, P. D. (1979). 'Jackal helpers and pup survival', *Nature, London*, **277**, 382–383.
Ojasti, J. (1973). *Estudio biologico del chigüire a capibara*, Fondo Nacional de Investigaciones Agropecuiarias, Caracas.
Rood, J. P. (1978). 'Dwarf mongoose helpers at the den', *Zeitschrift für Tierpsychologie*, **48**, 277–287.
Savage-Rumbaugh, E. S., Rumbaugh, D. M., and Boysen, S. (1978). 'Linguistically-mediated tool use and exchange by chimpanzees (*Pan troglodytes*)', *The Brain and Behavioral Sciences*, **1**, 539–554.
Sperlinger, D. (1979). 'Scientists and their experimental animals', in *Animals' rights—a symposium* (D. Paterson and R. D. Ryder, eds), Centaur Press, Fontwell, Sussex, pp. 196–200.
Stettner, L. J. and Matyniak, K. A. (1968). 'The brain of birds', *Scientific American*, **218**, 64–76.
Taylor, W. D. and Hardy, A. R. (1980). 'Movement of common rats (*Rattus norvegicus*) in farmland determined by radio tracking', in *A handbook on biotelemetry and radio tracking* (C. J. Amlaner and D. W. Macdonald, eds), Pergamon Press, Oxford, pp. 657–666.
Thorpe, W. H. (1965). 'The assessment of pain and distress in animals', appendix to *Report of the Technical Committee to Enquire into the Welfare of Animals Kept under Intensive Livestock Systems* (F. R. Brambell, Chairman), *Cmnd 2896*, HMSO, London.
Tinbergen, N. (1958). *Curious naturalists*, Country Life, London.
Tinbergen, N. and Perdeck, A. C. (1950). 'On the stimulus situation releasing the begging response in the newly hatched herring gull chick (*Larus argentatus argentatus* Pont.)', *Behaviour*, **3**, 1–39.
Tinbergen, N., Broekhuysen, G. J., Feekes, F., Houghton, J. C. W., Kruuk, H., and Szulc, E. (1962). 'Egg shell removal by the black-headed gull, *Larus ridibundus* L.; a behaviour component of camouflage', *Behaviour*, **19**, 74–117.
Van Iersel, J. J. A. and Bol. A. C. A. (1958). 'Preening of two tern species. A study of displacement activities', *Behaviour*, **13**, 1–88.
Wells, M. J. (1978). *Octopus—physiology and behaviour of a advanced invertebrate*, Chapman and Hall, London.
Wyman, J. (1967). 'The jackals of the Serengeti', *Animals*, **10**, 79–83.

Animals in Research
Edited by David Sperlinger
© 1981 John Wiley & Sons Ltd.

Chapter 10

The Use of Animals in Schools in Britain

DAVID PATERSON

INTRODUCTION

Most very young children respond to animals with spontaneous, instant, and concentrated fascination. Unprimed by adult fears and whimsies, the family cat presents itself as a 'furry', moving at a child's own level; an object to be reached, touched, and cuddled. In its turn, puss usually tolerates advances which, if offered by an adult human, would deservedly earn sharp and instant retribution. Dogs are also tolerant to a limit, beyond which the wiser ones generally remember urgent appointments elsewhere!

Whether babies see cats as fellow beings or vice versa, or both, is another issue: what it beyond doubt is that there is a quite remarkable degree of mutual interest and friendly tolerance between these very disparate species.

As a child matures, this pristine and innocent interest changes but does not wane. At heart, other things being equal, most children and many adults find animals fascinating to a quite extraordinary degree. Witness the present plethora

Figure 10.1

of 'wildlife documentaries' on television, and the increasing certainty with which even hardened news-editors find prime space for cruelty stories which involve animals.

It is this certainty of being able to gain interest through introducing animals to young children in the classroom that still leads both wise and unwise to attempt to combine the vocation of an animal keeper with that of a classroom teacher. After all, what better teaching aid than one which *infallibly* grips a child's mind? Of course, its ultimate effectiveness, to say nothing of the fate of the animal, depends upon the effectiveness and wisdom of the teacher concerned.

It is undeniable that in most elementary schools a balanced education programme was ensured by combining the 'Four R's', reading, writing, religion, and arithmetic, with nature study. Nor is it merely as a 'carry-over' from the early 1900s that today's primary school finds a justifiable place for animals in an establishment built for educating children. In nature's scheme of things the two go, and so should stay, together.

PRIMARY SCHOOLS

There are 23 341 primary schools in England and Wales, 'housing' 5 014 198 pupils and, at a conservative estimate, between 250 000 and 500 000 animals. We have already seen one reason why animals are there—as 'attention getters'. *Explanations* given for their presence vary, but seldom (for obvious reasons) include this factor. Thus, we are told that:

(a) Animals are kept as 'pets', serving thereby as 'educational aids' from birth to death in human communities where contact with the undoubted reality which they bring would otherwise be denied.

(b) The keeping of animals helps foster the 'finer emotions' of love, affection, empathy, respect, and even awe for the various forms of life.

(c) The involvement of young children in the keeping of animals helps to instil in them a sense of personal responsibility. They relate their own subordinate position in life to that of the small animals in their care, which depend wholly upon *them*. Realizing the animal's needs increases the childrens' sense of personal responsibility, thereby elevating their own status.

(d) In some cases (especially with emotionally disturbed children) animals may become 'transitional objects', offering a stable base in times of stress.

All these explanations have at least an element of theoretical validity in them—but one thing is all too often forgotten in practice.

Just as the best lessons are taught by good example, so the worst are learned through *bad* example. The lesson is still learned; it is simply the *wrong* lesson. For this reason a school or classroom with *no* animals in it at all is really far preferable to one where any animals are badly looked after. It matters little whether bad management is culpable or 'merely' due to ignorance or indifference.

Thus, rough or excessive handling of animals, a lack of clean bedding, inadequate or incorrect caging, or a poor diet not only have a bad effect on the animals; they are also totally counter-productive educationally speaking. Animal keeping of this nature is all too common; it *reduces* respect for life instead of instilling it; bad example counteracts whatever verbal lesson is attempted.

It is for this reason that advance planning, common sense, and a knowledge-able approach to the educational use and care of animals are so very important. With them, both children and teachers develop a rational, balanced, caring, and understanding attitude towards the animal kingdom of which they are a part.

A classroom animal flourishes under these conditions; otherwise it may all too readily be neglected and languish in a far corner, where it no longer sets up in active competition with the teacher. Under such circumstances its presence is worse than useless; a new approach is needed.

An interesting, basic, and informative new point of departure for any scheme of work with animals involves looking more deeply into their origins. In this way children can deduce (and so, it is hoped, provide for) their physical and behavioural needs. This scheme can, of course, be adapted for use at almost any educational level.

The gerbil, for instance, is a comparatively recent arrival in the classroom. Its relatives still live in colonies in the desert regions of Mongolia, Libya, and Egypt. Admittedly, much still needs to be discovered about its normal life style, but sufficient is probably already known for most school purposes. Even a superficial look at its natural life patterns will, particularly when undertaken in conjunction with realistic classroom observations, reveal many of its basic needs.

Gerbils, children will learn, are colonial animals, living in extensive burrows, which they make in dry soil. They feed on vegetable material, particularly grain; they drink little, and pass little urine—so appearing to be particularly well adapted to their natural environment. In nature they seem to live in 'extended families', but in captivity thrive best as monogamous pairs. (Perhaps because of reduced territorial availability?) They are diurnal.

With such background information, children should be able to deduce that gerbils would do best if *not* put into a 'standard cage'. To simulate their natural environment they can be kept in a reasonably large and dry area which has a deepish floor-material into which they can burrow. A peat/hay mix perhaps could simulate dry and fibrous soil. They will need fresh water, vegetable matter for chewing and bedding, and seed-type foods.

In the classroom, in fact, they therefore do best (and create least mess) when kept in a 'gerbilarium'—a well ventilated aquarium tank topped with a lid of fine mesh and with a deep layer of slightly dampened peat/hay in the base. In such an environment they not only thrive, but their behaviour patterns are readily observable for further study without disturbing them. This system also ensures one particularly desirable element: it provides its own built-in safeguard to protect the animals from excessive handling or other disturbance. It also means that far less cleaning-out is required than is needed for normally caged animals. Lastly, of course, it is educational; whilst still preserving territorial integrity for the animals it teaches children much of the respect due to them—including the important concept of 'personal space'.

Indeed, all too few teachers remember that in a classroom situation they, the children, and the classroom itself are *all* part of the animals' environment. For this reason, noise, vibration, excess heat or cold, etc., to say nothing of over-handling, are marks of insensitivity to the needs of others, including animals.

Pet-facilitated psychotherapy is a modern and complex term for a well established practice. Children who do not relate well to adults or even to each other will often readily form a useful, stable, and out-going relationship with an animal; through this, they then relate meaningfully to their own kind. Frances Farrer (1977) cites a classic example of an Asian child who could, apparently, not originally communicate with his teachers; he learned to do so by telling long stories in English to his rabbit. I myself have seen maladjusted children with a total aversion to any form of teaching come willingly to school, mixing effectively with their peers for the first time, when accompanied by a beloved pet animal (who found the classroom a good place to sunbathe or to wait for milk and biscuits). In such extreme cases, a vital link between the real world and a child is created, and subsequently extended, through the medium of a (classroom) animal.

In a similar way, units for the subnormal (adults or children) have found for many years that pets and/or 'farm' animals often provide common ground for setting up meaningful and permanent relationships.

In perhaps lesser ways, then, a meaningful relationship with animals in the primary school can undoubtedly be beneficial to the children involved. It will also, of course, be of benefit to the animals themselves, who come to be seen and treated as fellow creatures and 'friends'.

Having said that, though, it is *in* the primary school that children all too often begin to develop and establish permanent 'double standards' in their relationships with animals.

Many, if not most, are now 'exposed' to some animals in school during their formative years. Many are also exposed to them at home. The effects of this (double) exposure have never, apparently, been seriously assessed. Animals are treated (or mistreated) as 'pets' both at home and later, in a different context, when a child first enters the educational system.

As he learns to read (if not beforehand) the first dichotomy emerges, for pictures in elementary readers often show animals in 'idealized' situations. Hens are seen scratching about in farmyards, cows roam freely in beautiful green meadows. No reference is made to where they are *really* kept and above all no reference is made to our ultimate purpose in keeping them there. Similarly, monkeys and lions are seen behind bars in zoos with no reference to the fact that this is not their natural habitat, nor why we put them there. In a word, text, pictures, and reality do not correlate. The underlying assumption is, moreover, that our dominion over animals is absolute.

Potentially awkward situations are, therefore, treated by avoiding the real issues from a child's earliest years; wild animals are shown in captivity while intensively reared ones roam in (comparative) freedom. All are 'happy', and contented with their enviable lot.

Of course, whether or not one sympathizes with it, one can see the point. School dinners would, if children were shown what they involved for animals, go uneaten. Primary schools would become hotbeds of vegetarian revolution—though whether that would be altogether a bad thing or not is a different issue. Certainly, though, *no* issue at all can fairly be decided by a child who is given only one viewpoint.

A similar conflict between the basic needs of animals in their *natural* environment and as a part of our complex society emerges when children visit zoos, safari parks, and similar 'wildlife areas'. It must be said, though, that the advent of popular, accurate, realistic, and interesting television programmes showing animals in their true habitats at least provides a standard of comparison in the light of which few commercial establishments can reasonably be justified. When their sensitivities are aroused, even comparatively young children are very sensible, and extremely sensitive, critics.

SECONDARY SCHOOLS

A Schools' Council survey (Kelly and Wray, 1971) showed that some 40 species or groups of living organism (i.e. including plants) were then being used in primary schools and that more than 100 were being used in secondary schools (70 per cent of which kept living organisms on their premises).

Since the tendency has been for syllabus revisions in the past to result in considerably *more* animals being used than previously, this figure has probably increased quite substantially since then, and a reasonably educated guess shows that at least 250 000 (non-human) animals are now to be found in our 4 988 secondary schools. In fact a cursory survey shows that, as in primary schools, the minimal figure could most probably be doubled.

Examinations virtually dictate the secondary school syllabus, whether we like it or not. Of the numerous published syllabuses which were examined in detail in 1974–75, the majority (over 60) suggested, and 20 absolutely demanded, the use of living organisms. This figure remains substantially unchanged today. Bear in mind, too, that of the 5 million or so young people who are 'involved' in secondary education, some 255 000 take public examinations yearly, about 25 000 of these being 'A' Level candidates.

It is while a child is at secondary school that final dichotomies in their attitude towards animals become established. Some animals are 'nice', some are not; some are for mans' use in clothing, as food, or for other purposes such as entertainment; some are not. Some are furry 'cuddlies' at one moment and in one context, confusingly becoming transformed into 'vermin' in another. A pattern for the avoidance of future difficulties is set. Certain issues have socially acceptable answers and may be discussed; others may be debated only in an academic way, but are never to be really questioned.

In the secondary school itself, though, some issues constantly *do* arise. They include dissection, the use of living animals as experimental subjects, and the application of the Law itself.

THE DISSECTION ISSUE

Despite early experience of animals in primary school and in the home as being 'pets', a young person who wishes to study subjects such as 'A' Level Biology will have little or no option at present than actually to dissect similar animals.

Of course, it could be presumed that students at this level are more mature, emotionally speaking, than are many of the candidates for lower level examinations. Consequently, the adoption of certain 'learning by discovery' techniques could seemingly be excused. For instance, the dissection of various organisms, otherwise ruled out on diverse grounds, including possible emotional shock, could be permitted here—*provided only that there was a readily demonstrable necessity for dissection in the first place*, and that other safeguards were taken. The teacher who proceeded to dissect his kitten before a class of 14-year-olds was *asking* for trouble.

In fact, though, the whole dissection issue is a worrying one which could profitably be examined in more detail at this stage since the principles involved have a much wider application.

A recent government minister observed that the immediate objectives of dissection must be the achievement of knowledge which would otherwise be more difficult, if not impossible, to obtain. Otherwise it should not be carried out at all. . . . Remember this 'official' position in what follows.

Dissection can certainly be used to teach candidates the uniqueness of the individual living animal of which it is but a dead exemplar, thereby giving a certain reality to standard text book diagrams, in which all animals are stereotyped for simplicity. Any biologist worth his or her salt will therefore (or so the argument goes) come to have *more* regard for life after having undergone this somewhat traumatic procedure. Certainly he will no longer be able to regard animals as 'envelopes for the unknown', like the beloved 'black boxes' of the theoretical physicist. Dissection, then, is ostensibly undertaken to help a student understand animals, whether he likes the process or not. It has become a sort of 'technical purgative' which for his own good he has to suffer.

To come to a sensible understanding of animals, though, some sensible and sensitive guidance is needed and the carrying out of many, if not most or even all,

Figure 10.2

school dissections should most certainly be questioned. The prime route to an understanding of life is not a study of death.

In fact, though, dissection is all too frequently still *demanded* by schools themselves, through science departments or individual teachers—even for Nuffield 'Combined Science' or similar courses (which are, remember, designed for children aged between 11 and 12 years old). At the very least, many schools demand some teaching of dissection for those undertaking 'O' Level courses (age 14–15 years). This despite the fact that many objections are received nationally on grounds of principle, ethic, or 'simple' revulsion, from children, their parents, or even 'outside' organizations which are concerned at the effects which this process is having in desensitizing the next generation.

Why, incidentally, has any form of revulsion at suffering or at the appearance of suffering (a natural and human feeling) now become a thing to be scoffed at and systematically ignored? Why are children who appear to be overblessed with a feeling of empathy towards living things the very ones who are automatically thought to be unfit to study life's wonders because they don't like its antithesis? Educational systems are not necessarily logical, and nor are those who enforce them! Science still has far too many 'sacred cows', particularly where training techniques are concerned. Teachers all too readily teach as they were taught, without question.

As far as moral revulsion is concerned, it is perhaps ironic that precisely analogous feelings would be backed up by authority if they more plainly flowed from what the Act calls 'matters of *religious* belief', or even so-called 'secondary religious practices', such as diet (*Education Act 1944*, Section 25 (3)). A personal moral attitude (an eminently desirable quality, one would have thought) apparently rates as having little or no actual significance for society when/if it leads to an ethical stand being taken on the ground of beliefs, however deeply held. To put it mildly, that's illogical.

Perhaps it is worthy of note that the *Education Act 1944 does* in fact impose obligations in matters of 'moral development'. Thus: 'it shall be the duty of the local education authority for every area to contribute towards the moral, mental and physical development of the community' (Section 7). Note that *morality* comes first in this list; physical education and recreational facilities (valuable as they are) come last: would that there were some parity of provision even between these two!

Furthermore, parents *do* seem to be given at least a sound basis for fighting for their rights in this matter (as in so many others) by the Act itself: 'the Minister and the local education authorities shall have regard to the general principle that ... *pupils are to be educated in accordance with the wishes of their parents*' (Section 76). Surely such 'wishes' include the desire to avoid the imposition of teaching methods which conflict with ethical principles?

Quite apart from the fundamental morality of the issue, *dissection is not necessary anyway*. Thus, reports from the RSPCA (Paterson, 1973), the Royal Society, and the Institute of Biology in the mid-1970s emphasized that *no* dissection was needed prior to the sixth form 'A' Level Biology/Zoology courses, if then. Such reports are all too commonly ignored save where invoked in the interests of economy; money is often more powerful than ethics!

Of course, relatively few children actually *take* 'A' Level Biology or Zoology courses anyway, and *they* can reasonably be assumed to be emotionally more mature than younger children (besides having at least an element of control over what they are studying). But dissection remains wholly unjustified for the remaining 90% of schoolchildren.

Nor does the matter end there either: are we justified in assuming that dissection *is* desirable even for *senior* pupils, particularly since those who proceed to university or higher education courses *begin again* anyway? In fact, it is frequently only upon actual entry to a university that today's students finally choose a given course. Biologists in Oxbridge are certainly accepted from training backgrounds which include history, mathematics, or physics—never having actually studied biology at all, let alone dissection. So it *can't* be essential in schools!

Why, then, do so many schools and Examination Boards continue to demand this largely useless, desensitizing, and often counter-productive practice?

LIVING ANIMALS, SCHOOLS AND THE LAW

No-one who has taught a Nuffield 'Combined Science' course, and seen the sometimes devastating and often unpleasant effects which witnessing the dissection of a recently killed pregnant rat, with its fetuses at an advanced stage of development, can have upon a class of 12- or 13-year-old children, can doubt the un-wisdom of extending the principle of 'direct experience teaching' to this area. But what can be said of practices which actually involve the use of *living* animals as experimental subjects in secondary schools?

Far too many biology text books imply that whereas one has to be a little careful of what one does to vertebrates, invertebrates don't really matter. This is because they assume that, since 'experiments' seem to be involved, only the *Cruelty to Animals Act 1876* needs be considered. In fact, of course, there is *no way* in which the present Act can apply to the school situation. *No* school is licensed by the Home Office (a pre-requisite for being covered by the Act). Indeed, the only relevance of the 1876 Act in the school situation is that it renders anyone who experiments in any way upon a living vertebrate animal liable to prosecution.

What does apply though is the *Protection of Animals Act 1911* (1912 in Scotland), to say nothing of the various other animal protection Acts. The 1911 Act is much more extensive in many ways than is that of 1876. Violations of its requirements have resulted in a considerable number of teachers being pro- secuted for causing 'unnecessary suffering' to animals. Thus:

> 'if any person shall . . . cause (or permit) any unnecessary suffering . . . (or) . . .shall wilfully, without any reasonable cause, or excuse, administer . . . any poisonous or injurious substance to any animal . . . or . . . subject any animal to any operation which is performed without due care and humanity, such persons shall be guilty of an offence of cruelty within the meaning of this Act'

Furthermore,

> 'Any person who impounds or confines . . . any animal . . . shall . . . supply it with a sufficient quantity of wholesome and suitable food and water . . .'.

Teachers who effectively abandon animals in schools without adequate provision being made for all their needs, and particularly those who leave animals without proper food and/or water, are, therefore, liable to prosecution, as many have discovered to their cost.

Similarly, those (particularly primary school) teachers who leave 'their' school animals in an unshaded greenhouse or an unventilated shed for the sake of easy access during the summer holidays, or send them 'for a holiday with the children' without giving adequate training or ensuring supervision, must also beware. School folklore abounds with stories of gerbils taken home unbeknown to a child's parents and consequently (and lethally) hidden in the airing cupboard or given to the cat. Too many animals suffer and die during school holidays through ill-treatment or sheer neglect.

It must of course also be remembered that the term 'animal' in the 1911 Act refers to *any domestic or captive animal*, of whatsoever kind, including not only mammals but also fish, reptiles, amphibia and birds, where these are in any way held in captivity. Thus, it undoubtedly applies to wild animals which have been brought into the school to be kept for 'observation', to say nothing of more normal populations of assorted 'pets'.

Responsibility for Holiday Care

In these and other ways, holiday care poses many problems for teachers who have responsibility for school animals (and, ultimately, for the school head-master or the head of department concerned, since these are usually held to be legally responsible).

To avoid all argument, where animals *have* to be sent to childrens' homes, explicit parental permission must be obtained in advance, written instructions on regular and daily care must be given, and correct food, bedding, and litter ought also to be provided. Local veterinary care should also be available, at the expense of the school itself if the education authority makes no formal provision for payment.

An alternative suggestion, which can work quite well, is for secondary schools to help primaries, providing a centre to which other schools in their (catchment) area can send animals during holiday periods. In fact there is all too little practical co-operation between schools over this matter: science advisers might perhaps be more active in advising here than appears to be customary?

Of course, it goes without saying that, if the 'school centre' method of provision is adopted: (a) animals from different schools should be segregated to prevent any cross-infection, and (b) animals should not be left in the 'general care' of school cleaners or caretakers.

A minor extension of the laboratory technician's hours on a daily basis, with or without a defined rota system, is all that is needed to provide adequate and professional care for all school animals in an area at minimal cost.

Figure 10.3

Three other approaches to this problem could perhaps be mentioned here; the 'animal bank', the 'city farm', and the itinerant lecturer.

An animal bank is generally organized like a lending library, from which animals can be *borrowed*, rather than as a bank, from which they could simply be obtained. Its advantages are obvious; a range of animals can be made available to schools for 'short-term loans'. No school holiday problems. No excessive population of school animals which depend upon the Rates. The possibility at least of constant and expert health-checks.

But *dis*-advantages are present too. Animals kept in an individual school for only a short time certainly don't foster much sense of responsibility, personal or otherwise. Nor do they empicture much of the continuous cycle of birth and death. In fact, all in all, there appears to be very little that they usefully *can* do! On the contrary, their transitory passage confirms children in their growing suspicions that animals are disposable assets, without feelings or needs of their own.

The animals themselves are all too liable to be inadequately provided for, both as regards accommodation and diet. Moreover, they are constantly exposed to the maximum of disturbance both in school (being new), during loading and unloading, and in transit—to say nothing of the compounded cross-infections which they take back to give to their companions in the 'bank', and subsequently all pass on to children or animals in other schools. Ringworm? Psitticosis? Salmonella?

Bear in mind too that the simplest way for most 'laymen' to recognize that an animal is ailing is the perception of abnormal behaviour patterns. If a teacher is not familiar with what is 'usual' and what is not, then how can *variations* be noticed? The animal may well die before it is even seen to be ailing.

Generally, then, an animal bank, even a well run one, is a veterinarian's nightmare and an epidemiologist's delight. Why do people who have what are apparently such good ideas not look into their ultimate but all too possible consequences?

City farms are a different proposition; here the animals are kept in a stable (no pun!) environment, where they can be well looked after under expert supervision. This obviates many of the difficulties found in the 'animal bank'. On the other hand, they are constantly exposed to more and more groups of excited children who are vociferously determined to touch, cuddle, and squeeze each animal in turn.

Some 'farms' go further than others, though, and transport van loads of assorted livestock around local schools, attempting to extend their sphere and range of influence on an almost daily basis. They explain this as being their way of 'keeping town children in contact with nature', but such practices are ill thought out and, I would maintain, not very educational either: a 'gimmick', in other words, to which these establishments ought to be opposed in virtue of every principle behind their foundation. . . .

Some schools encourage 'itinerant lecturers' in a totally non-discriminatory fashion. These are brought in to occupy the children rather than to help educate them—and paid a fee to do so. Fair enough, though totally unqualified visitors from 'outside' are all too readily listened to and accepted as authorities who speak wisdom beyond that of the school's own teachers.

Where the fee which these lecturers are paid comes from (often being raised as a levy on the children themselves, or extorted from 'school funds') is another matter. Another matter also is the fact that these 'lecturers' often bring with them an assorted collection of birds, rats, mice, and even snakes. *Their* lives are certainly not worth living, to say nothing of the bad example given to children and the whole legal question of captive species (even those which are theoretically protected—birds especially). Remember the chaos inadvertently caused by all the children who fell for 'Kes', and robbed nests galore in emulative consequence? Bad example, remember, is all too readily accepted, imitated, and learned from. . .

Responsibility for Health and Hygiene

Health and hygiene, incidentally, are all too often matters more honoured in theory than in practice as far as general animal care in schools is concerned. Few teachers remember that: (a) children can give diseases to animals, (b) animals can (re-transfer) diseases to children and, perhaps above all, (c) animals brought in to the school or laboratory from the wild can all too readily provide the most rapid method that *I* have ever seen to kill off every other animal in the school!

In this matter, as in others, teachers have a legal obligation to ensure that children in their charge are guarded from possible harm. Despite 'popular' impressions to the contrary, a badly run biology laboratory is potentially *far* more lethal than is, say, a physics laboratory using radio-active sources. Bacteria can kill too—and more often do so!

EXPERIMENTAL WORK ON ANIMALS

As far as any actual experiments upon living animals are concerned, of course, experimental work upon at least warm-blooded vertebrates is forbidden by law (those who use American text books, beware). Confusingly, though, the *Cruelty to Animals Act 1876* is usually construed by the Home Office as not applying to work in schools or colleges—on the grounds that 'experiments' here are more in the nature of 'procedures', since their outcome is foreseen. Since the outcome of most experiments in most research is equally foreseen, the logic of this argument escapes me for the moment; either way, though, the 1911 Act certainly still applies and is enforced upon occasion.

Personally, I would like to see some test cases taken to clarify the apparently differing interpretations which are put upon the application of the law by the

Home Office itself and between the Home Office and the Department of Education and Science, but the present fluid state of the law in the area of experimentation makes such an exercise untimely. One could well ask, though, why *any* one seems to be allowed to 'pith' a frog by inserting a needle into its brain (and spinal cord), subsequently exposing its peripheral nervous system by dissection? A similar operation on a warm-blooded vertebrate could lead to prosecution. Does the Home Office effectively classify frogs as non-vertebrates in a similar way to that in which a Scottish Court recently made a solemn decision that prawns were 'insects and not animals'? Apart from showing that 'the law is an ass', such unprofessional decisions demonstrate very clearly that the application of the law is biased and irrational. Zoologists should at least be cynically amused to note that the Home Office is equally irrational in attempting to apply this irrational law, for they extended the full protection of the law to that classically primitive animal amphioxus in 1977–78. At the same time, of course, *no* protection is ever given to (for instance) the octopus, which is neurologically far, far, more highly advanced. Yet the 1876 Act which they administer ostensibly sets out to prevent cruelty to living animals in laboratories. Ah well!

More to the immediate point; how is it that teachers and students are not only allowed but are actually encouraged to carry out injection 'procedures' on animals such as the *Xenopus* toad? Their outcome is no more 'certain' than in much other scientific work; the work in question is plainly an *experiment* and, if not, then surely it can be classified as a 'treatment', and so is against the *Veterinary Surgeons' Act*?

Similarly, and even more seriously, work such as is described for 'A' Level students in the Nuffield Biology Course, in which a litter of young mice are deliberately stressed by being tossed from hand to hand some 40 times daily until they are 12 days old *in order to determine the effects of stress*, would seem to me to rank as an experiment, If it isn't, then many industrial 'procedures' in which rodents are exposed to various 'conditions' in order to determine their effects have results which are just as predictable, and should therefore not be covered by the 1876 Act either (which means that all involved in this work could, at least technically, be prosecuted). All that I am saying is that at least the same rule should apply in similar ways in either situation.

Be that as it may, such procedures in schools certainly expose many animals to 'unnecessary suffering' within the meaning of the 1911 Act. Work which produces only predictable results *and* causes suffering, almost by definition, brings about suffering which is *unnecessary*.

Injecting toads with gonadotrophic hormone would also seem to mean either inflicting pain (because the operator is inexperienced) or at the very least to be forbidden as 'the administration of a poisonous or injurious substance', since it interferes with what would otherwise be the animal's normal bodily state and function. Induction of ovulation (or labour) in more advanced vertebrates is no task for the unskilled.

Even more outstanding cases for actual prosecution arise in so-called 'rural study units', where animals are not only kept and bred by the children, but often seem to be killed by them too, 'for practice'. I know from personal and bitter experience of at least one wilfully blind headmaster who presides over a school where unfortunate cockerels and geese survived several successive neck-wringing attempts, only finally being put out of their misery by a child who was stronger than his peers.

What *are* our schools coming to? They were most decidedly never meant to be places for this type of tutelage, to say nothing of carnage. Official protection should be extended to animals and to children—not to those teachers who inflict suffering, and who should know better.

Similar practical insensitivity, at diverse levels, is preached in many areas of most biology and rural science syllabuses, though how this appears in practice depends largely upon the individual teacher. A mentor who is unthoughtful, callous, and insensitive can turn any lesson into at least an emotional disaster for his captive audience.

Invertebrates, for instance, are all too often regarded as being automata without nervous systems and are treated like toys—to be dismantled. Fairly advanced vertebrate fetuses (eggs) are exposed and examined 'to see the heartbeat', the child who manages to keep his 'preparation' alive longest being praised for his diligence, while the embryo dies before him on the laboratory bench. Whatever are we coming to?

Yet biology teachers are not *really* more insensitive than the rest of us, even if they do seem to forget what should be the logical and ethical outcome of the evolutionary theory which they teach. They love their families, and their family pets, like anyone else. But that's not enough: not in their position.

Those who have been taught to be 'objective' must never equate this with being *insensitive*. They must learn (or learn again) to *show* compassion. That's what teaching is all about: *leading* children towards a better and a more humane world. Teaching insensitivity towards the needs of living things is the sure way to end up with—nothing.

REFERENCES

Farrer, F. (1977). 'Animal magnetism', *Times Educational Supplement*, 28 October.
Kelly, P. J. and Wray, J. D. (1971). 'The educational use of living organisms', *Journal of Biological Education*, **5**, 213–218.
Paterson, D. A. (1973). 'The use of invertebrates in schools' and 'The use of experimental animals', unpublished papers, RSPCA Education Department, Horsham, Sussex.

Animals in Research
Edited by David Sperlinger
© 1981 John Wiley & Sons Ltd.

Chapter 11

Live-Animal Science Projects in US Schools

MICHAEL W. FOX AND HEATHER MCGIFFIN

There is a widespread use of live animals for teaching purposes in the United States today, both in high (or secondary) schools and at the undergraduate and graduate college levels. Actual figures of the numbers of animals used annually, and the range of species, are not available. Biological supply houses provide teachers with a wide range of organisms from *Hydra* to frogs and hamsters. Dissected, embalmed, and colour-dyed cadavers—fetal pigs, rats, and cats—are also widely used. Since the most readily accessible information about the kinds of experiments that are encouraged and undertaken in high schools on live animals can be obtained from local and national high school science fair exhibits (open to the public), the main focus of this chapter will be on these science fair studies.

The International Science and Engineering Fair (ISEF) is held annually with approximately 400 contestants who are the finalists from over 200 affiliated fairs in the US and abroad. The fair is sponsored primarily by Science Service Inc., a non-profit organization founded in 1921, supported by substantial grants from General Motors and Westinghouse Corporation. While Science Service sends their ISEF rules to local high school science fairs, there is no overall supervision of these local fairs. At the local level therefore there is a great need for more rigorous supervision and guidance especially for those students using live animals in their projects. A recent state science fair in Kansas included such school and home experiments as injecting rats with insecticides, dosing hamsters with valium, and inducing lead poisoning in infant rats.

The range of studies that will be described gives an accurate picture of what kinds of experimental interventions on vertebrates are accepted and encouraged in secondary schools in America. The 'hands on' approach of learning by doing is perhaps one cultural idiosyncrasy, not necessarily unique to, but certainly characteristic of, the US. Great value is placed upon students acquiring technological skills and manual dexterity, and this does have its virtues. Only too often, however, such utilitarianism may do more to develop technologists who are trained to 'do' rather than scientists who are educated to think, to ask questions, to test hypotheses, etc.. Again reinforcing the utilitarian approach, taking animal life or subjecting animals to physical pain or other suffering is considered acceptable if the students acquire new knowledge or skills. Significantly, those who have publicly criticized such utilitarian insensitivity towards animals in the classroom and who have questioned the ethics of such utility frequently evoke this conditioned response from *Genesis*: that man has dominion over animals and therefore a God-given right to do whatever he may wish to animals for his own benefit.

The International Science and Engineering Fair regulations clearly reflect and support the underlying values behind the rationale that condones school children experimenting upon live animals. The 1979 regulations of the ISEF state that

> 'The legitimate use of animals in the classroom, in the laboratory, or in science fair projects presupposes two postulates. First, the use of animals for learning, as it is for testing and research, is morally acceptable; and second, that man has a responsibility to grant the animals he uses for his benefit every humane consideration for their comfort and well being.'

On the surface, this statement seems acceptable; but it places, in terms of legitimacy and moral acceptability, the use of animals for teaching purposes in the same category of essentiality as for those animals used in research and for testing purposes.

The second segment of the above statement concerning man's responsibility to give experimental animals every humane consideration for their comfort and well-being hides one over-riding factor: namely, that if some degree of pain or suffering is unavoidable as a consequence of experimental intervention, then within limits set within the guidelines, it is justifiable. A similar loophole exists in the US *Animal Welfare Act* where animals need not be given the recommended analgesics or anaesthetics to alleviate pain or suffering if such treatment directly interferes with the experiment.

How are school children going to learn the full significance of humane ethics towards sentient animals if they are allowed to subject animals to any form of suffering, or to kill them, not in the search for new information, but for learning scientific skills? Such utilitarian and dominionistic lip service to humane ethics could well condition students to accept more readily the cost/benefit rationalizations of animal experimentation in the biomedical and commercial fields later in life. The 'costs' to animals (in terms of life and/or suffering) are invariably outweighed by the many and diverse 'benefits' to man, and more so when one's attitudes were formed initially in high school where student learning is valued above animals' lives and/or suffering.

Some educators and scientists believe that if school children do not experiment upon living animals, society will have fewer biomedical researchers and doctors in the future because the students won't have been sufficiently encouraged in the formative years of their training. This is a short-sighted if not self-serving belief that lacks both objectivity and supportive data. It also reflects a very biased view of what education entails in the full sense of the word. (This will be discussed in more detail subsequently.)

The following statement from the 1979 ISEF regulations attests to this short-sighted belief system:

'The moral responsibility that we all have toward animals means that we cannot give free rein to students in projects involving animals. Consequently, those of us who would nurture a healthy curiosity in youngsters are placed in a delicate position. To "turn off" a prospective biologist or physician by excessive limitations would be a serious mistake. For we know that through science fair work, thousands of today's physicians, dentists, veterinarians, scientists, engineers, and science teachers were given an important impetus toward their careers.'

But 'an important impetus toward their careers' need not entail students subjecting sentient animals to painful or stressful studies. Although the new ISEF regulations require student supervision for animal experimentation and suggest that students utilize 'lower' forms of life such as protista and other invertebrates, direct, interventive experiments on sentient animals are still condoned (including the use of anaesthetics, drugs, thermal procedures, physical stress, pathogenic organisms, ionizing radiation, carcinogens, and surgical procedures). No suggestions are given as to what non-interventive behavioural-type studies might be done. There is still a rather widespread contention that ethological studies are too subjective and not 'scientific'. Yet students could learn all the basic scientific principles of data collection and analysis, hypothesis testing, and so on simply from direct observation of animals without having to subject them to any physical stress or traumatic manipulation.

The following statements from the 1979 ISEF regulations show what kinds of experiments are acceptable for the fair, and what experiments, therefore, are being produced for local and national fairs in high school biology classrooms, and by students at home.

'An experiment in nutritional deficiency or ingestion of hazardous or reputedly toxic materials may proceed only to the point where symptoms of the deficiency or toxicity appear. Appropriate measures shall then be taken to correct the deficiency or toxicity, if such action is feasible, or the animals shall be killed by a humane method.'

Determining the point at which disease symptoms appear requires sophisticated and continued monitoring of the animal—a difficult requirement to fulfil under such varied conditions.

'Experiments involving stress will be permitted only when such stress does not produce pathological lesions.'

Stress-aversive electrical shock, conditioned avoidance, starvation, cold exposure, etc., only rarely produce pathological lesions, which could probably not be recognized by a school child or biology teacher. Psychological stressors even more rarely result in pathological lesions. This totally inadequate statement clearly accepts and condones high school students performing stressful experiments upon animals.

The following variety of interventive experiments are permitted, students doing these either in class, at home or in laboratories on summer internship programs.

'No experiment may be undertaken that involves anaesthetics, drugs, thermal procedures, physical stress, organisms pathogenic to man or other vertebrates, ionizing radiation, carcinogens, or surgical procedures, unless these procedures are performed UNDER THE DIRECT SUPERVISION OF AN EXPERIENCED AND QUALIFIED BIOMEDICAL SCIENTIST OR DESIGNATED ADULT SUPERVISOR...

'... A BIOMEDICAL SCIENTIST IS DEFINED AS ONE WHO POSSESSES AN EARNED DOCTORAL DEGREE IN SCIENCE OR MEDICINE, AND WHO HAS A WORKING KNOWLEDGE OF THE TECHNIQUES TO BE USED BY THE STUDENT IN THIS RESEARCH PROTOCOL.

'A designated adult supervisor is defined as an individual who has been properly trained in the techniques and procedures to be used in the investigation. The biomedical scientist must certify that the designated adult supervisor has been so trained...

'... A teacher's certification indicating familiarity, understanding, and compliance with current ISEF rules must be signed and dated PRIOR TO the initiation of experimental work of any kind involving live vertebrate animals.'

Students are not allowed to euthanize their experimental animals however. This must be done by the qualified scientist, designated adult supervisor, or animal care supervisor. ISEF rules also state which humane techniques they consider to be acceptable.

Few high school teachers are trained in laboratory animal care. A special mini-course was run for teachers at the 1976 ISEF. Yet even if the science teachers have some knowledge after attending a mini-course, how adequate are the facilities at the school? Certainly non-existent in many cases where experiments, often entailing surgery and anaesthesia, are conducted at home. Unless the student has conducted his or her experiments at an approved scientific institution or school laboratory, competition in any state science fair should be prohibited if such experiments entail any of the elements listed in the above paragraph of the *ISEF Guiding Principles*

ISEF GUIDING PRINCIPLES

Setting up appropriate facilities with well trained high school science teachers and graduate biomedical and scientific consultants or student 'internships' in research laboratories are only half-way steps towards the more general improvement that is needed in the biology classroom and in future science fairs. A more rigorous scrutiny of student projects before completion, i.e. at the protocol stage, based upon objective, scientific, and humane values, may help in screening out a number of experiments hopefully before they are undertaken, rather than when they are put up for exhibition.

There is a marked difference between gaining biomedical knowledge in a laboratory and the pursuit of a learning project at school or at home. There is absolutely no way that the two can be equated. A school science project is not likely to produce new findings that would justify injuring an animal. Therefore, sharp limitations should be placed on the way animals can and should be used if the student is not working in an established laboratory under qualified supervision. If animals are caged, they should be laboratory species already adapted to such conditions. (The use of wild animals or those species such as dogs, cats, monkeys, etc., whose instinctual and socio-emotional needs cannot be satisfied under confinement, should be prohibited.) Experiments where animals are observed in a natural state (or semi-natural, e.g. in zoos or farms) for normal behavioural purposes and not exposed to any injury, pain, fear, or anxiety should be encouraged.

There is no ethical or scientific justification for conducting classroom or home experiments which cause pain or suffering to any type of animal. It is fair to say that the child who is conditioned by either parent or teacher to indifference and apathy toward the pain and suffering of any sentient createre is being taught to repress his or her emotions, affection, and compassion, reaction formations which have been implicated in the development of antisocial and psychopathic personality disorders. Inhumane science fair projects tend to discredit true science; and, further, they inhibit the aim of education which, in its fullest sense, entails the development of the whole person and human nature involving ethics and sentiments as well as the intellect.

The following range of experiments should not be performed or allowed to enter a science fair if they have not been conducted in a well recognized laboratory facility: any form of surgery; the injection of or exposure to micro-organisms which can cause disease in man or animal; exposure to ionizing or other radiation; treatment with cancer-producing agents; injection or other treatment with any chemical agent; exposure to extremes of temperatures; electric or other shock; excessive noise; noxious fumes; exercise to exhaustion; overcrowding or other distressing stimuli; deprivation of food, water, or other factors essential for survival (including deficient diets).

Alternative guidelines for the study of animals in classroom of elementary and secondary schools have been drawn up by the Humane Society of the US. The Animal Welfare Institute, and the Canadian Council on Animal Care (see Appendix 2) which are in sharp contrast to those of ISEF. Recently the Scientists' Center for Animal Welfare, Washington, DC prepared the following guidelines:

'IV Experimental Studies

1. All experiments should be carried out under the supervision of a competent science teacher. It is the responsibility of the qualified science teacher to ensure the student has the necessary comprehension for the study to be undertaken.
2. Students should not be allowed to take animals home to carry out any studies. All animal projects must be performed in a suitable area in the school.
3. Projects involving vertebrate animals (mammals, birds, reptiles, amphibians, and fish), will normally be restricted to measuring and studying *normal* physiological functions such as *normal* growth, activity cycles, metabolism, blood circulation, learning processes, *normal* behaviour, ecology, reproduction, communication, or isolated organ tissue techniques. None of these studies requires infliction of pain.

All students carrying out projects involving *vertebrate animals* must adhere to the following rules:

A. No experimental procedures shall be attempted on a vertebrate animal that causes pain or distinct discomfort, or interferes with its health.
B. Students shall not perform surgery on vertebrate animals.
C. Experimental procedures shall not involve the use of:
   a. micro-organisms which can cause disease in man or animals
   b. ionizing radiation
   c. cancer producing agents
   d. drugs or chemicals at toxic levels
   e. alcohol in any form
   f. drugs that may produce pain
   g. drugs known to produce adverse reactions, side effects, or capable of producing birth deformities
D. Experimental treatments should not include any of the following: electric shock; exposure to extremes of temperature, noxious fumes, excessive noise or other distressing stimuli; exercise to exhaustion; overcrowding of living quarters.
E. Behavioural studies should use only reward (positive reinforcement) and not punishment in training program.
F. Diets deficient in essential foods are prohibited. Food shall not be withdrawn for periods longer than 12 hours. Water shall be available at all times and shall not be replaced by alcohol or drugs.
G. If egg embryos are subjected to experimental manipulations, the embryo must be destroyed humanely 2 days prior to hatching. If normal egg embryos are to be hatched, satisfactory humane provisions must be made for disposal of the young birds.'

The following is a report on one recent International Science fair to illustrate the range of experiments with live animal subjects that high school students are permitted to do under present regulations. The greatest concern by far are the kinds of experiments that are done at the local level before they reach state and national finals. Some projects that do go through to the national finals and are accepted by the ISEF judges are ethically highly questionable, as this report will demonstrate.

Prizes are awarded to students from a variety of industrial companies and such professional organizations as the American Medical, Dental, Veterinary,

Psychological, and Hearing and Speech Associations. The top two finalists also have an expenses-paid visit to the annual Nobel Prize ceremonies in Sweden.

At state and national levels it is obvious from the following examples from the 27th International Science and Engineering Fair, held in Denver, 10–15, May 1976, that much more regulated scrutiny is necessary and more important still that existing criteria governing the use of animals *in any kind* of experiments are inadequate.

(a) *Medicine and Health Division*   The following are some of the finalists' projects that we found particularly questionable. One, entitled 'Effects of caffeine on albino rats' involved injecting caffeine in high doses which caused considerable physical and mental distress in the rats. (This is needlessly repetitive and nothing more than an exercise in applied skills for the benefit of the student.) Another project, 'Studies of the heart in the guinea-pig' was 'supervised' at home by the student's father, an MD. The student injected sodium tetradecyl into the heart muscle of the guinea pigs to simulate a heart attack. (This chemical causes acute and undoubtedly extremely painful damage.) A study in which mice were exposed to ultraviolet and infrared rays after areas of their skin had been removed was entitled 'The influence of phototoxic rays on the skin'. Radiating these animals cannot be justified as the project demonstrated nothing new and contributed nothing to the future mitigation of suffering in man or animal. Another experiment (on mice) 'Prevention of the bends by breathing oxygenated liquid' had been approved by a professor of psychology and a professor of surgery and anesthesiology. Ironically, when it reached the national finals, the judges decided the exhibit should be dismantled and also not shown to the public. All animals in this study died. (This was testimony to inadequacy of design, lack of basic knowledge, and, above all, humane concern.) An unsupervised surgery project done at home and entitled 'Homotransplantation of skin in mice' turned out to be a repeat of one of a series of studies conducted by Drs Burnet and Medawar 25 years previously. Another finalist's contribution was entitled 'Possible hearing loss in cats due to hair spray'. This was a mimic of the type of safety test conducted by the cosmetics industry. It was a poorly designed experiment in which the student used four of her kittens at home, spraying the ears of two for 30 and 60 days and testing how well they responded to her call to them on a tape recorder at various distances and volumes. In yet another disturbing study, 'Nitrite: a comparison of nitrite content in cured meat and its effects on mice', the student (who actually used rats, not mice) described findings already well documented in scientific journals. Effects were not obtained until 100 times the FDA limitation level of nitrite was given.

(b) *Zoology Projects*   There were some particularly impressive non-interventive studies in this area, including *in vitro*, histochemical tissue culture and micro-organism experiments. Two highly original and useful projects, a survey of calf mortalities due to various livestock practices on different farms, entitled 'Environmental factors causing stress in dairy calves' and a study, 'Dry poultry waste: an economical and efficient supplement for sheep', illustrate the potential utility of such investigations. More traditional studies on ecology, ecosystems, and animal behaviour were in general well designed and presented. 'Study of symbiotic relationships between three shrimps and two anemones', 'The paper builders' (a study of wasps), 'Do birds have color preferences?' 'Lead shot controvery', 'Architectural study of orb webs and identification of spiders', and 'Chirping patterns of crickets' are the kinds of research that students can do well and often fortuitously contribute to scientific knowledge, using their natural talents of observation and enquiry without being distracted by technical skills and other self-limiting influences which, at an early age, may be detrimental to the development of an integrative and creative mind. Such studies should automatically be given a high rating by judges since the student interferes minimally with the animals he studies and does not have to resort to surgery, drugs, or other 'technique and interference oriented' questioning.

Repetitive studies were also found in the Zoology division of project exhibits, some of which would have been placed more appropriately under Medicine and Health, e.g. 'Chloramphenicol-concentration for anemia production in the rabbit'. Blood anomalies are well documented for this drug; the student merely acquired haematological skills—which could have been developed without having to make animals ill in the process.

Some studies were clearly without merit, one being the 'Behavioral effects of microwaves'. Rabbits, which were placed in a microwave for 10 minutes each day received radiation exposure for a period of 6 weeks. The student's conclusions were 'Their memory was affected'. This conclusion was one of a number of possible interpretations and the student, who was interviewed, obviously lacked not motivation but supervision and constructive criticism from a scientist.

(c) *Behavioural and Social Sciences* This 'mixed' class of animal and human studies included several well designed and executed studies with animals, such as biorhythms of crickets, sowbugs, and rats, vocalizations of green frogs, and tarantula behaviour. Again, some exhibits might have been more suitably placed in Medicine and Health—or in a new future category of Environment and Conservation, such as a survey of steel versus lead shot in ducks, cadmium toxicity in Florida crabs, and effects of food additives on mutation rates in fruit flies. A few studies were disconcerting. In one, a toxicological study of the effects of benzaldehyde on *Euglena* (a one-celled organism), the student was questioned and was totally unaware of a major well-know phenomenon of such organisms—the higher the population, the greater is the resistance to various poisons: such lack of background knowledge is a reflection of inadequate science instruction and supervision.

There was only one project in this division that was questionable, entitled 'Psychological stress in relation to physical activity in rats'. This was a typical animal psychology experiment, 20 years ago; therefore, little was required of the student other than the physical design and construction of the experiment. An electric shock (physical stress) was given to one rat who could learn to avoid being shocked over a 24 hour period. Another rat 'yoked' to the first by an electrical circuit could do nothing to avoid shock and always received a shock if the first rat ever made a mistake. This 'yoked' rat was the more active of the two, after experimentation; this was attributed to it 'being more nervous or having more muscular tension'. This is a wholly unnatural and meaningless experiment, relevant to no real-life or real-rat situation but only to rats under such confined and defined conditions. Students of obvious talent and motivation should not be misled into believing that studies of this type are of any scientific or humanitarian value, they are simply an intellectual exercise of questionable significance.

The purpose here is not to criticize individual projects *per se* but rather to select those projects that are indicative of the values and attitudes towards animals that are prevalent today in the scientific establishment and which are being perpetuated through the educational system.

In a time when the scientific community itself is questioning the ethics and scientific value of the use of animals in biomedical research, the educators of the youth of America must also question the fact that inappropriate and inhumane animal projects are commonly conducted within their own realm of jurisdiction. It is still often considered necessary for a student on the elementary or high school level to conduct an experiment which causes pain, suffering, or discomfort if it can be justified as a 'learning experience'.

The high school science fair is a well established and continuing tradition that few seem to question when, for example, animals are used in experiments that are unnecessary repetitions of work already well documented in great detail in scientific literature. Other than gaining manipulative and basic experimental skills, the reiteration of well researched areas seems pointless and should be questioned when the lives or suffering of animals is involved.

The acquisition of such skills in the young 'scientist' is surely less important than developing ethical and humane values as well as the powers of observation and quantitative and qualitative analyses of data; the latter can be derived from more humane experimentation. For example, there is a wealth of experiments that the student can design in the field of animal behaviour which would provide an excellent learning experience and familiarize them with scientific methodologies without necessitating animal suffering or death. Also Russell (1978) has compiled a series of physiological experiments which students may perform upon each other, thus eliminating the need for live animal subjects for classroom or science fair projects in many areas of basic physiology. Certainly the science fair gives an accurate indication of the values and attitudes toward research and the use of animals in science and medicine in general.

There is great need for improvement in values and attitudes after a rigorous scrutiny and objective evaluation of the goals, purposes, whys, and hows of biomedical research. At the high school level this is indeed critical, since the unthinking acquisition of values, skills, and methods of experimentation with animals may be a self-limiting traditionalism. The student may be unable to conceive of alternative methods and subjects or to explore untrodden areas if his modelling or programing of and by the graduate biomedical scientist is narrow in terms of values and attitudes. Today students can tell *how* they performed their experiments, but few can clearly articulate why. Hopefully over the next few years, the motivation and intelligence of these potential scientists will be complemented by the highest ethical, humane, and purposive values acquired from their instructors and supervisors.

Expertise and sophistication of technique are valid accomplishments, but are only a means, and yet too often they become an end in themselves. The animals used are then mere tools. For many students, this has become a major factor in their disillusionment with biomedical research. The answers to human problems are sometimes best answered by studying man himself. That is not to say that all research on vertebrates should be abolished, only carefully scrutinized for its validity. Great pressure to the other extremes—continuation of unthinking experimentation with convenient, familiar, and standardized animals (rats, mice, rabbits, guinea-pigs, etc.) is being constantly exerted. The words of the National Society for Medical Research have a ring of irony—'study life to protect life'; an edict which may lead one to justify an animal's suffering under an all-encompassing and protective umbrella of relevance to man. There should be concern for the perpetuation of such unthinking usage. With enhanced scientific and humane education, and more rigorous regulations, this vicious circle may be broken. Intellectual potentials (as distinct from manual/technical aptitude) of young students may then be freed from the narrow confines of mechanistic reductionism and their creativity add to their understanding of life processes.

The science fair can help to foster such awareness in future years as an annual focal point for young people interested in the many areas of science and medicine, as it has in the past—but more so once the intrinsic values, motives, attitudes, and goals underlying biomedical and scientific research are recognized, understood fully, . . . and changed when necessary. A rational, objective, ethical, and humane appraisal is needed each year, for each exhibit, and for each and every person, field of science, and area of humane interaction with the natural world. Only then will science and humanity advance and science fairs and high school biology projects be a consummate expression of the highest human qualities emergent in our children.

BIBLIOGRAPHY

Fox, M. W. (1975). *Concepts in ethology: animal and human behavior*, University of Minnesota Press, Minneapolis. (For senior high schools.)

Gates, W. D. and Fox, M. W. (1977). *What is your dog saying?*, Coward, McCann, and Geoghegan, New York. (An introduction to animal behaviour for elementary school.)

Mills, T. K. and Keeling, J. A. (1973). *Animals for schools*, Harrap, London.

Orlans, F. B. (1970). 'Better nutrition studies', *The American Biology Teacher*, **32**, 484–486

Orlans, F. B. (1974). 'Biology students as experimental subjects', *The American Biology Teacher*, **36**, 401–406.

Orlans, F. B. (1977). *Animal care from protozoa to small animals*, Addison-Wesley, Reading, Mass. (Examples of science fair projects.)

Price, E. C. and Stokes, A. (1975). *Animal behavior in laboratory and field*, W. H. Freeman, San Francisco. (Many projects for high school.)

Rowsell, H. C. (1974). 'Canada's experience with student use of living animals', *The American Biology Teacher*, **36**, 31–36.

Russell, R. J. (1978). *Laboratory investigations in human physiology*, Macmillan, New York.

Scott, J. P. (1974). *Animal behavior*, University of Chicago Press, Chicago. (An introductory text for senior high school level.)

Scott, J. P. (1980). *Humane biology projects*, Animal Welfare Institute, Washington, DC. (Examples of science fair projects.)

APPENDIX 1

The following is a selection of ISEF student projects in the 28th ISEF in Cleveland, Ohio, 1977.

Behavioral and Social Sciences Division

114—'Some biological effects of organic mercury administered topically and orally'. Coleen Truax (Junior) 17

This experiment utilized 80 rats. For the initial training shocking chambers were used. The rats were given the maximum legal dose of 2 p.p.m. of mercury. It is known that mercury taken orally can cause abdominal pain, diarrhoea, tremour, irritability, and apprehension. It is also known that human beings can be intoxicated by mercury simply from exposure to mercury vapour or even prolonged vaginal douching so the need for further study of this agent's toxicity is questionable. (Ref. *Harrison's Principles of Internal Medicine*, McGraw-Hill, New York, 1970.) (This project won second place.)

115—'High protein diet effects on the memory of a rat'. Frank Troyka (Senior) 17

In this experiment rats were given a total of 150 protein tablets. Although this probably did not cause any pain to the animals the need for this kind of study is repetitious and its necessity questionable.

801—'Effects of cigarette smoke on libido and reproduction of mice'. Sherry Dicharry (Junior) 16

Eight mice were exposed to cigarette smoke and compared to non-exposed mice. The mice which were exposed to the smoke showed decreased libido, greater numbers of still births, and 'weakened' sperm. The reproductive organs were removed for study but there was no mention as to how this surgical procedure was accomplished or if and how the animals were euthanized. This experiment was no more than a copy of the same old theme, smoking affects those who come into contact with it physiologically. All this information is available in any textbook. (This project won third place.)

803—'Effects of lead on blood, heart, liver, kidneys, and brain'. Precia A. Boudreaux (Senior) 18

This experiment utilized 36 New Zealand white rabbits to study the over-studied problem of lead poisoning. This study was relatively comprehensive and confirmed the results of previously reported studies. However, it is unfortunate that the supervising science teacher did not instruct the student as to a proper method of killing the animals. The animals were killed by cervical dislocation, a method not recommended for large species such as the rabbit. Blood specimens were drawn from live rabbits by cardiac puncture.

Medicine and Health Division

805—'Evaluating the effects of atmosphere on surgical cuts'. Bruce Dingeman (Sophomore) 15

Incisions were made along the backs of rats to study the effect of different amounts of oxygen or ozone on healing. Although the results of this mutilative study demonstrated faster healing of wounds in an atmosphere of ozone, the usefulness of this information is totally obscure as it is certainly not practical to place a patient in an environment of ozone so his wounds can heal 'faster'. Again, highly sophisticated equipment was necessary to carry out this experiment. (This project won third place.)

816—'"C" stands for cholesterol: is it really bad?'. Linda Blakely (Senior) 17

This blatantly duplicative study simply verified the redundant fact that a high cholesterol diet causes abnormalities of blood vessels. Rats were the experimental animal.

817—'Physiology and behaviour of white mice as affected by caffeine'. Yvonne Harmon (Junior) 17

This experiment which used 45 mice was performed in the high school classroom, a needlessly repetitive study.

820—'Effect of capsicum on the internal organs of hamsters'. Maria Belen **Rios** Matos (Sophomore) 15

There may be justification for finding out if capsicum (chili pepper) has a harmful effect but this should be studied relative to the usual ingested dose. This study investigated an unrealistically large dose of chili pepper in 14 hamsters. The student gave each hamster a daily dose of 2–5 ml of chili pepper per day for a period of 65 days. If a hamster weighs 0.25 lb, then a dose of 2 ml to the hamster would be equivalent to 1200 ml to an average (150 lb) human being. It is not surprising that the student found the hamster's stomachs and intestines to be ulcerated. (This project was awarded fourth place.)

829—'If you have tired blood will geritol help?'. Neil Spencer (Senior) 17

Twenty female, Sprague–Dawley rats were force-fed Geritol tablets then blood was drawn by a cut in the tail twice per week. This experiment was highly unimaginative and repetitious. The student had never been informed that the rat is one species which does not need supplemental feeding, i.e. the rat was the wrong model to use in the first place.

831—'Weight reducing pills, are they safe?'. Joan Brenchley (Senior) 17

In this experiment 16 mice were given Dexamyl, which is a compound containing dextrose amphetamine sulphate better known as 'speed' and phenobarbital. This was not only repetitious but it was a poorly designed experiment as the dose was based on human doses for two generations of mice instead of weight to weight ratios between the experimental animal and man.

   In summary, high school science fair projects fall into four major categories, as they do in animal research in general: (1) original, well-designed, and meaningful contributions to a given research area (few were of this category); (2) unoriginal, repetitive replications of earlier studies by other investigators; (3) exercises in manual skill or technique *per se*, lacking any other purpose and being a means that justifies a meaningless or obscure end; (4) inhumane, illogical, ill conceived, and of neither value nor purpose. While few ISEF projects were in this final category, examples have been given of the kinds of experiments that clearly fit into these four categories.

   The overall trend of animal experimentation at the science fairs is basically *technique oriented* (e.g. surgical or drug interference or histochemical analysis of brain) and *replication/repetition* of many research topics that have already been widely studied and published (i.e. categories 2 and 3 of the above classification). One may wonder what the latter 'cookbook copying' does to a student's creativity and originality. It is also an unnecessary waste of animals, a criticism that is also often leveled against full-time 'professional' research in many areas, where replication/repetition and minor variations in an over-worked field are commonplace.

APPENDIX   2

The Humane Society of the United States: Guiding Principles for Use* of Animals in Elementary and Secondary Education

It is the policy of The Humane Society of the United States that any study of animals shall foster a humane regard for the animal kingdom and a respect for life. The Society believes all live animal experiments (other than those for the purpose of behavioral observations and ecological studies that involve no direct manipulations) should be prohibited in elementary and secondary schools. Learning experiences that entail animal suffering cannot be justified or add positively to a child's character development.

The Society recognizes, however, that although this goal may not be reached for some time, it is a humane imperative to minimize and eliminate where possible all animal suffering and has, therefore, developed the following guiding principles with this goal in mind.

Choice of Subject

In biological procedures involving living organisms, species such as plants, bacteria, fungi, protozoa, worms, snails, or insects should be used wherever possible. (Phyletically 'lower' life forms should be sought as alternatives to 'higher' warm-blooded forms). Their wide variety and ready availability in large number, the simplicity of their maintenance and subsequent disposal makes them especially suitable for student work. In mammalian studies, non-hazardous human experiments are often educationally preferable to the use of species such as gerbils, guinea-pigs, or mice.

Procedures

Procedures to be performed on any warm-blooded animal that might cause any extreme physiological or psychological reactions; i.e. pain, suffering, anxiety, discomfort, or any interference with its normal health should be avoided. (Warm-blooded animals include man, other mammals such as gerbils, guinea-pigs, mice, rabbits, hamsters, and rats. It also includes birds, such as hens, quail, and pigeons. This means that a student shall do unto other warm-blooded animals only what he can do to himself without pain or hazard to health). Birds'

---

* 'Use' implies preferentially the non-manipulative study of life forms, i.e. observation, which is no less educational and scientifically valid than direct manipulations, and other treatments which should be discouraged.

eggs subjected to experimental manipulations should not be allowed to hatch; such embryos should be destroyed 2 days prior to the normal hatching time. If normal egg embryos are to be hatched, satisfactory arrangements must be made for the humane disposal of chicks. No lesson or experiment shall be performed on a vertebrate animal that involves: micro-organisms which can cause disease in man or animal; ionizing radiation; cancer-producing agents; extremes of temperatures; electric or other shock; excessive noise; noxious fumes; exercise to exhaustion; overcrowding; or other distressing stimuli. No alien substance (drug or chemical) should be exposed to, given in the food or water, or injected into the animal, and no surgery shall be performed on any living vertebrate animal (mammal, bird, reptile, or amphibian).

Diets deficient in essential foods are prohibited. Food shall not be withdrawn for periods longer than 12 hours. Clean drinking water shall be available at all times (and shall not be replaced by alcohol or drugs).

Projects involving vertebrate animals will normally be restricted to measuring and studying *normal* physiological functions such as *normal* growth, activity cycles, metabolism, blood circulation, learning processes, *normal* behavior, ecology, reproduction, communication, or isolated organ/tissue techniques. None of these studies requires infliction of pain.

Supervision

Observations must be directly supervised by a competent science teacher who shall approve the student's protocol before the study is initiated. Students must have the necessary comprehension and abilities for the work contemplated. The supervisor shall oversee all experimental procedures, shall be responsible for their non-hazard nature, and shall personally inspect experimental animals during the course of the study to ensure that their health and comfort are fully sustained.

Care

Vertebrate studies (except natural field or zoo or farm studies) shall be conducted only in locations where proper supervision and adequate animal care facilities are available; either in a school or an institution of research or higher education. No vertebrate animal studies shall be conducted at a home (other than observations of *normal* behavior of pet animals such as dogs or cats).

In vertebrate studies the housing, feeding, and maintenance of all subjects should accord with established standards of laboratory animal care, established under the *Animal Welfare Act* of 1970. Palatable food shall be provided in sufficient quantity to maintain normal growth.

The comfort of the animal observed shall receive first consideration. The animal shall be housed in appropriate spacious, comfortable sanitary quarters.

Adequate provision shall be made for its care at all times, including weekend and vacation periods. The animal shall be handled gently and humanely at all times.

In rare instances when killing of a vertebrate animal is deemed necessary, it shall be performed in an approved humane (rapid and painless) manner by an adult experienced in these techniques.

Respect for life shall be accorded to all animals, creatures, and organisms that are kept for education purposes, and such respect shall take precedence over any inhumane experiment which might be otherwise rationalized and justified as a valid learning experience.

(These guidelines were substantially revised in the autumn of 1979 after this chapter had been completed.)

ADDENDUM

In 1981, the rules of the International Science and Engineering Fair will require those students who wish to perform surgical procedures on sentient animals to conduct such experiments in biomedical facilities.

Part III

Animal Experimentation: Some General Issues

Animals in Research
Edited by David Sperlinger
© 1981 John Wiley & Sons Ltd.

Chapter 12

Alternatives and Laboratory Animals

ANDREW N. ROWAN

INTRODUCTION

Over the years, the use of animals in biomedical research has resulted in much impassioned and emotional argument from both anti-vivisectionists and biomedical research workers. Back in the 1800s when vivisection meant experimentation on live, unanaesthetized animals, many medical men and leaders of society supported the anti-vivisection movement (French, 1975). Nowadays, there is relatively little sympathy from medical quarters for anti-vivisection arguments, presumably because of the perceived improvements in human health that have resulted from animal research (Lapage, 1960). In addition, nineteenth century vivisection is no longer the mainstay of biomedical advance since little research involves cutting open live and conscious animals. However, many animals still endure pain and suffering in modern laboratories.

Although the anti-vivisection movement has lost a lot of its power compared to the late 1800s, the widespread use of animals in biomedical programmes has been challenged in the last 10 years by the development of the new concept of 'alternatives' and the re-examination by philosophers of a scholarly argument in favour of animal 'rights' (Morris and Fox, 1977). The beginning of the alternatives concept may be traced back many years but received its first formal enunciation in 1959 when two scientists, sponsored by the Universities Federation for Animal Welfare in Britain, published a book called *The principles of humane experimental technique* (Russell and Burch, 1959) in which the authors argued the case for the three R's (replacement, reduction, and refinement).

For several years, this was the only available material on the subject, but then an American anti-vivisection organization, the United Action for Animals, published a few leaflets on the subject and introduced the term 'alternatives'. In addition, the Lawson Tait Trust for Humane Research, an offshoot of the National Anti-vivisection Society (UK), had been supporting a few laboratory projects which did not involve animal research. In 1969, FRAME (Fund for the Replacement of Animals in Medical Experiments) was established in London to promote, on a technical and scientific level, the concept of alternatives. This activity prompted some response from the biomedical establishment and from organizations dedicated to a defence of the use of animals in biomedical research. For example, Sir Peter Medawar made the following moderate statement in 1969 at the Research Defence Society's annual meeting:

'The use of animals in laboratories to enlarge our understanding of nature is part of a far wider exploratory process, and one cannot assay its value in isolation—as if it were an activity which, if prohibited, would deprive us only of the material benefits that grow directly out of its own use. Any such prohibition of learning or confinement

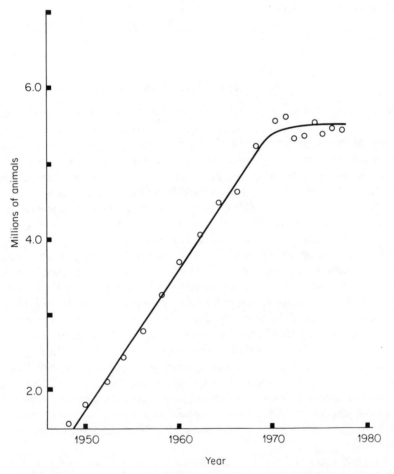

Figure 12.1 Number of experiments every year in Great Britain (Home Office, 1978)

of the understanding would have wide-spread and damaging consequences; but this does not imply that we are for evermore, and in increasing numbers, to enlist animals in the scientific service of man. I think that the use of experimental animals on the present scale is a temporary episode in biological and medical history, and that its peak will be reached in ten years time, or perhaps even sooner. In the meantime, we must grapple with the paradox that nothing but research on animals will provide us with the knowledge that will make it possible for us, one day, to dispense with the use of them altogether.' (Medawar, 1972).

There are few reliable statistics available on the growth in demand for laboratory animals, but those that have been collated and published indicate that Medawar was correct when he predicted a peak in demand within 10 years. For example, the Home Office returns indicate that experimental activity in Great Britain has stabilized at around $5\frac{1}{2}$ million animal experiments every year (Figure 12.1). On a world scale, Table 12.1 provides some indication of the demand for laboratory animals. While demand has stabilized for the moment, it may well increase again in the next few years as testing requirements under the United States' *Toxic Substances Control Act 1976*, the United Kingdom's *Health and Safety at Work Act 1974*, and other similar legislation is put into effect. Rodents comprise the vast majority of laboratory animals. For example, the 1978 USA figure of 90 million can be broken down into the following approximate figures: 50 million mice, 20 million rats, 5 million amphibians, 5 million birds, 4 million hamsters, 3 million guinea pigs, 2 million rabbits, $\frac{1}{2}$ million dogs, 200 000 cats, and 30 000 monkeys. This breakdown is based on interviews with breeders and on a survey conducted in 1965 (Anonymous, 1966).

Several factors are probably responsible for the present stabilization of

Table 12.1 Annual number of laboratory animals used in various countries (000's)

Country	Years		
	1956–59	1969–71	1976–77
Australia	400	800	
Austria	120		805
Finland	140	165	
France	1 250	4 420	
Holland	660		3 000
India	270	1 066	
Israel	350	545	
Japan	1 600	13 155	
Norway	21	93	
Sweden	170	875	
United Kingdom	2 500	5 580	5 386
United States	20 000	64 000	90 000

Compiled from ICLA (1959, 1962), Home Office (1978), Tajima (1975), and personal communications demand, including a slow-down in the real growth of biomedical research funding (Burger, 1975) and the development of systems of research and testing which require fewer animals to produce the same amount of data—or in other words, the development and wider use of alternatives.

ALTERNATIVES—A DEFINITION

One of the main problems with the use of the term 'alternatives' is that few people take the trouble to define it properly and hence fail to provide clear parameters for further discussion and debate. Some anti-vivisection groups use it to refer only to *replacement* techniques and as a result the scientific community has avoided using the term, preferring to talk about 'complementary' methods. However, last year the Research Defence Society published a book entitled *Alternatives to animal experiments* (Smyth, 1978) which represents the first significant endorsement of the concept by the biomedical establishment. In addition, the UK Prime Minister affirmed in Parliament at the end of 1977 that it was his Government's intention to encourage the speedy development and application of alternatives. In the United States, the concept has yet to be accepted by the biomedical community, although there are many signs of a softening of attitudes.

The definition which has the most widespread acceptance is as follows. An alternative is any technique which could: (a) replace the use of animals altogether; (b) reduce the number of animals required; and (c) reduce the amount of stress suffered by the animal through suitable refinements in the techniques used.

At the same time, and this is a most important point, any alternative system must provide data which leads to the same conclusion with at least the same degree of confidence as that obtained from the system being replaced.

REPRESENTATIVE ALTERNATIVES

The relatively broad definition of 'alternatives' means that a wide variety of technical improvements could qualify—including improved capture and quarantine methods for rhesus monkeys destined for the laboratory. Such improvements resulted in fewer wild monkeys dying before or during experimentation, and this was one of the contributory factors for the fall in the number of non-human primates imported into the United States after 1959–60 (see Figure 12.2). However, certain techniques have received particular attention when 'alternatives' are discussed, and some of the more important are outlined below. More detailed descriptions are available in Smyth (1978) and FRAME (1979).

(a) *Physico-chemical techniques* The physical and chemical techniques now available to the biomedical research worker are much more sensitive and

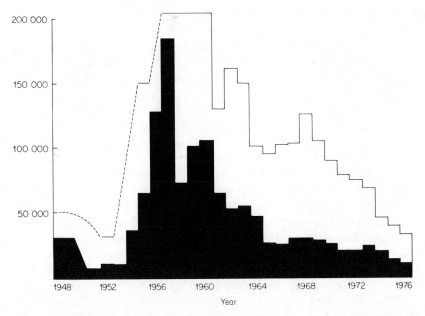

Figure 12.2 Number of non-human primates imported into the USA every year: open bars, total primates; solid bars, rhesus macaques (LeCornu and Rowan, 1979)

powerful than those used before the last war and this has resulted in the total replacement of animals in some areas. For example, laboratory animals used to be necessary for the assay of the fat-soluble vitamins but now, vitamins A, D, and E can be assayed by gas – liquid chromatography and mass spectroscopy techniques (Wiggins, 1976). However, one still has to use the rat antirachitic assay for the vitamin D content of cod-liver oil because not even modern GLC techniques can satisfactorily separate the vitamin from the other fats in the oil (British Pharmacopoeia, 1973a). The greater sensitivity of modern machines also means that fewer animals need to be used in other research areas since more assays can be done on smaller amounts of material.

(b) *Computer mathematical analysis and modelling*   The number of animals required for a series of studies can sometimes be reduced by effective use of statistics and by prior analysis using a computer model (Newton, 1977; Harrison and Harrison, 1978). However, it is not correct to state (as has been done on occasion) that such systems can replace all animal experimentation.

(c) *Microbiological systems*   Microbiological systems have, in a few cases, totally replaced the use of animals (as in vitamin assays, Gyorgy and Pearson, 1967) but more commonly are employed to reduce the total number of animals required. For example, the Ames test for detecting mutagens (and perhaps carcinogens?) which employs *Salmonella* detector strains does reduce the need for animal studies in that initial screening studies can be done in the bacterial system (Ames, 1979; McCann and Ames, 1975, 1976).

(d) *Tissue culture*   Tissue culture is the technique for which most has been claimed with respect to the development of satisfactory and practical alternatives. For example, Professor Sergey Fedoroff of Saskatchewan University, a well-known member of the Tissue Culture Association, has stated that 'the application of tissue cultures to biomedical research is limited only by the imagination of the scientists employing them' (S. Fedoroff, personal communication). This may appear to be a very radical statement at first glance, but not if one interprets it as meaning that tissue culture can be applied to a very wide range of research problems without necessarily replacing the need for animals.

Tissue culture systems have replaced the need for animals in a few cases, notably virus vaccine production and some vaccine potency and safety tests (Petricciani *et al.*, 1977), but in general the technique's main potential lies in its ability to reduce demand. For example, a pharmaceutical virology laboratory reduced its demand for mice from 13 000 to 1600 per annum between 1963 and 1975. At the same time, they increased the number of compounds screened from 1000 to 22 000 per annum. These changes were entirely due to the employment of cell and organ culture systems as preliminary screening systems before testing the most promising chemicals in the mouse system (Bucknall, 1980).

(e) *Clinical and epidemiological studies*   The use of human patients, volunteers, or populations is often the final step in the application of a new cure after extensive animal and *in vitro* studies. There is no doubt that results obtained with human subjects are more relevant than those from animal studies, but there are strict ethical guidelines controlling such experimentation and many research projects are not possible for ethical reasons (National Commission for the Protection of Human Subjects of Biomedical and Behavioral Research, 1978).

(f) General refinements in techniques may lead to reduced stress and can contribute significantly to a decrease in the number of painful procedures. This would qualify as an alternative and Smyth (1978) has identified reduction of the number of painful experiments as the main area in which advances need to be made. For example, the British Pharmacopoeia (Addendum) (1977) now specifies a paralytic rather than a lethal end-point (British Pharmacopoeia, 1973*b*) for the tetanus antitoxin potency test. The paralysis is mild, involving only minor loss of movement in the hind limb of a mouse followed by recovery. There is no doubt that this procedure is considerably less stressful than use of the lethal end-point.

TYPES OF BIOMEDICAL PROGRAMMES AND RESEARCH MODELLING

When discussing alternatives, it may be helpful to recognize that the scope for developing and using an alternative varies according to the objectives of the particular laboratory activity using animals. For example, the arguments in favour of using live animals in an educational exercise may well be considerably less pressing than those for using animals in research to develop a vaccine against poliomyelitis. At the very least, it is probable that the public will respond in a very different way to the two situations. Animal use in the laboratory can be divided into a number of relatively distinct categories including (but not necessarily limited to) education, diagnosis, toxicity testing and safety evaluation, and biomedical research involving the generation of new data and the testing of hypotheses.

Table 12.2 provides a breakdown of the United Kingdom use of laboratory animals in 1977. There is no education category (live animals may not be used in painful experiments to improve manipulative skills) but diagnosis, toxicity testing, and research are all represented. Approximately 3 percent, of the animals are used in diagnosis, about 30 percent are used in toxicity testing (not shown in Table 12.2) and the remainder are used for the development of therapeutics (c. 40 percent) and biomedical research.

Biological and Medical Education

At least 3 million animals per annum (out of about 90 million) are used in the United States for secondary school and university educational purposes (ILAR, 1970), including frogs in high school biology classes and pound dogs for practice surgery in medical and veterinary curricula. There is little objective evidence that manipulative exercises on live animals are necessary in many of these educational courses. Certainly, all animal experimentation involving significant intervention could be excluded from high school and many undergraduate programmes without jeopardizing the development of critical, yet imaginative, scientific intellects nor the flow of young scientists into research laboratories. In most cases, adequate alternatives either exist or could be developed.

Diagnosis

A certain number of animals are used every year to diagnose diseases. For example, in 1956 animals were used to diagnose a wide variety of diseases including tuberculosis, diptheria, brucellosis, and anthrax (Russell and Burch, 1959). Nowadays, tuberculosis can be diagnosed using *in vitro* culture techniques and guinea pigs are no longer required (Marks, 1972). However, the number of recognized pathogens has increased considerably and therefore the demand for animal diagnosis has not declined significantly despite many advances in culture technology (Austwick, 1977; Meier-Ewert, 1977; Raettig, 1977; Storz *et al.*, 1977).

Toxicity testing

Toxicity testing is conducted to determine whether chemicals are 'safe' for

Table 12.2 Laboratory animal (live) use in the United Kingdom in 1977: organizations and objects of experiments (Home Office, 1978)

Organizations	Diagnosis	Development of therapeutics	Legislative requirements	Other testing	Other experiments	Unclassified	Total
Universities, medical schools, and polytechnics	66 710	110 235	5 874	3 808	696 259	25 463	908 349
Government and quasi-government organizations	43 958	185 237	45 650	12 630	343 243	9 583	640 301
Hospitals and public health laboratories	53 709	28 677	7 378	13	124 524	3 975	218 276
Other non-profit organizations	4 828	248 602	440 165	5 614	145 193	13 859	858 261
Commercial concerns	1 790	1 513 995	943 055	103 042	170 055	28 451	2 760 388
Totals	170 995	2 086 746	1 442 122	125 107	1 479 274	81 331	5 385 575

general human use or to identify the 'safe' limits of use for chemicals known to be hazardous. Tests are carried out on a wide range of chemicals and biological substances (including drugs, vaccines, pesticides, food additives, industrial chemicals, cosmetics, and household products) and involve millions of animals every year (Table 12.3). Standard toxicity tests for new drugs include procedures to detect certain specialized effects such as the induction of genetic damage, the induction of cancer, or the induction of deformities in the developing fetus. The design of an overall toxicity protocol will depend on the proposed use of the new substance and may incorporate only a few tests. The cost of testing varies from $ 20 000 per substance (using a limited range of tests and short-term screens) to over $ 1 000 000 for an in-depth toxicological evaluation.

There are a number of problems involved in attempting to promote the development of alternatives for toxicity testing programmes. First, relatively little is known about the mechanisms underlying toxic reactions in mammalian systems and it is, therefore, difficult to design non-animal tests which reliably mimic the human response. Second, despite the fact that animal testing is crude, cumbersome, and expensive, many of the techniques have become entrenched in regulatory guidelines throughout the world. The resulting 'bureaucratic inertia' discourages innovation. Third, consumer groups are becoming more vocal about the unknown hazards represented by the many industrial chemicals in our environment and are consequently pressuring government agencies to widen the

Table 12.3 Numbers of experiments for toxicity testing in Great Britain—1977 (Home Office, 1978)

	Study of normal and abnormal structure and function	Development and testing of:			
		therapeutic products	other substances	Other purposes*	Total
Acute toxicity tests	53 647	290 629	121 984	93 657	559 917
Sub-acute and chronic toxicity tests	12 547	134 799	58 460	22 310	228 124
Distribution and metabolism studies	44 896	90 392	11 560	13 911	160 759
More than one of above	5 643	7 756	26 795	14 161	54 355
Totals	116,733	523,576	218,799	144,039	1,003,155

* Includes experiments in which the animal was used for more than one of the purposes specified in the original tables.

scope of toxicity testing (usually involving more intensive testing on more animals).

Partly as a result of the burden which these increased demands have placed on toxicity testing programmes, steps are now being taken to develop, validate, and apply alternative systems which are less expensive and more efficient. For example, several years ago, a group of toxicologists estimated that a suitable battery of short-term tests, involving fewer animals than the present procedures, could result in a tenfold reduction in cost and fivefold reduction in time. Furthermore, they argued that these tests would involve little or no sacrifice of safety (Muul *et al.*, 1976). However, there have been few systematic approaches to review the problem so far. This question is explored in more detail later in this paper.

Basic and Applied Research

In all of the three above categories the animal is being used as a *surrogate* for man. In research, because of the reduced options of employing a human subject in a research project, an appropriate *model* (usually animal) of the human response is used. The classic approach has involved the use of an animal to model a particular disease or normal function. A perfect model would be indistinguishable from the original and, therefore, animal or *in vitro* models depart, to a greater or lesser degree, from the properties of the original. For the purposes of the present discussion, there are two important ways in which a model differs from the original—namely, in its *fidelity* and its *discrimination* (Russell and Burch, 1959).

For example, Tinbergen and Perdeck (1950) studied the type of stimuli which produced food-begging behaviour in the herring gull chick. They presented the chicks with two models (Figure 12.3). One was a high-fidelity model of the parent bird's head and the other was a thin red rod with three sharply etched white bands

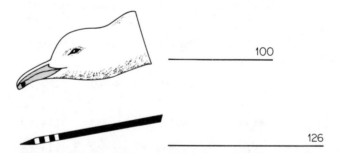

Figure 12.3 Food-begging behaviour in the herring gull chick—responses elicited by a high-fidelity (model of adult's head) and low-fidelity (thin red rod) model (Tinbergen and Perdeck, 1950)

at the tip. The relative number of food-begging reactions released by the two models were 100 (for the model of the head) and 126 (for the rod). Thus, the high-fidelity model had a lower discrimination capacity than the rod with the three key stimuli (redness, colour contrast, and elongation). This example demonstrates that a low-fidelity model may be even more satisfactory in discriminating between two states than a high-fidelity model. Therefore, when selecting a model, the researcher needs to be aware of whether or not the problem requires a high-fidelity model (usually a mammal) or not. If not, it will frequently be possible to use or develop a suitable non-sentient system.

Scientists should not only be careful in selection of the appropriate model, their approach to the research problem itself also requires detailed consideration. For example, a simplified outline of the 'scientific method' used in research involves the formation of an hypothesis or theory, the deduction of certain consequences from this hypothesis, the design of an experiment to test these deductions, and then the acceptance, modification, or rejection of the hypothesis depending on the results of the experiment. However, the theoretical basis for research and standards of critical analysis vary widely from one project to another. For example, Bernard Dixon has stated:

'Where the blind empiricist will stumble from one question to the next, and the scientist with the accountant mentality will deploy a massive induction-based experiment (which may somehow throw up the result he is looking for), the top-class creative thinker designs a single, crucial experiment that decides absolutely between one hypothesis and another.' (Dixon, 1976).

It is not possible to legislate against lack of insight or to draw up regulations which will abolish mistakes, but more could be done to reduce the amount of wasteful research involving animals. Universities could institute lecture courses and seminars on alternatives in order to raise the general level of consciousness and sensitivity among biomedical students so that animal research is initiated only after careful thought.

The induction-based research mentioned in the quotation refers to an approach which is widespread in the pharmaceutical industry—namely, the screening of thousands of chemicals in the hope that one or more will have therapeutic properties. The researcher is guided by theoretical considerations only to a relatively limited extent and will try anything which is not patently unreasonable if there is some likelihood of success. A standard technique for investigating chemicals for drug development involves designing a screening system using an animal model of the human disease. Thousands of chemicals are then tested to determine if they have any effect on the disease. In a number of cases, it is possible to develop and use a non-animal system as a preliminary screen before proceeding on the animal model. The number of negative tests conducted in an animal system, and the overall number of animals required, will thus be considerably reduced. As an added advantage, alternatives are often less expensive and more efficient when used in screening systems. One pharmaceutical company which has taken advantage of this feature now uses animals for only about 30 percent of its drug screening programme as opposed to about 80 percent 10 years earlier (Spink, 1977).

ALTERNATIVES AND TOXICOLOGY TESTING

Because of the extremely varied nature of basic biomedical research, it is difficult to put forward both specific and realistic proposals for the development of alternatives. Each research problem requires a different approach and frequently a complete answer can only be obtained from a variety of methods, some of which involve research on live animals. However, there are other areas where the animal is being used as a surrogate for man to answer relatively well-defined questions such as what is the effect of chemical X on a whole organism or on a particular organ. (In these cases, the use of an animal is more a function of the investigator's ignorance of how particular phenomena occur.) This is particularly true of toxicity testing where the number of effects being sought (or guarded against) is relatively small (see Table 12.4). At present, knowledge of the mechanisms is scanty and, as a result, regulatory agencies are unwilling to take a chance with an apparently satisfactory *in vitro* screening system since there is no absolute certainty that such a system can detect all those chemicals which are harmful (to man). Of course, the whole animal system will also fail to detect some harmful chemicals—compared to modern biological technology, animal testing is crude, cumbersome, and expensive (Kennedy, 1978)—but at least it has a

Table 12.4 Sample toxic effects investigated in safety/hazard evaluation studies

Acute toxicity (general)	Oral
	Dermal
	Inhalation
	Subcutaneous
	Intravenous
Acute toxicity (specific)	Ophthalmic irritation
	Dermal irritation
	Dermal sensitisation
Sub-acute toxicity (30–90 days)	Oral
	Dermal
	Inhalation
Chronic toxicity (specific)	Carcinogenicity
	Mutagenicity
	Teratogenicity
	Reproductive toxicity

The number and type of tests done varies according to the substance under investigation and the particular requirements of the regulatory authority. For example, specific neurotoxicity studies or wildlife studies may have to be conducted for environmental chemicals.

better chance of detecting unexpected effects. However, toxicological knowledge is being augmented at an accelerating rate and many new techniques are now potentially available to reduce the reliance on animal testing. Some suggested alternatives in toxicology testing are discussed below.

The LD_{50} Test

LD_{50} is an acronym for 'lethal dose—50 percent' and it indicates the amount of a substance which, when administered in a single dose to a group of animals, will result in the death of half the group within 14 days. The figure is usually determined to a relatively high degree of statistical precision (95 percent confidence limits). The LD_{50} of a wide variety of substances, including drugs, industrial chemicals, household products, and cosmetics is determined as a matter of course and the following arguments are usually presented to justify the practice (Morrison et al., 1968). First, the LD_{50} determination is claimed to be a useful quality control parameter. Second, it is argued that the LD_{50} is required as a guide for toxicological and pharmacological studies. Third, clinicians need an idea of the human lethal dose of those chemicals which may be ingested and it is claimed that this is provided by animal LD_{50} data. Fourth, the LD_{50} measure is an implicit (if not explicit) requirement of regulatory agencies around the world. The question which is raised here is not whether *some* measure of acute toxicity is required, but rather how precise that measure needs to be. For example, is there any value to knowing the LD_{50} together with its 95 percent confidence limits or would a rough estimate suffice. An examination of each of the proposed justifications for determination of the LD_{50} leads to the conclusion that there are, in most cases, alternatives to the LD_{50} which require the use of substantially fewer animals.

(a) *Quality control* The LD_{50} determination was originally developed as a potency test for drugs such as digitalis which could not be standardized in other ways (Trevan, 1927). However, there are many sources of variation in the test (Morrison et al., 1968) and the end-point (death) is, to say the least, extremely crude. Nowadays, its use as a quality control device should be limited to those few biological therapeutics (vaccines, antitoxins, and the like) which cannot be tested by other means (British Pharmacopoeia Commission, 1977). Therefore, the LD_{50} measure (or a relatively precise acute toxicity test) does serve a limited function in the quality control of biologicals.

(b) *Toxicology and pharmacology* The LD_{50} test is usually the first test performed in a toxicological protocol, but the quantitative information provided is far less important than the qualitative information (which could be obtained using fewer animals). The quantitative data is subject to many errors including interspecific differences (see Table 12.5) and differences between acute and chronic toxic reactions which make a nonsense of the determination of precise,

Table 12.5 Human LDLo and animal LD_{50} values (mg/kg body weight)

Chemical	Human LDLo*	LD_{50}			
		Rat	Mouse	Rabbit	Dog
Acetyl salicylic acid	—	1750	1100	1800	80
Aminopyrine	—	1380	1850	160	150
Amiline	350	440	—	—	—
Amytal	43	560	—	575	—
Binapicryl	—	161	1600	1350	—
Bisacodyl	—	4320	17 500	—	—
Boric acid	640	2660	3450	—	—
Caffeine	192	192	620	—	—
Carbofuran	11	5	2	—	—
Cycloheximide	—	3	133	—	65
Diallate	—	395	—	2250	510
Fenoflurazole	—	238	1600	28	—
Lindane	840	125	—	130	120
Thiopental	—	—	350	600	150

*The human LDLo is the lowest lethal dose of the chemical which has been recorded for humans.

Compiled from Christensen and Luginbuhl (1975) and Sunshine (1969).

statistical figures. A rough idea of acute toxicity would be quite satisfactory for most (if not all) purposes.

(c) *The LD_{50} and clinical requirements*   The interspecific differences in LD_{50} values for a wide variety of chemicals (see Table 12.5) indicate clearly that a clinician can only use animal LD_{50} data as a very rough guide for human lethal doses. As such, it would be quite sufficient if the animal data provided only a rough estimate (i.e. semi-quantitative) of lethality. In addition, few textbooks of clinical toxicology pay much attention to animal LD_{50} data; much more significance is attached to the qualitative toxic effects. Therefore, a rough figure for the lethal dose in animals would be satisfactory.

(d) *Regulatory agencies and the LD_{50}*   A wide variety of regulatory agencies use animal LD_{50} data. For example, transport regulations for chemical substances base safeguards for chemical transport on a variety of parameters, including an LD_{50} measure. If no LD_{50} is available, then the chemical is usually placed into the most toxic category, which naturally increases the costs of transport. In another example, the Environmental Protection Agency has issued proposals for hazardous waste management in which waste is established as being hazardous if, *inter alia*, the value of the human LD_{50} rises above a certain value (EPA, 1978). The human LD_{50} is calculated by multiplying the rat LD_{50} by a factor of 0.16 or the mouse LD_{50} by 0.06. This clearly illustrates the manner in which regulatory agencies have come to regard LD_{50} values in the same way as they perceive density—that is, as fundamental property of the chemical (or mixture of chemicals). Unfortunately, the LD_{50} is far from being an immutable figure and regulatory agencies would do much better to rely on qualitative and semi-quantitative (i.e. rough) acute toxicity data than let themselves be lulled into a false sense of security by the apparent precision of LD_{50} determinations.

Given that the LD_{50} measure is too precise, what are the alternatives. At this stage, it is not altogether possible to do away with animals in acute toxicity tests though some interesting *in vitro* studies have recently been published (Autian and Dillingham, 1978). In the following proposals animals are still required, albeit not in the same quantity as before (see Table 12.3—acute toxicity tests account for half the number of animal tests performed and 10 percent of the total use of animals in biomedical programs).

First, in quality control work, every effort should be made to replace tests involving lethal endpoints with tests in which either fewer animals are used or where the endpoint is less stressful. The LD_{50} measure should not be used in the quality control of chemicals and other substances which can be tested adequately by other means. Second, in other studies, the LD_{50} test should be replaced by determination of the 'approximate lethal dose' (this technique involves the use of only 6–10 animals (Deichmann and Leblanc, 1943)) or by the use of the 'limit' test when the chemical is essentially non-toxic. The approximate lethal dose (ALD) test involves administering increasing doses to single animals. The doses are increased by approximately 50 percent at a time and the ALD is the lowest dose which results in the death of the animal. The technique does not give a statistically precise figure but it should be more than adequate for the toxicologist, the pharmacologist, the clinician, and the regulator. For example, Table 12.6 list a few results from a comparative trial carried out by Deichmann and Mergard (1948). In tests involving a number of different substances and modes of administration (87 in all), the largest difference found between an LD_{50}

Table 12.6 Comparison between LD_{50} values and approximate lethal dose (ALD)—oral (Diechmann and Mergard, 1948)

Compound	Species	LD_{50} (mg/kg)	ALD (mg/kg)	% Deviation
Pentachlorophenol	Rat	78	80	+ 3
Thioglycolic acid	Rat	152	120	− 26
2-Aminothiazole	Rabbit	370	280	− 24
Phenol	Rat	1500	1400	− 7
o-Nitrodiphenyl	Rat	2230	2100	− 6
p-Nitrodiphenyl	Rat	1230	2100	+ 70
Methocyclohexanol	Rat	2900	3200	+ 10

Of 87 ALDs determined, 88 per cent ranged between ± 30 per cent of the LD_{50}. The largest deviation was + 70 per cent for o-nitrodiphenyl.

and the ALD was 70 percent for *o*-nitrodiphenyl. Considering that far larger interspecific differences occur all the time, this is relatively unimportant. Third, the upper limit for acute toxicity test should be 2 g/kg body weight (oral administration). If *no* toxic effects are seen at this level then it is extremely unlikely that the LD_{50} will be below 20 g/kg, which is well above the arbitrary figures established by regulatory agencies to differentiate between relatively toxic and non-toxic substances. (One end-point which could be developed as a possible replacement for lethality is weight loss—loss of weight usually being one of the earliest signs of toxicity.)

The General Chronic Study

In chronic toxicology tests, animals are given several doses of the chemical, at least one of which is large enough to produce some toxic effects. This study is the safety net for toxicity evaluation in that the toxicologist will be looking for a very wide range of effects, often without having any idea of which might appear. The length of the study may vary from 3 months (for a drug which will only be administered for a short period of time) to 7 years in non-human primates (for birth control substances). Because of the general purpose of the study and its non-specific end-points, it is difficult to suggest any particular alternatives. However, the animal chosen as a surrogate for man should be the species which is not only of a suitable size and life span but should also absorb, metabolize, and excrete the chemical in a manner similar to man. If the metabolic profile is very different to man's, then the results of the trial will be of questionable relevance to human exposure.

Carcinogenicity and Mutagenicity

There is a body of informed and expert opinion which considers that short-term *in vitro* systems could be as satisfactory as animal (rat or mouse) bioassays in the detection of carcinogens. For example, Brusick (1977) of Litton Bionetics has said

'I see no scientific reason why mouse or rat studies are necessarily any better at determining the carcinogenic potential of chemicals for man than would be a short-term test using *Salmonella* or cultured animal cells. There can be arguments, I am sure, about this, but I think that there are arguments in both directions that are quite valid'.

In the regulatory arena, there are also persons who are prepared to endorse these sentiments. For example, Hutt (1978) states

'Although they [*in vitro* short-term carcinogenicity predictive tests] are presently too unpredictable to justify, by themselves, regulatory decisions on the safety of food ingredients, it is only a matter of time before their deficiencies are corrected and they become at least as reliable in predicting human carcinogenicity as animal studies. It is likely, indeed, that in time they will be perfected to a point where they are able to mimic human response far more accurately than animal testing.'

It is this last point which is so contentious. Nobody now denies the value of short-term carcinogen tests as *preliminary screens* but few people are, as yet, prepared to use them as a basis for regulation.

In the next few years, this may well change because the demand for toxicity data is far greater than can be satisfied by the facilities currently available. It has been estimated (Maugh, 1978) that there are 63 000 chemicals in common use and that approximately 1000 new substances are introduced every year. Only about 3500 chemicals have been adequately tested for carcinogenicity (Saffiotti, 1976) and, with current resources, no more than 500 chemicals could be started on bioassays each year. It is interesting to note that about 7000 chemicals have been tested for carcinogenic properties, but that half the studies were totally inadequate (Saffiotti, 1976). The result has been a waste of time, money, and animals. Unfortunately, the situation with the current bioassay program appears to be little better (Smith, 1979). In addition, the full National Cancer Institute bioassay now costs over $350 000 per chemical, takes 3–4 years to complete and requires a minimum of 800 animals. Given these figures it is not surprising that so much attention is now being given to the very much quicker and cheaper short-term tests.

Any regulatory programme would have to be based on a suitable battery of short-term tests (Muul et al., 1976) rather than one or two alone. A list of some of the short-term tests which have been (or are being) developed is provided in Table 12.7. As can be seen, the tests are based on a variety of end-points, including DNA damage, chromosome damage, and the 'transformation' of mammalian cells. There are other test systems, not listed in the table, which are also currently under development (for example, the National Cancer Institute *in vitro* carcinogenesis programme is currently evaluating three mammalian cell systems which are not mentioned in the table). In proposing the use of a battery of short-term tests in regulatory decision-making, the following caveats *must* be borne in mind. First, if there are problems about extrapolating from animals to man, there will be even more concern about extrapolations from bacteria to man. Second, many of the systems are still in the development stage and, as such, have not been standardized for use in a wide range of different laboratories. Their reliability as regulatory tools is, therefore, still in question. Third, the tests are being used at the moment to *guide* regulatory agencies in their decision making, but they are only guides. Fourth, the estimated costs in the table are given as rough estimates only and refer to the cost of testing one chemical. If more than one chemical is tested at a time, the cost per chemical would be reduced.

In mutagenicity studies, consideration must be given to mutations caused to both somatic and germinal cells. The primary concern in somatic cell mutation is the potential for carcinogenesis while in germinal cell mutation the effects will be seen in the offspring. Three live animal test systems are available for the

Table 12.7 Short-term tests for carcinogenicity and mutagenicity (see Hollstein *et al.*, 1979, for review)

Endpoint	System	Duration of test (weeks)	Cost per test ($)	Comment
A. *DNA damage*				
Mutation	*Salmonella*[a]	1	1000	90 + % predictive accuracy
	E.coli polA⁺/PolA⁻[b]	1	1000	
	Drosophila[c]	2		
	Mammalian cells[d] (CHO, Mouse lymphoma L5178Y)	2		
DNA repair	Human cells[e]	2	1000	29/34 positives with carcinogens; 0/24 positive with non-carcinogens
Chromosome damage	Mammalian cells[f]	2	c.2500	
B. *Cell transformation*				
Altered morphology	Syrian hamster cells[g]	2–3	c.3000	90% predictive
Colony growth in soft agar	BHK 21 cells[h]	4–6	c.3000	90 + % predictive accuracy

C. Miscellaneous				
Degranulation of rough endoplasmic reticulum	Rat liver[i]	1	c.1000	71% predictive accuracy
Biphenyl-2-hydroxylation enhancement	Rat liver[j]	1	c.1000	15/16 positive with carcinogens; 1/15 positive with non-carcinogens

[a] Ames (1979), McCann and Ames (1975, 1976), and Purchase et al. (1978).
[b] Rosenkranz et al. (1980).
[c] Vogel and Sobels (1976).
[d] Arlett (1977), Clive and Spector (1975), O'Neill et al. (1977).
[e] San and Stich (1975), H. F. Stich (personal communication).
[f] Brewen (1977), Latt et al. (1977).
[g] Pienta (1979), Pienta et al. (1977).
[h] Purchase et al. (1978).
[i] Williams and Rabin (1971).
[j] D. V. Parke (personal communication).

evaluation of mutagenic effects on germinal cells—the dominant lethal test, the specific locus test, and the translocation test. Of these, the dominant lethal test is the most widely used because it is less expensive and simpler to perform. However, even so, one is talking of an expenditure of over $20 000. The dominant lethal test involves the dosing of males who are then mated with untreated partners throughout a period of several weeks to cover the full spermatogenesis cycle. The females are killed and autopsied and the number of live and dead embryos counted. Deaths are assumed to result from a dominant lethal mutation. The advantages of the test include its relative cheapness, the potential to employ a wide variety of treatments, the relatively short time required (8 weeks for a subchronic study), and the fact that the sensitivity of the germinal cells at all stages of the cycle can be evaluated (Maxwell and Newell, 1973). The disadvantages include the very real possibility of pre-implantation loss due to non-genetic causes, its relatively low sensitivity for point mutation effects, the fact that direct genetic analysis of the mutation is not possible, and the fact that chromosome breakage (the cause of dominant lethals) does not necessarily result in a mutation (Maxwell and Newell, 1973). In addition, there are the usual problems of sensitivity and extrapolation to other species.

By contrast, all the genetic damage tests, with the exception of *Drosophila*, listed in Table 12.7 utilize somatic cells exclusively. This is one of the problems in using these *in vitro* systems to predict germ cell effects, but there are counterbalancing advantages. These include greater speed and sensitivity, the ease with which point mutations can be detected, and the potential for using human cell systems to provide data for the determination of human health hazards. Once again, the *in vitro* systems are very useful as preliminary screens but further development and validation is required if they are to reduce significantly the use of animals in mutagenicity testing.

Teratogenicity

Prior to the thalidomide tragedy of 1961, nobody routinely screened new chemicals for teratogenicity, although it was already well known that certain chemicals could produce malformations in experimental animals (Baker, 1960). (*Note*: Thalidomide was *not* adequately tested prior to marketing. When it was tested properly, the drug produced characteristic lesions in the rabbit and monkey fetus and caused pregnant rodents to resorb damaged embryos, albeit at much higher doses than those required for human effects.) After 1962, new testing requirements were established, typically requiring the use of at least two species—usually a rat or mouse and the rabbit, and specifying that dosing should occur during organogenesis. In general, established teratogens in humans are teratogenic in other mammals, but many agents which act as teratogens in other mammals do not affect man. One of the reasons for this may be the differences in placental structures between rodents and the 'higher' mammalian forms.

There are a number of shortcomings in the currently employed animal test systems. For example, there are the usual problems of interspecific differences in the handling and metabolism of foreign chemicals and the consequent difficulties about extrapolating from animal to man. However, not only are there variations in maternal metabolism, but one also has differences in placental and fetal metabolism. This adds several additional levels of complexity to the problem. Other factors which also add to the difficulties of extrapolation include the question of just what constitutes a malformation and the variation in response of developing tissues and organs in different morphogenetic situations and gestation periods.

However, if there are problems in the reliable detection of teratogens using whole animal systems, there are even greater problems in attempting to develop an *in vitro* technique which is sufficiently reliable to be of use, even as a screening technique. Ordinary cell culture is of little use at the present time because the cells do not develop in the progressive manner seen in differentiation. Organ culture has been extensively investigated (Saxen, 1976) and a batch of *in vitro* techniques have been reviewed recently (Wilson, 1978). One of the most interesting systems is being developed at Edinburgh (Clayton, 1980) using techniques which may be applicable to general toxicity testing as well as teratogenesis. The method is based on the premise that cell biochemistry, behaviour, and morphology are directly related and that the protein profiles of different cell populations will consequently be different. High-resolution protein separation in gels coupled with autoradiography and staining forms the basis of the test and the results so far indicate that the quantitative balance of the protein contents in cells in culture is indeed modified by abnormality. The technique is simple, it can be automated, and it does not require highly experienced pathologists.

Fetal cells are employed as the target cells and several different organ systems are represented—namely, lens epithelium, neural retina, kidney fibroblasts, and limb fibroblasts. The test substance is not added directly to the cell cultures. Instead it is administered to female animals from whom samples of amniotic fluid are taken for application to the cultures. There are still many problems which must be investigated and much more validation is required. However, using this test system, the investigators not only identified which of two unknowns was teratogenic, they also correctly identified that the teratogen produced phocomelia and cataract formation (Clayton, 1980). In the words of the author, 'We felt reasonably encouraged to continue with the work and test many more substances . . .'.

Other

There are many other types of tests for specific toxic effects such as those for sensitization/allergenic effects, dermal irritation, mucous membrane irritation, and ophthalmic irritation. In testing for topical effects, the potential for developing and using cell culture systems is high because metabolism of the compound is not as critical a factor (although dermal and ophthalmic toxicity may involve a significant systemic response).

One of the topical tests which has produced a great deal of public reaction is the Draize ophthalmic irritancy test (Draize et al., 1946) which appraises irritancy on the basis of corneal opacity and conjunctival inflammation, chemosis, and discharge. The test was developed in 1944 but, despite many modifications, it is still not satisfactory. Weil and Scala (1971) conducted a study of inter- and intra-laboratory variation in the use of the Draize eye and skin tests which involved 24 major industrial and government laboratories. The results demonstrated a remarkable degree of variation (one laboratory would identify a substance as being non-irritant while another would identify the same substance, using the same test procedure, as irritant). As a result the same authors concluded that 'the rabbit eye and skin procedures currently recommended by the Federal agencies for use in the delineation of irritancy of materials should not be recommended as standard procedures in any new regulations. Without careful re-education these tests result in unreliable results.' In addition, results of ophthalmic toxicity in rabbits may bear little relation to the human situation (Davies et al., 1972). In 1974, the Food and Drug Administration (FDA) lost a court case in which they attempted to prosecute a manufacturer for marketing an eye irritant shampoo. The court ruled in favour of the manufacturer in part because the FDA had failed to show that the results of tests on rabbit eyes can be extrapolated to human beings (General Accounting Office, 1978).

Relatively few attempts have been made to develop an alternative to the Draize test. However, a pilot project, funded by the Walter Hadwen Trust (a division of the British Union for the Abolition of Vivisection), which involved cell viability determinations following treatment with shampoos of known irritancy, produced some promising results (Simons, 1980). As a result, a group of British cosmetic companies are investigating the system further in a collaborative research effort.

The above analyses and examples illustrate some of the problems and potential in attempting to reduce the number of animals used in toxicology testing. However, the calls for increased consumer safety apply pressure in the opposite direction—namely, for more animal testing with no guarantee that this will in fact lead to greater safety. Recent improvements in in vitro techniques demonstrate that the potential conflict between the desire to reduce animal testing as well as human health hazards can be avoided. However, both imagination and more research and validation is required.

CONCLUSION

This chapter has explored the growth in demand for laboratory animals, the areas in which the animals are used, the concept of alternatives, and the application of the concept to toxicity testing techniques. The potential for developing and using alternatives in biomedical research and in education programmes is not addressed or assessed. In research, more than in any other activity involving laboratory animals, the development and application of alternatives depends heavily on habit, the educational background and training of the investigator, the investigator's sensitivity to and interest in the concept of alternatives, and easy access to suitable facilities, equipment, and information. Any programme to promote the wider use of alternatives must therefore include the following features.

First, biomedical students must be made aware of the issues at an early age. The use of animals in primary and secondary education must not involve procedures which will result in desensitization. At higher educational levels, the student will need active instruction in ethical concerns and appropriate technologies (e.g. tissue culture) and should be taught that good science and animal experimentation are not necessarily synonymous. Undoubtedly, animal research is a necessary *part* of biomedical advance, but too often the two appear to be perceived as being equivalent. The example set by authority figures will be another vital ingredient in the educational process. (Journals, books, and instruction manuals also contribute significantly to the 'authority' deferred to by the student.)

Second, every research institution using animals should provide suitable facilities to encourage the use of alternatives. At present, the facilities available favour the scientist using laboratory animals, but this could be changed relatively easily. Instead of establishing only laboratory animal service facilities, relevant institutions could include other features and provide a broad range of research support services. These would include tissue culture banks and associated expertise, computer and statistical expertise, and a current data information service on research methods.

In toxicity testing, the situation becomes more complicated because of the additional inertia built into the system by bureaucratic requirements. As long as a particular test system appears to be functioning well, there is very little incentive in industry or among the regulatory agencies to change it. Where a new test is introduced, it is usually added to an already long list of tests rather than being used as a replacement. As a result, the development of a quick, efficient screening test is often of little advantage to industries which are closely regulated. Regulatory agencies need to encourage innovation in testing systems with the emphasis on producing less costly (in time, money, and animal suffering) tests without reducing safety factors.

Development and application of alternatives is therefore governed as much by attitude as by technical constraints. There are those who have never tried cell culture because they 'know' it is too difficult. Changing attitudes will not be easy.

However, the concept of alternatives, provided it is articulated in a clear, practical, and scholarly fashion, is a promising approach to the whole ethical dilemma of laboratory animal use.

REFERENCES

Ames, B. N. (1979). 'Identifying environmental chemicals causing mutations and cancer', *Science, New York,* **204**, 587–593.

Anonymous (1966). 'Survey of laboratory animal numbers', *Information on Laboratory Animals for Research,* **9**, 10.

Arlett, C. F. (1977). 'Mutagenicity in cultured mammalian cells', in *Progress in genetic toxicology* (D. Scott, B. A. Bridges, and F. H. Sobels, eds) Elsevier/North-Holland, Amsterdam, pp. 141–154.

Austwick, P. (1977). 'The use of experimental animals and *in vitro* systems in the diagnosis of fungal diseases', in *International symposium on experimental animals and in vitro systems in medical microbiology,* WHO Collaborating Center for Collection and Evaluation of Data on Comparative Virology, Munich, pp. 203–213.

Autian, J. and Dillingham, E. O. (1978). 'Overview of general toxicity testing with emphasis on special tissue culture tests', in *In vitro toxicity testing, 1975–1976* (J. Berky and C. Sherrod, eds), Franklin Institute Press, Philadelphia, pp. 23–49.

Baker, J. B. E. (1960). 'The effects of drugs on the foetus', *Pharmacological Reviews,* **12**, 37–90.

Brewen, J. G. (1977). 'The application of mammalian cytogenetics to mutagenicity studies, In *Progress in genetic toxicology* (D. Scott, B. A. Bridges, and F. H. Sobels, eds), Elsevier/North-Holland, Amsterdam, pp. 165–174.

British Pharmacopoeia (1973a). *Biological assay of antirachitic vitamin (vitamin D),* British Pharmacopoeia Commission, London, pp. A113–A114.

British Pharmacopoeia (1973b). *Biological assay for tetanus antitoxin,* British Pharmacopoeia Commission, London, P. 104.

British Pharmacopoeia (Addendum) (1977). *Biological assay for tetanus antitoxin,* British Pharmacopoeia Commission, London, p. A8.

British Pharmacopoeia Commission (1977). *The LD50 test,* Statement to the Advisory Committee to the Home Office (UK) on the *Cruelty to Animals Act 1876,* Home Office, London.

Brusick, D. (1977). 'New developments and applications of microbial mutagenesis techniques', in *In vitro toxicity testing, 1975–1976* (J. Berky and C. Sherrod, eds), Franklin Institute Press, Philadelphia, pp. 172–191.

Bucknall, R. A. (1980). 'The use of cultured cells and tissues in the development of antiviral drugs', in *The use of alternatives in drug research* (A. N. Rowan and C. J. Stratmann, eds), Macmillan, London, pp. 15–27.

Burger, E. J. (1975). 'Science for medicine—time for another reappraisal', *Federation Proceedings. Federation of American Societies for Experimental Biology,* **34**, 2106–2114.

Christensen, H. E. and Luginbuhl, T. T. (eds) (1975). *Registry of toxic effects of chemical substances,* National Institute of Occupational Safety and Health, Rockville, Md.

Clayton, R. M. (1980). 'An *in vitro* system for teratogenicity testing', in *The use of alternatives in drug research* (A. N. Rowan and C. J. Stratmann, eds), Macmillan, London, pp. 153–173.

Clive, D. and Spector, J. F. S. (1975). 'Laboratory procedure for assessing specific locus mutations at the TK locus in cultured L5178Y mouse lymphoma cells', *Mutation Research,* **31**, 17–29.

Davies, R. E., Harper, K. M., and Kynoch, S. R. (1972). 'Interspecies variations in dermal reactivity', *Journal of the Society of Cosmetic Chemists*, **23**, 271–281.

Deichmann, W. B. and Leblanc, T. J. (1943). 'Determination of the approximate lethal dose with about six animals', *Journal of Industrial Hygiene and Toxicology*, **25**, 415–417.

Deichmann, W. B. and Mergard, E. G. (1948). 'Comparative evaluation of methods employed to express the degree of toxicity of a compound', *Journal of Industrial Hygiene and Toxicology*, **30**, 373–378.

Dixon, B. D. (1976). *What is science for?*, Penguin, London.

Draize, J. H., Woodard, G., and Calvery, H. O. (1944). 'Methods for the study of irritation and toxicity of substances applied topically to the skin and mucous membranes', *Journal of Pharmacology and Experimental Therapeutics*, **82**, 377–390.

EPA (1978). 'Hazardous waste: Proposed guidelines and regulations and proposal on identification and listing', *Federal Register*, 18 December, p. 59023.

FRAME (1979). *Alternatives to laboratory animals*, Fund for the Replacement of Animals in Medical Experiments, London.

French, R. D. (1975). *Antivivisection and medical science in Victorian society*. Princeton University Press, Princeton, NJ.

General Accounting Office (1978). *Lack of authority hampers attempts to increase cosmetics safety*, US General Accounting Office (HRD-78-139), Washington, DC, p. 33.

Gyorgy, P. and Pearson, W. N. (eds) (1967). *The vitamins*, Vol. 7. Academic Press, New York.

Harrison, R. J. and Harrison, M. J. (1978). 'Computer simulation as an aid to the replacement of experimentation on animals and humans', *ATLA Abstracts*, **62**, 22–35.

Hollstein, M., McCann, J., Angelosanto, F., and Nichols, W. (1979). *Mutation Research*, **65**, 133–226.

Home Office (1978). *Experiments on living animals, Statistics—1977, Cmnd 7333*, HMSO, London.

Hutt, P. B. (1978). 'Unresolved issues in the conflict between individual freedom and government control of food safety', *Food Drug Cosmetic Law Journal*, **33**, 558–589.

ICLA (1959). *International survey on the supply, quality and use of laboratory animals*, Supplements I, II, and III, International Committee for Laboratory Animals, Oslo.

ICLA (1962). *International survey on the supply, quality and use of laboratory animals*, Supplement IV, International Committee for Laboratory Animals, Oslo.

ILAR (1970). 'Laboratory animal facilities and resources supporting biomedical research, 1967–1968', *Laboratory Animal Care*, **20**, 795–869.

Kennedy, D. (1978). 'Animal testing and human risk', *Human Nature*, May.

Lapage, G. (1960). *Achievement: some contributions of animal experiments to the conquest of disease*, W. Heffer and Sons, Cambridge.

Latt, S. A., Allen, J. W., Rogers, W. E., and Juergens, L. A. (1977). '*In vitro* and *in vivo* analysis of sister chromatid exchange formation', in *Handbook of mutagenicity test procedures* (B. J. Kilbey, M. Legator, W. Nichols, and C. Ramel, eds), Elsevier/North-Holland, Amsterdam, pp. 275–291.

LeCornu, A. P. and Rowan, A. N. (1979). 'The use of non-human primates in the development and production of poliomyelitis vaccines', *ATLA Abstracts*, **7**, 11–19.

McCann, J. and Ames, B. N. (1975). 'Detection of carcinogens as mutagens in the *Salmonella*/microsome test: Assay of 300 chemicals', *Proceedings of the National Academy of Science, USA*, **72**, 5135–5139.

McCann, J. and Ames, B. N. (1976). 'Detection of carcinogens as mutagens in the *Salmonella*/microsome test: assay of 300 chemicals: discussion', *Proceedings of the National Academy of Science, USA*, **73**, 950–954.

Marks, J. (1972). 'Ending the routine guinea-pig test', *Tubercle*, **53**, 31–34.

Maugh, T. H. (1978). 'Chemical carcinogens: the scientific basis for regulation', *Science, New York*, **201**, 1200–1205.

Maxwell, W. A. and Newell, G. W. (1973). 'Considerations for evaluating chemical mutagenicity to germinal cells', *Environmental Health Perspectives*, December, 47–50.

Medawar, P. B. (1972). *The hope of progress*, Methuen, London.

Meier-Ewert, H. (1977). 'The use of experimental animals and *in vitro* systems for proving virus diagnosis', in *International symposium on experimental animals and in vitro systems in medical microbiology*, WHO Collaborating Centre for Collection and Evaluation of Data on Comparative Virology, Munich, pp. 236–242.

Morris, R. K. and Fox, M. W. (eds) (1978). *On the fifth day: animal rights and human ethics*, Acropolis Books, Washington, DC.

Morrison, J. K., Quinton, R. M., and Reiner, H. (1968). 'The purpose and value of LD50 determinations', in *Modern trends in toxicology*, Vol. 1 (E. Boyland and R. Goulding, eds), Butterworth, London, pp. 1–17.

Muul, I., Hegyeli, A. F., Dacre, J. C., and Woodard, G. (1976). 'Toxicological testing dilemma', *Science, New York*, **193**, 834.

National Commission for the Protection of Human Subjects of Biomedical and Behavioral Research (1978). *Ethical guidelines for the protection of human subjects of research*, DHEW Publication No. (05) 78–0012, US Government Printing Office, Washington, DC.

Newton, C. M. (1977). 'Biostatistical and biomathematical methods in efficient animal experimentation', in *The future of animals, cells, models and systems in research, development, education and testing*, National Academy of Sciences, Washington, DC, pp. 152–164.

O'Neill, J. P., Brimer, P. A., Machanoff, R., Hirsch, G. P., and Hsie, A. W. (1977). 'A quantitative assay of mutation induction at the hypoxanthine – guanine phosphoribosyl transferase locus in chinese hamster ovary cells (CHO/HGPRT system): development and definition of the system', *Mutation Research*, **45**, 91–101.

Petricciani, J. C., Hopps, H.E., Elisberg, B.L., and Early, E. M. (1977). 'Application of *in vitro* systems to public health', in *The future of animals, cells, models and systems in research, development, education and testing*, National Academy of Sciences, Washington, DC, pp. 240–254.

Pienta, R. J. (1979). '*In vitro* transformation of cultured cells', in *Proceedings of the XIIth Internationl Cancer Congress, Buenos Aires*, Pergamon Press, Oxford, in press.

Pienta, R. J., Poiley, J. A., and Lebherz, W. B. (1977). 'Morphological transformation of early passage golden Syrian hamster embryo cells derived from cryopreserved primary cultures as a reliable in vitro bioassay for identifying diverse carcinogens', *International Journal of Cancer*, **19**, 642–655.

Purchase, I. F. H., Longstaff, E., Ashby, J., Styles, J. A., Anderson, D., Lefevre, P. A., and Westwood, F. R. (1978). 'An evaluation of six short-term tests for detecting organic chemical carcinogens', *British Journal of Cancer*, **37**, 873–903.

Raettig, H. (1977). 'Bacteria: the use of experimental animals and *in vitro* systems in investigating scientific diagnostic problems', in *International symposium on experimental animals and in vitro systems in medical microbiology*, WHO Collaborating Centre for Collection and Evaluation of Data on Comparative Virology, Munich, pp. 214–220.

Rosenkranz, H. S., McCoy, E. C., Briuso, L., and Speck, W. T. (1980). 'Short-term microbial assays in the assessment of carcinogenic risk', in *The use of alternatives in drug research* (A. N. Rowan and C. J. Stratmann, eds), Macmillan, London, pp. 103–145.

Russell, W. M. S. and Burch, R. L. (1959). *The principles of humane experimental technique*, Methuen, London.

Saffiotti, U. (1976). 'Validation of short-term bioassays as predictive screens for chemical carcinogens', in *Screening tests in chemical carcinogenesis* (R. Montesano, H. Bartsch, and L. Tomatis eds), International Agency for Research on Cancer, Lyon, pp. 3–13.

San, R. H. C. and Stich, H. F. (1975). 'DNA repair synthesis of cultured human cells as a rapid bioassay for chemical carcinogens', *International Journal of Cancer*, 16, 284–291.

Saxen, L. (1976). 'Advantages of organ culture techniques in teratology', in *Tests of teratogenicity in vitro* (J. Ebert and M. Marois, eds), North-Holland, Amsterdam, pp. 262–284.

Simons, P. J. (1980). 'An alternative to the Draize test', in *The use of alternatives in drug research* (A. N. Rowan and C. J. Stratmann, eds), Macmillan, London, pp. 147–151.

Smith, R. J. (1979). 'NCI bioassays yield a trail of blunders', *Science, New York*, 204, 1287–1292.

Smyth, D. H. (1978). *Alternatives to animal experiments*, Scolar Press, London.

Spink, J. D. (1977). 'Drug testing', in *The welfare of laboratory animals: legal, scientific and humane requirements*, Universities' Federation for Animal Welfare, Potters Bar, London, pp. 44–50.

Storz, J., Moore, K. A., and Spears, P. (1977). 'The use of experimental animals and in vitro systems for the detection and identification of chlamydial and rickettsial agents', in *International symposium on experimental animals and in vitro systems in medical microbiology*, WHO Collaborating Centre for Collection and Evaluation of Data on Comparative Virology, Munich, pp. 221–235.

Sunshine, I. (ed.) (1969). *Handbook of analytical toxicology*, CRC Press, Boca Raton.

Tajima, Y. (1975). 'Present status of experimental animals in Japan', *Experimental Animals*, 24, 67–77.

Tinbergen, N. and Perdeck, A. C. (1950). 'On the stimulus situation releasing the begging response in the newly hatched herring gull chick', *Behaviour*, 3, 1–39.

Trevan, J. W. (1927). 'The error of determination of toxicity', *Proceedings of the Royal Society B*, 101, 483.

Vogel, E. and Sobels, F. H. (1976). 'The function of *Drosophila* in genetic toxicology testing', in *Chemical mutagens: principles and methods for their detection*, Vol. 4 (A. Hollaender, ed.), Plenum Press, New York, pp. 93–142.

Weil, C. S. and Scala, R. A. (1971). 'Study of intra- and inter-laboratory variability in the results of rabbit eye and skin irritation tests', *Toxicology and Applied Pharmacology*, 19, 276–360.

Wiggins, R. A. (1976). 'Replacement of biological methods by chemical and physical methods. II. Chemical analysis of vitamins, A, D and E', *Proceedings of the Analytical Division of the Chemical Society*, 13, 133–137.

Williams, D. J. and Rabin, B. R. (1971). 'Disruption by carcinogens of the hormone-dependent association of membranes with polysomes', *Nature, London*, 232: 102–105.

Wilson, J. G. (1978). 'Survey of in vitro systems—their potential use in teratogenicity screening', in *Handbook of Teratology*, Vol. 4 (J. G. Wilson and F. C. Fraser, eds), Plenum Press, New York.

Animals in Research
Edited by David Sperlinger
© 1981 John Wiley & Sons Ltd.

Chapter 13

The 'Defined' Animal and the Reduction of Animal Use*

MICHAEL F. W. FESTING

INTRODUCTION

About 6 million animals are used annually in research laboratories in the UK. Although this is numerically insignificant compared with approximately 398 million chickens, 15 million pigs, 13 million sheep, and 4.5 million cattle slaughtered for meat each year, the public are concerned that some of these animals may suffer unnecessarily. The aim of the chapter is to outline the current status of laboratory animal science, and to consider some of the ways in which the scientific validity of the work may be improved, and the suffering minimized, by the use of high quality, defined animals maintained by trained professional staff, and used in well designed experiments with the correct statistical analysis (Festing, 1977).

The Use of Animals in Research and Safety Evaluation

The health of modern society relies heavily on laboratory animals, both in research and in the production of medicines and vaccines. In fact, in 1978 a fifth of all laboratory animals used in the UK were used in mandatory tests for evaluating the safety of drugs and standardization of sera, vaccines, medicines, or materials required under the *Medicines Act 1968,* the *Health and Safety at Work Act 1974*, the *Poisonous Substances Act 1952*, and the *Food and Drugs Act 1955*, or equivalent overseas acts (Home Office, 1978). Laboratory animals are also widely used in cancer research, immunology, toxicology, biochemistry, experimental surgery, animal behaviour, and a wide range of other disciplines in medicine, veterinary science, and agriculture. Although the value of using animals as models of man has been questioned, recent research has tended to

---

* The views expressed in this paper are the personal views of the author and do not necessarily correspond to those of the Medical Research Council.

emphasize the similarities between man and animals, and hence the value of careful extrapolation from animals to man. The theory of evolution and the universal nature of the genetic code, emphasizes the relationships between all living things. Within the vertebrates, many of the biochemical pathways, immunological responses, spontaneous diseases, susceptibilities to toxins, and even certain aspects of behaviour are similar in a range of species including man. Research with animals continues to advance the state of biological knowledge leading to the development of new drugs, and the discovery of unsuspected toxic environmental hazards. Past discoveries have led to dramatic saving of life. According to a recent television documentary, the discovery of insulin alone has saved over 31 million lives, many of them children. The development of vaccines against smallpox, diptheria, and polio and the development of antibiotics and other drugs have made a substantial contribution to reducing the child mortality rate from 3 deaths per thousand in 1931–35 to only 0.4 deaths per thousand today. Similarly, deaths from tuberculosis have declined from 27 600 in 1935 to less than 1000 in 1974, and deaths during childbirth have declined from 3000 per annum to only 132 during the same period. These reductions are largely due to advances in medical science, though there have of course been some important changes in general hygiene and nutrition which deserve some of the credit. The development of new drugs, which has depended on animal research, has made great progress during the last two decades.

Yet many diseases remain unconquered. Cancer and heart disease still cause nearly half a million premature deaths each year in the UK alone, and the numbers have nearly doubled since 1935. These, and other 'diseases of affluence' may be closely associated with changes in life-style, exercise, diet, and exposure to drugs and pollutants associated with modern western civilization. Illness continues to cause considerable suffering in people of all age groups, quite apart from loss of earning and costs to the health services. The need to use animals in research will continue for many years, though the scientist and the animal welfare societies have a common interest in reducing the total numbers of animals used wherever possible.

The Organization of Laboratory Animal Science

Unfortunately, the husbandry, breeding, care, and use of laboratory animals—laboratory animal science—is not one of the more glamorous scientific disciplines, yet it is of vital interest to a range of institutions and scientific societies, which spend several million pounds annually on animals, food, equipment, animal houses, and staff.

Unlike many other disciplines, however, the universities in the UK have played a relatively small part in the development of the subject. There are no university departments of laboratory animal science, and no full-time courses on the subject in the UK, although a number of universities in the USA and the Federal Republic of Germany have chairs of laboratory animal science with postgraduate courses in laboratory animal medicine usually for veterinary graduates.

The main British institute with a direct interest in the subject is the Medical Research Council Laboratory Animals Centre (LAC), which was founded in 1947 after a meeting of the major institutional users of laboratory animals who were concerned about the poor quality of animals available at that time. The LAC now has a total staff of over 80, of whom about 18 are graduates, with specialists in laboratory animal nutrition, husbandry, genetics, microbiology, physiology, and pathology. The LAC supplies high quality breeding stock to breeders and institutions in the UK and acts as a World Health Organization reference centre for the supply of defined laboratory animals to users abroad wishing to set up breeding colonies. It also administers a voluntary Accreditation and Recognition Scheme for commercial laboratory animal breeders, conducts research, and acts as an international centre for information on laboratory animals. The LAC has had a strong influence on the development of the subject throughout the world.

The Laboratory Animal Science Association (LASA) is the main British scientific society concerned with the subject, and publishes a journal, *Laboratory Animals*. The main support for this society comes from the staff of animal divisions in the research councils, pharmaceutical industry, and universities, though some of the more enlightened animal users are also members. However, there are over 19 000 licence holders under the *Cruelty to Animals Act, 1876* yet the total membership of LASA is less than 400. There are also a number of specialist scientific societies, such as the British Laboratory Animals Veterinary Association, and trade associations, such as the Laboratory Animal Breeders Association, with an interest in laboratory animals in the UK. There are equivalent societies in a number of other countries.

Although there is no formal training at the graduate or postgraduate level in the UK, animal technician training is organized by the Institute of Animal Technicians, whose qualifications are widely recognized by employers. Animal technology offers a good career with excellent employment prospects for those with suitable qualifications.

THE DEFINED ANIMAL

Although it is possible for a research worker to pay little attention to the health and quality of the animals used in research, the best results are undoubtedly obtained by those research workers who use only microbiologically and genetically 'defined' animals which have been properly acclimatized to the animal house environment.

Microbiological Definition

Animals, like humans, suffer from infectious diseases ranging from those causing only mild symptoms to acute disease outbreaks with high mortality. Although some of these diseases may be partially controlled by medication, both the disease and the medication may seriously interfere with experimental work. Fortunately, most of the infectious disease of laboratory animals may be controlled simultaneously by the use of 'specific pathogen-free' (SPF) techniques (Bleby, 1976).

SPF stocks were initially developed by hysterectomy of the pregnant females just prior to parturition using aseptic surgical techniques. The young were resuscitated in a building designed to prevent the entry of disease-causing organisms, or in closed isolators, and were hand-reared using a sterile milk substitute. This procedure immediately eliminates a wide range of pathogenic organisms which are normally transmitted from mother to offspring, because young within the uterus are usually microbially sterile. SPF animals are normally free of all parasites, many viruses, and most pathogenic bacteria. Once established, such colonies breed well and can be used to supply high quality breeding stock which does not carry the risk of introducing disease when introduced into an existing colony. SPF mice, rats, guinea-pigs, and cats are available from commercial breeders in the UK. Although some other species have been derived into SPF conditions, they are not freely available at present.

In the past, the term 'SPF' caused confusion, as it was not clear exactly which pathogens an SPF colony was free from. This led the Laboratory Animals Centre to introduce a grading scheme for Accredited commercial laboratory animal breeders (Townsend, 1969; LAC, 1974). The lowest grade of animals that it is permissible for an accredited breeder to supply are graded as one-star. Such animals must be free from all zoonoses, i.e. all diseases that can be transmitted to man. Samples of animals from each breeder are routinely tested for a range of potential human pathogens, and the results are published in the *Register of Accredited Breeders*. Such animals should be safe for use in schools and for dissection and teaching. At the other extreme, animals free from all demonstratable microorganisms ('germ-free') and maintained within a closed isolator system, are graded as five-star. Four-star animals, i.e. animals which are free of all higher parasites, all pathogenic bacteria, and a number of specified viruses are what are more loosely termed 'SPF' animals, except that these four-star animals are rigidly defined and tested by the LAC according to published protocols. Such animals, though more expensive to produce than conventional and one-star animals are extremely valuable in research for the following reasons.

(a) A colony of animals carrying infectious pathogenic organisms (e.g. category one-star) is likely to be more variable than a group of SPF animals and therefore more animals will be needed to achieve the same degree of statistical precision. For example, it has been shown that underweight mice carried five times as many parasitic nematodes as their normal weight littermates (Eaton, 1972). Although in this case cause and effect cannot be separated, it seems reasonable to assume that such an uneven parasite burden will lead to increased variability among the experimental animals. This means that for the same statistical precision, more parasitized animals will be needed than if the animal were free of such parasites.

(b) In conventional animals there is the danger that the effects of disease may be mistaken for the effects of the experimental treatment. Vitamin A deficiency, for example, has been recorded as causing pneumonia and lung abscesses in rats, when in fact all that the deficiency is doing is to increase the severity of the infectious chronic respiratory disease found in all non-SPF rats (Lindsey et al., 1971). Pneumonia does not occur in SPF rats which are vitamin A-deficient. The activation of latent disease through the experimental treatment can be extremely misleading as the control animals not subjected to the same degree of stress may be unaffected (Baker et al., 1971).

(c) In some cases, mild infections may mask the results of an experimental treatment. For example, rats are widely used in inhalation toxicology, yet the lesions of chronic respiratory disease may completely obscure the effects of the experimental treatment. In one case, chlorine gas caused lung lesions, but as the animals grew older the differences between the treated and control groups were completely obscured by the chronic respiratory disease, so that the two groups were histologically indistinguishable (Elmes and Bell, 1963).

(d) Many long-term studies require a substantial number of animals to reach old age. As fewer SPF animals die as a result of infectious disease, fewer animals need to be started in each experimental group (Lindsey et al., 1971). Thus, for long-term studies SPF animals may be substantially more economical than conventional ones.

(e) In a number of studies animals are used as models of some human disease. However, infections can drastically alter the characteristics of such models. A striking example of this was recorded by Sabine et al. (1973) and Heston (1975). The inbred mouse strain C3H-A^{vy} develops a 100 per cent incidence of mammary tumours at an early age under normal laboratory conditions. Yet, when such mice were transferred to a laboratory in Australia, the incidence of such tumours declined to about 10 per cent (Sabine et al., 1973). Subsequent investigation of the cause of this change showed that the mice transferred to Australia had developed a severe mite infestation which was reducing growth rate. It is well established that growth rate and tumour incidence are directly correlated (Heston, 1975), and that this mite infestation was probably responsible for the reduction in tumour incidence. Obviously, if infection can alter the characteristics of a strain so dramatically, it can lead to highly confusing research results. The use of SPF mice would reduce such confusion.

Genetic Definition

Inbred Strains

Laboratory animals should also be genetically defined. Behaviour, response to drugs, size, weight, and shape of many organs, numbers, and types of spontaneous tumours, and response to antigens depends not only on the species but also on the strain of animal (Festing, 1979a). Inbred strains of mice, rats, hamsters, and guinea-pigs produced as a result of at least 20 generations of brother × sister mating, are readily available. They are much better experimental subjects than the more widely used outbred 'white' mice and rats for most studies. As early as 1942 Dr L. C. Strong wrote that:

> 'It is the conviction of many geneticists that the use of the inbred mouse in cancer research has made possible many contributions of a fundamental nature that would not have been made otherwise. Perhaps it would not be out of place to make the suggestion that within the near future all research on mice should be carried out on inbred animals or on hybrid mice of known (genetically controlled) origin where the degree of biological variability has been carefully controlled' (Strong, 1942).

Grüneberg (1952) even went so far as to state that 'The introduction of inbred strains into biology is probably comparable with that of the analytical balance into chemistry'.

The main characteristics of genetically defined inbred strains are:

(a) All individuals of a strain are genetically identical (isogenic). The genetic uniformity of inbred strains means that each strain can be genetically typed for characters such as their blood group in the knowledge that all animals within that strain will be the same. Such data is essential in many immunological and cancer research studies, and cannot be gathered in outbred stocks, where each individual is genetically unique. Isogenicity also leads to phenotypic uniformity for all highly inherited characters and this means that the statistical precision of an experiment using these animals is increased.

(b) Inbred strains are genetically stable. Once a strain has been developed it stays genetically constant for many years. Non-inbred strains may change as a result of selective forces, but such forces cannot act on inbred strains, which can only change as a result of the accumulation of mutations—a slow process. This stability means that background data on strain characteristics remains constant for long periods—allowing for the use of such information in planning experiments.

(c) Inbred strains are internationally distributed. This means that experiments conducted on some of the commoner inbred strains, which are maintained in laboratories throughout the world, may easily be confirmed at laboratories in entirely different parts of the world. Moreover, if many laboratories are working with the same strain, background data on that strain is accumulated much faster.

(d) Each strain has a unique set of characteristics which may be of value in research. In some cases a strain may have a disease which in some way models a similar condition in man. The best known of such models are the strains with a high incidence of a particular type of cancer. The inbred mouse strain C3H develops a very high incidence of breast tumour, strains AKR and C58 get leukaemia, SJL gets reticulum-cell sarcoma (Hodgkin's disease), and some sublines of strain 129 develop teratomas. Other strains develop autoimmune anaemia (NZB), amyloidosis (YBR and SJL), congenital cleft palate (A and CL), hypertension and heart defects (BALB/c and DBA/2 mice, and SHR and GH rats), obesity and diabetes (NZO, PBB and KK mice), and even a preference for alcohol when given a free choice of 10 per cent alcohol or plain water (strain C57BL mice). Each of these strains may be studied in order to obtain a better understanding of the disease in the mouse or rat. Once it is understood in the animal, it will be easier to understand in the human, even though it is unlikely that the conditions are exactly comparable in animals and man. In fact, it is clear from a study of a disease such as hypertension in the rat that the cause of the hypertension in SHR and GH is entirely different (Simpson et al., 1973), emphasizing that diseases of this sort in humans can have several different causes. Obviously, in such cases some animal models may mimic a human disease relatively closely, while in other cases there is little resemblance. Table 13.1 lists some examples of models of disease found among inbred strains of mice.

Inbred strains do not need to model any human disease in order to be of value in research. Strains can usually be found to differ for almost every characteristic studied including many aspects of behaviour, response to a wide range of drugs and chemicals, response to antigens, response to infectious agents, incidence of spontaneous diseases, and even anatomical features. These differences can be of great value in research in a number of different ways. At the most trivial level, if a scientist is studying a response to some treatment effect, it is often possible for him to find, by surveying a number of inbred strains, a strain which is highly sensitive to his experimental treatment. In some cases this will mean that fewer animals are needed to achieve the same degree of statistical precision in future experiments. In other cases, the more sensitive strain may well show the effect sooner than resistant strains, and this may reduce the time and facilities needed to complete the experiment.

At a slightly more sophisticated level, a comparison of sensitive and resistant strains may give extremely valuable information about the mechanism of some treatment effect. For example, if two strains differ in sensitivity to a drug, it would be of great interest to know whether this is because of differences in absorption, metabolism, excretion, or target organ sensitivity. Such a study could give information that would be extremely useful in evaluating the likely effect of the drug in man. Preferably, such studies should be carried out on two or more sensitive and two or more resistant strains in order to show whether the results are uniquely strain dependent, or can be generalized.

Table 13.1 Examples of disease models and characters of interest in inbred strains of mice

Character	Strains(s)
Alcohol (10%) preference	C57BL, C57BR/cd
Aggression/fighting	SJL, NZW
Audiogenic seizures	DBA/2
Autoimmune anaemia	NZB
Amyloidosis	YBR, SJL
Cleft palate	CL, A
Chediak–Higashi syndrome	SB
Hypertension and/or heart defects	BALB/c, DBA/1, DBA/2
Hyperprolinaemia and prolinuria	PRO
Obesity and/or diabetes	NZO, PBB, KK, AY
Osteoarthropathy of knee joints	STR/1
Polydipsia	SWR, SWV
Resistance to myxovirus infection	A2G
Tumours	
Leukaemia	AKR, C58, PL, RF
Reticulum-cell sarcoma	
(Hodgkin's disease)	SJL
Lung tumours	A
Hepatomas	C3Hf
Mammary tumours	C3H, C3H-A^{vy}, GRS/A, RIII
Ovarian teratomas	LT
Induced plasmacytomas	BALB/c, NZB
Testicular teratomas	129/terSv
Complete absence of spontaneous	
tumours	X/Gf
Whisker eating	A2G

Reproduced from Festing (1978) with the permission of Hans Huber, Publishers.

Any two inbred strains will normally differ from each other at several thousand different genetic loci. However, sets of inbred strains which differ from one another at only one or a few loci have been developed in order to study in greater detail those loci which are of particular importance in biomedical research. These are known as sets of congenic strains, and most of them have been developed in order to study the major histocompatibility complex (MHC). This complex locus is responsible for a range of immunological reactions, including immune responses and graft rejection. Obviously, if two strains can be developed which are genetically identical apart from the MHC, it becomes possible to study the MHC in detail simply by comparing the two strains. Such strains can be developed either as a result of a fortuitous mutation within an inbred strain, or by deliberate breeding using conventional genetic backcrossing techniques. Several hundred strains of this type have been developed, and they are now widely used in immunology and cancer research. They have undoubtedly given much insight into the biology of the mouse MHC, which in many respects is very similar to the MHC in man.

Mutants

There are more than 500 known mutants and variants in the mouse, and a further 100 in the rat, though in the rat many of these have now been lost. Some of these mutants appear to mimic similar mutants in several species including man, and may therefore be regarded as 'models' of human disease. Other mutants lack an organ such as the thymus, spleen, tail, or eyes, or they suffer from some hormone deficiency or a developmental defect. Such mutants can be extremely valuable for certain types of research even though they may not resemble any human condition. A list of some of these mutants, classified into models of disease, genetic alterations and deficiencies, and biochemical and immunological polymorphisms is given in Table 13.2. There are, for example, a number of types of genetically determined obesity and diabetes which have been extremely useful as models of similar conditions in man (Festing, 1979*b*). Such animals help to show up the immense complexity of the regulation of body fat via the hormonal control of a range of metabolic interactions, each of which may be controlled by regulatory mechanisms which interact with one another. Thus, the finding of a particular biochemical abnormality is no guarantee that it is the cause of the observed obesity. It is much more likely that it is a secondary effect of the primary genetic defect. However, although many of these models of obesity may have no exact counterpart in man, they may still be useful in screening drugs with a potential for reducing obesity (Cawthorne, 1979).

One of the most important mutants causing genetic alterations or deficiencies

Table 13.2 Examples of mouse mutants of medical interest and of biochemical and immunological polymorphisms

A. Examples of mouse mutants of medical interest

	Gene	Name
Models of disease		
Anaemia	*sla*	sex-linked anaemia
	Sl	steel
	W	dominant spotting
Chediak–Higashi syndrome	*bg*	beige
Diabetes and/or obesity	A^y	yellow
	A^{vy}	viable yellow
	db	diabetes
	db^{ad}	adipose
	ob	obese
Inborn errors of metabolism	*his*	histidinaemia
	pro	prolinemia

Table 13.2 A continued

	Gene	Name
Models of disease		
Kidney disease	*kd*	kidney disease
Muscular dystrophy	*dy*	dystrophia-muscularis
	dy²ʲ	dystrophia-muscularis-2J
Neuromuscular mutants	*jp*	jimpy
	med	motor and plate disease
	qk	quaking
	Swl	sprawling
	Tr	trembler
Genetic alterations or deficiencies		
Embryonic defects	*t*-alleles	tailess alleles
Hair absent	*hr*	hairless
	hrʳʰ	rhino
	N	naked
Hair and thymus absent	*nu*	nude
Growth hormone absent	*dw*	dwarf
Resistance to androgen	*Tfm*	testicular feminization
Sex reversal	*Sxr*	sex reversal
Spleen absent	*Dh*	dominant hemimelia

B. Examples of biochemical and immunological polymorphisms

Name of polymorphism	Gene locus
Aromatic hydrocarbon responsiveness	*Ahh*
Pancreatic amylase	*Amy-2*
ß-D-Galactosidase activity	*Bgs*
Liver catalase	*Ce-1*
Erythrocyte antigens	*Ea-1* to *Ea-7*
Esterases (serum and kidney)	*Es-1* to *Es-7*
Friend virus susceptibility	*Fv-1, Fv-2*
G 6 PD regulators	*Gdr-1, Gdr-2*
Haemoglobin alpha chain	*Hba*
Haemoglobin beta chain	*Hbb*
Haemolytic complement	*He*
Histocompatibility	*H-1* to *H-38*
Immunoglobulin	*Ig-1* to *Ig-4*
Macrophage antigen-1	*Mph-1*
Major urinary protein	*Mup-1*
Phosphoglucomutase	*Pgm-1, Pgm-2*
Sex-limited protein	*Slp*
Thymus cell antigen-1	*Thy-1*
Thymus leukaemia antigen	*Tla*

Reproduced from Festing (1978) with permission of Hans Huber, Publishers.

is the athymic nude mutation in the mouse. A similar mutation has now been described in the rat (Festing *et al* , 1978). The thymus is essential for the full development of the immune system, and homozygous nude mice or rats are deficient in the cell-mediated type of immune response. They are of value in fundamental studies of immune mechanisms, as well as in applied cancer research. This is because, lacking the cell-mediated immune response, they are unable to reject transplanted foreign tissue, including transplants of human tumours. Such transplanted human tumours usually grow, but they retain all the characteristics of human tissue. Therefore, it is possible not only to study human tumours when growing in an animal, but it is also possible to study the effect of drugs on such tumours. This is obviously of more value than having to rely simply on the study of animal tumours in animals.

Recently, there has been a great increase in interest in mutants. This probably follows the successful development of the nude mouse as a research model.

TRAINED STAFF

The microbially and genetically defined laboratory animal is a highly sophisti-cated research tool, which must be carefully nurtured and thoroughly under-stood if the maximum amount of information is to be extracted from the minimum number of animals. At its best, animal experimentation is a team effort involving the research scientists, the skilled animal house staff, and appropriate specialists, such as animal surgeons, geneticists, pathologists, and statisticians.

Animal technicians holding qualifications approved by the Institute of Animal Technicians (or the equivalent overseas) should be able to maintain the required standards of animal care. The institute examinations are based on both practical competence and academic knowledge, and include instruction on the symptoms and control of all the common animal diseases. Animal technicians should also be trained to recognize suffering in animals, and know the most humane methods of destroying them either when they are suffering unduly, or at the end of an experiment.

Some animal houses are supervised by an animal curator with scientific or veterinary qualifications, and an academic status equal to that of most of the research workers. Animal houses shared by several departments are often unified under the control of the animal curator who reports to a committee of animal house users. Failure to employ qualified staff often leads to poor standards and outbreaks of disease, though even the very best veterinary supervision can not guarantee the absence of disease in the absence of good physical facilities.

ENVIRONMENTAL AND NUTRITIONAL STANDARDIZATION

The need to house the defined laboratory animal in defined and stabilized environmental conditions, with a nutritionally adequate and controlled diet, is now becoming recognized. Both diet and environment can drastically alter the physiology of the animal, and change its response to drugs and other experimental treatments. Moreover, animals bought in from a commercial breeder may well take two or more weeks to acclimitize to their new environment. During this period their physiological responses may be unpredictable, depending on the difference between the two environments (Grant et al., 1971).

FAILURE OF MANY RESEARCH WORKERS TO MAKE OPTIMUM USE OF LABORATORY ANIMALS

The microbiologically and genetically defined laboratory animal, maintained in a controlled environment, fed on a defined diet and cared for by trained animal technicians under the supervision of a scientifically qualified animal curator is clearly a very superior animal. Such animals have been compared with the pure chemical demanded by research chemists.

Unfortunately, many scientists are not making full use of such animals. Thus, less than 50 per cent of animals used in the United Kingdom in 1972 (the latest figures available) were microbiologically defined, and only 24 per cent of mice and less than 16 per cent of any other species were genetically defined. Moreover, only 35 per cent of the 1219 animal technical staff employed in 289 registered animal houses had qualifications recognized by the Institute of Animal Technicians (Dr M. Gamble, personal communication). No figures are available on the number of animal premises controlled by a specialized animal curator, but there are many institutes where the animal house is shared by a number of different departments each of which acts independently, reserving the right to bring in animals from any source at any time, thus exposing all the animals to the risk of epidemic diseases. Even allowing for some improvement since these figures were collected, it is clear that there is scope for improvement in the quality of animals used, the employment of qualified animal technicians, and the organization of animal houses under a single professionally qualified curator.

Many disease outbreaks are due to failure to provide adequate quarantine facilities for animals that need to be introduced into the colony. For example, at least five mouse colonies became infected by the ectromelia (mouse pox) virus in 1974. This was apparently caused by importation of infected mice from Czechoslovakia followed by inadequate quarantining of animals introduced into each colony. All infected colonies had to be destroyed, resulting in considerable disruption to research and financial loss.

Many research workers lack formal training in the statistical aspects of experimental design, leading to the failure to use formal designs such as randomized blocks and Latin squares. This may be responsible for some wastage of animals, and unrepeatable experimental results. Relatively few papers state that the animals were assigned to the treatment groups at random. It is not acceptable to assign a box of mice to the treatment and control groups on the basis of the first animals which are caught being assigned to one group and the remainder to the second group. Such animals would be stratified on the behavioural variable 'catchability'. Those caught first would almost certainly include any which are blind or sick, and there may well be important physiological differences between groups. Such differences would then be completely confounded with the treatment effect. Even worse cases of violation of the principles of experimental design are not uncommon, though it is often difficult to detect such faults from the resulting publication. Some research workers still fail to use contemporary controls even though uncontrolled environmental variation can profoundly effect experimental results. In other cases, the treated and control groups may be housed in different rooms, or on different shelves. As there is usually a temperature gradient from floor to ceiling, animals housed on the bottom shelf will have a different environment from those housed on the top shelf, so biases will occur if equal numbers of treated and control animals are not represented on each shelf.

The most striking example of the violation of the principles of experimental design, which may be responsible for the large-scale wastage of laboratory animals, is in the field of toxicological screening.

Toxicological screening is carried out using a number of different types of test: long-term toxicity and carcinogenesis tests, teratology, and in some cases multi-generation studies in which the substance is administered to animals over several generations.

The animal tests are usually carried out on genetically variable 'outbred' stocks. Two arguments are used to justify the use of such stocks. These are firstly that the animal is used as a 'model' of man, and man is 'outbred', therefore outbred laboratory animals should be used. Secondly, it is argued that the tests should include animals of a wide range of different genotypes in order to get a broad genetic base for extrapolation to man. For example, Grice *et al.* (1973) state that

> 'the use of highly inbred strains should be avoided since . . . while one inbred strain may be highly responsive to an agent, the same carcinogen may fail to induce tumours in animals of another inbred strain . . . use of animals with a more heterogeneous genetic constitution will, to some extent, overcome this problem even though the tumour incidence may not be particularly high within a given test.'

Unfortunately, neither of these arguments will withstand critical evaluation (Festing, 1975, 1979c), and they result in the use of genetically variable material which violates one of the first laws of experimentation, namely that the experimenter should control all relevant variables. An outbred stock is one in which there is an unknown, and uncontrollable, degree of genetic variation or 'noise' which in critical situations may well obscure any treatment effect.

The argument that outbred animals should be used because man is outbred fails to recognize the two-step nature of experimental inference. In any experiment the first step is to decide whether the experimental treatment has affected these particular experimental animals. This can be done most easily using inbred animals in which uncontrolled genetic variation is absent. Only after this question has been answered is it possible to interpret the importance of these results in human terms. There is no need for the experimental animals to model man in every respect. This would in any case be impossible, as the mouse differs from man in many respects. All that is required is that the animal should be able to give a clear-cut answer to the question posed by the experimenter. It is then up to the experimenter to interpret this result. Use of an animal incapable, because of 'noise', of distinguishing between two treatments is of no value at all.

The argument that the use of an outbred stock will give a wide range of genotypes, thus avoiding the possibility that all animals will be genetically resistant, is also invalid. Although the number of genetically determined types (genotypes) in an outbred stock may be very great, the range of observed types of animal (phenotypes) is much smaller. If a wide range of phenotypes is needed (and the arguments are persuasive), the best method is to include several different inbred strains in a factorial design of experiment (Festing, 1975). Differences between inbred strains are much greater than the differences between individuals of an outbred stock. This design has a number of statistical and biological advantages over the use of either a single inbred strain or a single outbred stock. The high statistical precision is obtained from the fact that the inbred strains are genetically uniform, while the broad basis for extrapolation to man is acheived from the differences between the strains.

The statistical assumptions behind such a design were studied by Haseman and Hoel (1979). After making a number of simplifying assumptions, they concluded that the factorial design was in all cases statistically more powerful than the use of a single outbred stock.

There are also a number of important biological advantages. If some strains should prove to be more susceptible to the test chemical than others, then studies of why this should be so may well yield useful data for extrapolation to man (Festing, 1979c, d).

It appears, therefore, that if toxicologists were to transfer their allegiance to inbred strains rather than outbred stocks, and if they were to use a factorial experimental design in screening experiments, it would be possible to reduce the number of animals used in toxicological screening without loss of statistical precision, and with an increase in biological data that could be used in the interpretation of the experiment. Unfortunately, at this stage toxicologists are not prepared to make such a change. This is largely because screening with the use of outbred stocks is currently the accepted method, and no one can be criticized for using the same method as everyone else. Moreover, governmental regulatory authorities such as the Federal Drugs Administration in the USA and the Medicines Commission in the UK are not at this time particularly concerned about the type of animal used in toxicological screening, and are quite content to accept results based on outbred stocks without giving much thought to the advantages of inbred strains. Eventually, the use of improved experimental designs based on inbred strains will become acceptable to the toxicology industry, but in the meantime there is considerable inertia which will take several years and some persuasive arguments to overcome. In the meantime animals are being wasted on experiments which could be much better than they are at present.

INADEQUATE STATISTICAL ANALYSIS

According to Anderson and Bancroft (1952) 'statistics is the science and art of the development and application of the most effective methods of collecting, tabulating, and interpreting quantitative data in such a manner that the fallibility of conclusions and estimates may be assessed by means of inductive reasoning based on the mathematics of probability.' Sterling (1971) reviewed the data on the toxicity of the compound 2, 4, 5-T to experimental animals, and made the following devastating criticism:

'Very serious, in view of the wealth of available statistical and mathematical techniques, is the relative naivety with which these studies were analysed. In many of these reports, the investigators did not subject their data to any statistical analysis. Simple summary figures, such as the arithmetic average, were in most instances not accompanied by measures of dispersion. The most sophisticated statistical analyses were multiple applications of Student's t-tests, comparing sets of measurements from each individual group (resulting from combinations of doses and other factors) with a control group. The numbers so analysed often were discrete rather than continuous and came from obviously non-normal distributions. Those values of t that met the famous criterion of statistical significant at $P < .05$ were duly starred with an asterisk, thus adding insult to injury. (It is a matter of curiosity why not a single one of these experimenters subjected his data to the robust and generally available technique of analysis of variance, let alone saying anything about taking out the effect of confounding variables by suitable modifications of that technique.)'

A very similar situation was reported by Gore et al. (1977), who studied 62 reports that appeared in the British Medical Journal from January to March 1976. They found that 32 (i.e. more than half) had statistical errors of some kind and in 18 (29 per cent) of cases the errors were serious. The summaries of five of the reports made claims that were unsupportable on re-examination of the data. Much of the data referred to human studies, but there is no reason to assume that statistical analysis of animal experiments is carried out at higher level than human experiments.

One result of the low level of statistical expertise of many research workers is that they are unable to extract all the information present in their data, and they are therefore likely to repeat experiments which really do not need to be repeated. Obviously this wastes time, money, and animals. It is really quite extraordinary that the science of getting information out of data, i.e. statistics, is so poorly understood by so many otherwise highly skilled and intelligent scientists.

POSSIBLE REASONS FOR FAILURE OF SOME SCIENTISTS TO MAKE THE BEST USE OF ANIMALS

Failure of some of the biomedical research community to make optimum use of the animal resources currently available is probably due to a number of causes. These include inadequate training, conservatism, interdepartmental rivalry, and unrealistic financial accounting. On the other hand, there are some institutes where the care and use of laboratory animals is exemplary, leaving only minor scope for improvement.

Inadequate Training

Many research workers working with laboratory animals have no formal training in laboratory animal science and some have little training in experimental design, statistics, or even biology.

No British university teaches laboratory animal science, though one or two lectures on laboratory animals are given at some veterinary schools, and a few lectures are given in some postgraduate courses. Thus, the average research worker starting animal experimentation must teach himself the biology of the species he favours (and to make a rational choice he needs to know the biology of several species). He must also become familiar with the properties of inbred, outbred, coisogenic, cogenic, and mutant strains; the significance of many individual infectious agents, including bacteria, viruses, and higher parasites; animal husbandry, breeding methods, methods of handling animals, and a whole range of experimental skills, such as anaesthesia, collection of body fluids, surgical techniques, post-operative care, and the use of therapeutic agents.

In the absence of formal course work many research workers simply do not have an opportunity to become familiar with these concepts, and treat their animals as 'white rats'. The humane care of the animal is, of course, governed by the 1876 Act which is administered by the Home Office inspectors, and there is no evidence of failure to comply with the provisions of this Act. However, the scientific quality of much work could be improved if research workers were more familiar with the wide range of techniques and knowledge which comprises the subject of laboratory animal science.

Ignorance of the principles of the design and analysis of experiments also has serious implications. It is still possible for a research worker to be awarded a doctoral degree for work on animals without attendance at any formal courses on statistics and experimental design. Even where formal courses have been given, they are often taught by theoretical statisticians who often teach at too high a level and are unfamiliar with the practical problems that arise in the collection and analysis of biological data. The over-emphasis by statisticians on highly sophisticated statistical techniques requiring great mathematical rigour, and the neglect of practical problems and 'dirty' data, is even causing concern among professional statisticians (Juvancz, 1976).

Although inadequate statistical analysis of the data is often apparent in published papers, inadequate experimental design may be a more serious problem in that it is often not apparent from the published work, and may result in entirely misleading conclusions. Thus experimental and control animals may be housed in different rooms, assessed at different times, and assessments with a subjective component may not be carried out 'blind'. Some pathologists, for example, claim that it is impossible to judge whether or not a histological section is normal unless it is known whether the animal is a control or a treated subject. This is a violation of correct experimental design. Such practices can easily lead to apparent differences between experimental and control animals which have nothing to do with the actual experimental treatment.

All research workers using animals should have sufficient statistical training to design their own experiments, or at least know when to consult a professional statistician.

Conservatism

Although scientists are often quick to jump on a new bandwagon, they are notoriously slow to accept new ideas in peripheral scientific fields such as laboratory animal science. The very gradual acceptance of the value of inbred strains in certain research areas, as shown in Figure 13.1 (Festing, 1979a), is evidence of this. Thus it took 30 years for the proportion of genetically defined mice used in papers published in the journal *Cancer Research* to rise from about 40 per cent to about 80 per cent in spite of their obvious advantages. There are, of course strong pressures on scientists to conform to currently accepted practices if they wish to have their papers published in the best journals. Thus, if the majority of research workers in the field are using undefined animals and poor statistical analyses (as in toxicological research at present) the use of defined animals and better statistical techniques may not be understood, and may be resisted by referees and editors.

A research worker who has used a particular stock of animals for many years will often attach great value to the 'background information' on that stock and will go to great lengths to ensure that he can continue to work with the same stock in the future. However, if his stock is very different from those used by other people, then his past results have little generality. If the stock is not very different from other strains, then 'background data' from these other strains will be perfectly adequate in helping him to design future experiments. He would of course have been better off if he had determined the extent to which his results were strain-dependent periodically during the course of his studies by the occasional use of several strains.

Even the use of disease-free animals has been resisted with arguments based more on an emotional resistance to change rather than a rational assessment of the benefits of such animals.

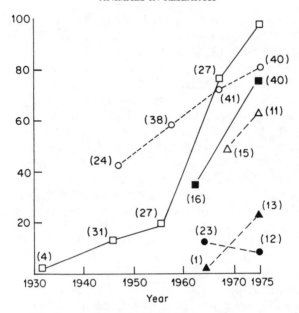

Figure 13.1 Use of genetically defined (inbred or mutant) mice in six journals. Vertical scale shows proportion of published papers using the mouse which used genetically defined mice. The actual number of papers is shown in parentheses. Note the general upward trend and the large variation between journals. ●, *Toxicology and Applied Phamacology;* ▲ , *Food and Cosmetics Toxicology;* △, *Physiology and Behaviour;* O, *Cancer Research;* ■, *Immunology;* □ , *Journal of Immunology.* (Redrawn from Festing (1978) by permission of Hans Huber, Publishers)

Lack of Interdepartmental Co-operation

In some large institutions, particularly universities, the different departments jealously guard the right to manage their section of the animal house as they want. This often involves bringing in animals of unknown health from outside suppliers. Unfortunately, disease does not respect departmental boundaries. Thus, in a shared animal house in which each department acts independently, the health status of the animals will be that of the worst. Animal houses must be under the control of a scientist with sufficient authority to ensure that it is managed in the best interests of all research workers, and animals.

Wherever possible different departments should collaborate in the management of the animal facilities, even if they are physically separated, so that they can afford to employ full-time animal curators and veterinary staff. A number of universities, including, for example, Oxford, Liverpool, Manchester, and

Cambridge, and a number of teaching hospitals, such as Charing Cross, The Royal Postgraduate Medical School, and the Royal Free Hospital have appointed such animal curators. The pharmaceutical industry has been a leader in this field for many years, possibly because it has to be commercially efficient and competitive—unlike universities and research institutes.

Unrealistic Financial Accounting

The quality of animals used in the pharmaceutical industry is usually very high. Expensive, but good quality animals are justified on the grounds that such animals give substantially better results than poorer quality ones, and the additional cost is insignificant in comparison with the total cost of research. For example, some years ago one contract research company calculated that the average investment in a rat on an 18-month toxicity trial was more than £ 65. The difference between a conventional and a disease-free (SPF) rat at that time was less than 50p, and SPF animals free of chronic respiratory disease are known to survive longer than conventional ones. Clearly, in these circumstances the SPF animal was a good investment.

In the universities, on the other hand, the higher cost of good quality animals means that fewer can be purchased from a fixed research budget which does not take account of all the costs involved in animal experimentation. Thus, the research worker is tempted to stretch his research funds by buying cheap animals from non-accredited sources, even though in strict accounting terms this could be totally uneconomical. For example, the financial advantage of using laboratory-bred dogs rather than conditioned pound (stray) dogs was emphasized by Fletcher et al. (1969) in a comparative study of dogs used in experimental open-heart surgery. They found that the 5-day survival in 79 laboratory-bred Labrador dogs was 93 per cent, compared with only 73 per cent in the pound dogs (only dogs of comparable body weight were used). They also calculated that each operation cost a total of $ 562, excluding the surgeon's time and the cost of the dog. Assuming that a new heart valve would require long-term evaluation in, say, 100 dogs, it would be necessary to carry out 136 operations with the pound dogs at a total cost of $ 76 432 or 107 operations on the laboratory-bred dogs at a total cost of $ 60 134. The saving of $ 16 298 through the use of the laboratory-bred dogs, amounting to $ 152 per operation, far more than paid for the difference in purchase price of the two types of animal. At that time, it was possible to breed a dog in the laboratory for only $ 80. Had the full cost of the operation, including the surgeon's time, also been taken into account, the value of the laboratory-bred dog would have been ever more overwhelming. Yet, research workers both in the USA and the UK continue to use these 'cheap' animals in the hope that it will preserve their research budgets.

Investment in the production and care of laboratory animals also varies widely. In general, the pharmaceutical industry considers the cost and quality of the animal in relation to total research costs, and invests heavily in producing high quality animals, because it is economically worthwhile. Universities are unable to quantify the value of research. Laboratory animals are much less glamorous than electron microscopes and computers, so lose out in the competition for funds. If the value of research papers could be quantified, the use of defined animals might be seen to be economically justified in universities as well as in the pharmaceutical industry.

CONCLUSIONS

(1) High quality, genetically and microbiologically defined animals may easily be produced with current technology. The more widespread use of such animals would give more valid scientific results and would reduce the number of laboratory animals that become sick from disease unrelated to the experimental treatment. Thus, on humanitarian and economic grounds it is highly desirable.

(2) The key to improving the quality of animals used in laboratories in the UK is probably better education of the animal users in all aspects of laboratory animal science, and the design and analysis of experiments. It is evident that such education is overdue.

(3) The education might overcome the conservatism of some scientists, and eventually would persuade the financial authorities that greater investment in good quality animals, facilities and staff is economically worthwhile.

(4) Education needs to be given at several levels. New Home Office licence holders might be required to attend 1- or 2-week short courses in laboratory animal science before being granted licences (this would not require any legislation under the present laws). Current licence holders should be encouraged to attend refresher courses, which could easily be organized at universities and technical colleges. Expert lecturers could be provided by universities and colleges through the various laboratory animal organizations and societies.

Education in experimental design and statistics presents more of a problem. These subjects should be compulsory for all undergraduate biologists and should be given strong emphasis in postgraduate education. No one should be awarded a research degree who has not attended formal courses on the design and statistical analysis of experiments, or has at least proved his competence in these subjects.

Finally, a postgraduate course in laboratory animal science should be established for those wishing to work full time with animals as animal house curators or animal specialists. Such courses are already well established in the USA.

REFERENCES

Anderson, R. L. and Bancroft, T. A. (1952). *Statistical theory in research*, McGraw-Hill Book Co., New York.

Baker, H. J., Cassell, G. H., and Lindsey, J. R. (1971). 'Research complications due to *Haemobartonella* and *Eperythrozoon* infection in experimental animals', *American Journal of Pathology*, **64**, 625–656.

Bleby, J. (1976). 'Disease-free (SPF) animals', in *The UFAW handbook on the care and management of laboratory animals*, 5th edn, Churchill Livingstone, Edinburgh.

Cawthorne, M. A. (1979). 'The use of animal models in the detection and evaluation of compounds for the treatment of obesity', in *Animal models of obesity* (M. F. W. Festing, ed.), Macmillan Press, London, pp. 79–90.

Eaton, G. J. (1972). 'Intestinal helminths in inbred strains of mice', *Laboratory Animal Science*, **22**, 850–853.

Elmes, P. C. and Bell, D. P. (1963). 'The effects of chlorine gas on the lungs of rats with spontaneous pulmonary disease', *Journal of Pathology and Bacteriology*, **86**, 317–326.

Festing, M. F. W. (1975). 'A case for using inbred strains of laboratory animals in evaluating the safety of drugs', *Food and Cosmetics Toxicology*, **13**, 369–375.

Festing, M. F. W. (1977). 'Bad animals mean bad science', *New Scientist*, **73**, 130–131.

Festing, M. F. W. (1978). 'Genetic variation and adaptation in laboratory animals, in *Das Tier in Experiment* (W. H. Weihe, ed.), Hans Huber, Bern, pp. 16–32.

Festing, M. F. W. (1979a). *Inbred strains in biomedical research*, Macmillan Press, London.

Festing, M. F. W. (ed.) (1979b). *Animal models of obesity*, Macmillan Press, London.

Festing, M. F. W. (1979c). *The value of inbred strains in toxicity testing*, Proceedings of an RSPCA symposium, London, 1978.

Festing, M. F. W. (1979d). 'Properties of inbred strains and outbred stocks, with special reference to toxicity testing', *Journal of Toxicology and Environmental Health*, **5**, 53–68.

Festing, M. F. W., May, D., Connors, T. A., Lovell, D., and Sparrow, S. (1978). 'An athymic nude mutation in the rat', *Nature, London*, **274**, 365–366.

Fletcher, S. W., Herr, R. H., and Rodgers, A. L. (1969). 'Survival of purebred Labrador retrievers versus pound dogs undergoing experimental heart valve replacement', *Laboratory Animal Care*, **19**, 506–508.

Gore, S. M., Jones, I. G., and Rytter, E. C. (1977). 'Misuse of statistical methods: critical assessment of articles in BMJ from January to March 1976', *British Medical Journal*, **i**, 85–87.

Grant, L., Hopkinson, P., Jennings, G., and Jenner, F. A. (1971). 'Period of adjustment of rats used for experimental studies', *Nature, London*, **232**, 135.

Grice, H. C., DaSilva, T., Stoltz, D. R., Munro, I. C., Clegg, D. J., and Abbatt, J. D. (1973). *The testing of chemicals for carcinogenicity, mutagenicity and teratogenicity*, Canada, Health and Welfare.

Grüneberg, H. (1952). *The genetics of the mouse*, 2nd edn, Nijhoff, The Hague.

Haseman, J. K. and Hoel, D. G. (1979). 'Statistical design of toxicity assays: role of genetic structure of test animal population', *Journal of Toxicology and Environmental Health*, **5**, 89–102.

Heston, W. E. (1975). 'Testing for possible effects of cedar wood shavings and diet on the occurrence of mammary gland tumours and hepatomas in C3H-A^{vy} and C3H-A^{vy}fB mice', *Journal of the National Cancer Institute*, **54**, 1011–1014.

Home Office (1978). *Statistics of experiments on living animals in Great Britain 1978, Cmnd 7628*, HMSO, London.

Juvancz, I. (1976). 'Is the Bradford Hill Principle still valid?', *Biometrics* **32**, 200–201.

LAC (1974). *The Accreditation and Recognition Schemes for Suppliers of Laboratory Animals*, Manual Series No. 1, Medical Research Council Laboratory Animals Centre, Woodmansterne Road, Carshalton, Surrey, UK.

Lindsey, J. R., Baker, H. J., Overcash, R. G., Cassell, G. H., and Hunt, C. E. (1971). 'Murine chronic respiratory disease', *American Journal of Pathology*, **64**, 675–716.

Sabine, J. R., Horton, B. J., and Wicks, M. B. (1973). 'Spontaneous tumours in C3H-A^{vy} and C3H-A^{vy}fB mice: high incidence in the United States and low incidence in Australia', *Journal of the National Cancer Institute*, **50**, 1237–1242.

Simpson, F. O., Phelan, E. L., Clark, D. W. J., Jones, D. R., Gresson, C. R., Lee, D. R., and Bird, D. L. (1973). 'Studies on the New Zealand strain of genetically hypertensive rats', *Clinical Science and Molecular Medicine*, **45**, 15s–21s.

Sterling, T. D. (1971). 'Difficulty of evaluating the toxicity and teratogenicity of 2, 4, 5-T from existing animal experiments', *Science, New York*, **174**, 1358–1359.

Strong, L. C. (1942). 'The origin of some inbred mice', *Cancer Research*, **2**, 531–539.

Townsend, G. H. (1969). 'The grading of commercially bred laboratory animals', *Veterinary Record*, **85**, 225–226.

Animals in Research
Edited by David Sperlinger
© 1981 John Wiley & Sons Ltd.

Chapter 14

The Fallacy of Animal Experimentation in Psychology

DON BANNISTER

INTRODUCTION

From its earliest days, modern psychology has accepted animal experimentation as a legitimate, respected, and substantial part of the discipline. Wilhelm Wundt, often cited as a founding father of scientific psychology, presented many of his early ideas under the title *Lectures on human and animal psychology* in 1863 and by the turn of the century Morgan in his *Introduction to comparative psychology* was effectively urging the introduction of experimental, as opposed to observational and anecdotal, methods into the field. For over half a century most academic institutions devoted to psychology have boasted an animal laboratory. The study of animal behaviour, with its own extensive range of journals and textbooks, has figured largely in the university psychology syllabus. The Norway rat probably outranks the American college student as the favourite object of study for psychologists. The single most influential theoretical framework in modern psychology, learning theory, used the navigational behaviour of the rat as its prime data and the rat, the cat, the pigeon, the monkey, the octopus, the lamb, the chicken, the cockroach, the amoeba, and a multitude of other species have been used as significant models of humankind. Countless animals have been surgically dismembered, drugged, starved, fatigued, frozen, electrically shocked, infected, cross-bred, maddened, and killed in the belief that their behaviour, closely observed, would cast light on the nature of humankind.

ASSUMPTIONS OF ANIMAL PSYCHOLOGY

Strangely, the theoretical assumptions underlying this vast undertaking have rarely been debated. The vast mass of animal work in psychology is published as if its fundamental value were unquestionable and only its detail needed to be examined. This may be, in part, due to a particular coincidence in the history of thought. Psychology's initial bid to establish itself as a science coincided with the growing dominance in biology and in social ideology of Darwinian theorizing. It seems that Darwin's central tenet, that higher forms of animal life have developed from more primitive forms, was taken as an unchallengeable justification for the development of work on animals as a central feature of modern psychology. Indeed psychology is frequently referred to as one of the 'biological' sciences, which seems to carry the implication that whatever holds fundamentally true for biology must be accepted as a basis for psychology.

This kind of approach was allied with, and strengthened by, two offshoots of positivism which seem to have dominated much of our thinking in this century. The first is the notion of reductionism which has promoted the idea that the grand strategy for science involves proceeding from the simple to the complex and correspondingly, in psychology, from what are argued to be simpler forms of behaviour (i.e. animal behaviour) to more complex forms (i.e. human behaviour). Thereby, we overlooked the counter-argument that 'simple' and 'complex' are statements about the nature of our thinking about phenomena, not statements about the phenomena themselves. A wooden stool is a simple object in terms of carpentry and an incredibly complex object when viewed by a physicist.

The second offspring of positivism, naive realism, a belief in the primacy of the touchable, has encouraged psychologists to flee from what are thought to be metaphysical concerns, such as 'mind' or 'person' and cling grimly to manipulating and observing what are deemed to be definable behaviours. Given beliefs such as these, what could be a more promisingly scientific basis for psychology than to manipulate and observe the speechless antics of the allegedly simple, i.e. animal experimentation?

PUBLIC DEFENCES OF ANIMAL EXPERIMENTATION

Public vindication of the enormous growth of comparative psychology is most often undertaken in the introductory chapters to standard text-books in the area. Such introductions usually present a brief and repetitive set of assumptions, so that almost any such textbook can be taken as representative. For the purposes of this essay the text by Waters *et al.* (1960) will be examined. Typically it opens with brisk confidence.

> 'The use of animals in psychological experimentation might well seem paradoxical to one not acquainted with the scope of contemporary psychology. "Why study animals?" is a question frequently asked of the psychologist. The question has a special significance for the comparative psychologist. It appears to raise the problem of the validity of his specific field of interest and study. Let us dispose of it at once.'

The text goes on to offer a number of justifications. Firstly it is stressed that the results of experimenting on animals yield information that can be applied to the training of animals as household pets, as circus performers, as hunting and working companions, and as more abundant sources of food. The text does not acknowledge that this is simply a justification for a *technology* of animal manipulation; it is no justification at all for animal experimentation as part of a *science* of psychology. The authors are proposing what is essentially an economic argument, not a scientific one. Long before psychology appeared as any kind of scientific discipline people had observed and manipulated animals to see how they could be more effectively hunted, trained, and bred. But to justify animal experimentation in terms of its economic pay-off is not to justify it as a part of a science of psychology, since it entirely begs the question of what psychology is about and how it is to be defined and in what terms it can be elaborated.

The text then goes on to justify the use of animals in the psychological laboratory on the grounds that we can do things to them which we would never consider it ethical to do to people. Animal experimentation is lauded as a solution for ethical problems. As the authors point out we can, by using animals, study problems which 'require the surgical removal of or injuries to the nervous system'; undertake investigations which 'call for the control of mating in the studies of heredity, for the application of intense and painful stimuli, or for prolonged subjection to hunger, thirst or other biological drives', and so forth. Here again, a question is begged. That we can, without personal distress, public condemnation, or legal penalty, experimentally inflict suffering on animals in no way proves that such activity helps to advance psychology as a science.

Finally, as part of an attempt to separate animal psychology from physiology, the text defines what it calls 'behaviour of interest to psychologists' and states that the characteristics of such behaviour are that it is: (i) in part autonomous, the product of conditions within the organism; (ii) persistent, tending to continue until some end goal is reached; (iii) variable, shifting, and changing until some appropriate way to the goal is discovered; (iv) docile or trainable. It is then alleged that these are the central characteristics of both animal and human behaviour and therefore psychology should and must encompass both. Ending the argument, as the text does, at this point, ignores the issue of whether there are characteristics of human behaviour which are not only additional to these but which cannot be properly encompassed or understood if we accept the four given characteristics as definitive. Human behaviour is reflective, as for example the behaviour of the authors of the text cited, in presenting arguments for their behaviour. Human behaviour is personal, in the sense that we distinguish our individual self from others and base our actions on the way in which we make this distinction. Human behaviour is mediated and negotiated through the enormous structure of language. Such characteristics are not only ignored by the text, they are *denied* if the argument is left at the quoted point.

ALTERNATIVE IMAGES OF PSYCHOLOGY

It seems fair, if we are going to question the assumptions of animal experimentation in psychology, that we make explicit our counter-assumptions. Clearly psychology *can* be defined (and has been defined by the authors quoted) in such a way that animal psychology is a legitimate part of it. Equally, psychology can be defined as being not centrally about 'behaviour' but as being centrally about 'experience'. Defined thus, psychologists require an access to experience which we cannot obtain by studying animals. Psychology can be defined in such a way that it accepts values and purposes as intrinsic to human undertakings, *including the undertaking of constructing psychology*, and an animal-based psychology precludes this acceptance.

The kind of animal experimentation practised in psychology *may* be seen as economically useful or as an extension of aspects of zoology. Equally the study of animals may provide stimulating *metaphors* for the study of humankind—as do Aesop's fables. But animal experimentation cannot be a valid content or method for a psychology which deliberately defines itself in terms of what we think *being human* is about.

It seems that the most powerfully *felt* argument for animal experimentation is rarely directly expressed. It may be something like the argument that *man is an animal* and that it follows, therefore, that the study of animals is a proper adjunct to the study of man. This does not seem a scientifically acceptable argument since sciences define themselves by what they *choose to see their phenomena as*. There is no *is* about it. A science is a language system and thereby each science not only studies phenomena but determines the 'nature' of the phenomena it studies. True, man is an animal and therefore zoologists who study animals may legitimately study man *as an animal*. Equally man is an organic system and therefore physiologists who study organic systems may study man *as an organic system*. Man is also, as the physicists see him (or anything else), *a mass of whirling electrons*. Equally, for an economist, man is seen *as an economic unit*. Since psychology, like any science, is an invention, not a discovery, we are at liberty to (and also obliged to) decide what we want to *see man as*. We do not have to see humankind as it is seen by other disciplines (cf. Bannister, 1970a). Psychologists are no more obliged to view man *as an animal* than they are obliged to mimic economists and see man *as an economic unit* or the physiologists who see man *as an organic system*. Indeed, we have created little besides a hybrid language and much confusion by hobnobbing with other disciplines and failing to meet the primary demand which inventing a science imposes—the demand for true novelty in thought.

INTELLECTUAL CONJURING TRICKS

If the assumptions underlying animal experimentation in psychology are, at best, highly questionable, then how does it come about that they have not frequently been questioned. It may be due to the fact that the massive stream of published work in the field of animal experimentation is presented in such a way that its assumptions are deeply inferred within it, are so implicit that they neither arouse, nor are they readily accessible to, questioning. To examine this possibility one volume of the *Journal of Comparative and Physiological Psychology* (1977, volume 91) was chosen at random and a selection of papers contained therein will be briefly commented on in terms of this suspicion that there are hidden assumptions.

Incidently, the combination title of the journal, 'comparative and physiological psychology', is of interest in that just as it is argued that animal psychology is essentially a paradoxical notion, similarly it can be argued that physiological psychology is a contradiction in terms (Bannister, 1968).

A cursory examination of the papers suggests that a large number of them sit so clearly within the boundaries of zoology and physiology that they raise no psychological issue and the reader is not tempted to question their psychological relevance simply because they do not claim to have any—except, that is, the whispered claim implied by their appearance in a psychological journal. Thus Phoenix (1977) presents a paper on 'Factors influencing sexual performance in male rhesus monkeys' which simply checks the effect of testerone on the sex performance of monkeys. It is not even hinted that similar effects would be found on the sexual performance of humans, nor indeed that the sexual performance of humans can be measured or regarded in the same way as that of monkeys. The issue is evaded, as in many of these papers, by simply not being mentioned. But the word 'sexual' is quietly left to carry its inevitable psychological overtones.

Many of the authors centre their work on what is clearly a *psychological* concept but are careful not to claim that their findings apply to humankind, or to say that their findings do *not* apply to humankind. Thereby challenge is evaded. Coover *et al.* (1977) present a paper on 'Conditioning decrease in plasma corticosterone level in rats by pairing stimuli with daily feedings' which shows that, in rats, corticosterone level is conditionable. True 'conditioning' is a psychological concept and we know much about how and what is conditionable in humankind, but here we are asked merely to contemplate the fact that corticosterone level in rats is conditionable and not to infer from that anything about either conditioning, as such, or the conditionability of such levels in humankind.

Another paper, which similarly restricts its message, but which uses an even more adventurous psychological concept, namely that of 'communication', is offered by Floody and Pfaff (1977) under the title 'Communication among hamsters by high frequency acoustic signals'. This paper seeks to demonstrate that the high frequency sounds emitted differentially by male and female hamsters serve a 'social function' between the two sexes. Again there is no suggestion that human beings do or do not communicate with each other sexually nor that they may or may not communicate by high frequency sounds—the parameters of the study are such that no statement is made beyond the range of a limited aspect of hamster behaviour. Yet again, phrases like 'social function' are inserted to carry psychological implications *which are not examined.*

A further form of question evading presentation involves demonstrating some pattern of animal behaviour which is so 'basic' that its equivalent in human behaviour is manifest. However, it is linked to a kind of question we would find it entirely unnecessary to ask in relation to human beings. An example is provided by Haskins (1977) in his paper on 'Effect of kitten vocalisations on maternal behaviour'. In this experiment he demonstrates that mother cats tend to 'investigate' their offspring when the kittens miaow. His conclusions are worth quoting verbatim since they demonstrate very clearly the ingenuous nature of the concealed implication.

'Nothing reported in this article is inconsistent with an evolutionary interpretation of the relation between kitten vocalisations and maternal behaviour . . . however, to claim that a behaviour was selected or is adaptive is certainly no explanation, and, from a psychologist's point of view, is even irrelevant. It remains the task of psychology to examine the stimulus–response relations—even, or especially, if the stimulus is the property of one organism and response a property of a second organism—and the situations under which these stimulus–response relations manifest themselves. The combination of experimental and observational methods used in this study reveal that kitten vocalisations were correlated with a variety of maternal behaviours and that kitten vocalisations reliably elicited approach and investigation of the sound source by mothers. Subsequent study may reveal similar effects for other infant-emitted cues. Taken together, such studies would yield a comprehensive understanding of the limited but important role of dyadic influences on maternal–infant behaviour.'

It is interesting to consider what kind of theorizing would lead us to set up experimental demonstrations of the fact that human mothers tend to investigate when their babies cry and where, intellectually, we would go from there.

A more sophisticated form of question begging in animal experimentation involves the use of some very broad psychological concept, the operational definition of this concept in terms of experimentally controlled animal behaviour, and a discussion of findings which extends them to humankind without further reference to the animal context. For example, Albert and Richmond (1977) conducted an experiment in which blocking agents were injected into the anterior septum of rats, the rats were subject to pushing and frustration and their degree of 'aggression' measured by the amount of responsive biting and so forth. The discussion following the description of the experiment drops all specific reference to rats and talks generally about 'aggression' in relation to highly defined aspects of the central nervous system, as if the meaning their experiment gives to 'aggression' will fairly transfer to its multiple connotations in human affairs. (This raises an interesting query: if I am showing 'aggression' in my comments on Albert and Richmond is my central nervous system undergoing the process they postulate?) Washburn (1978) comments cuttingly on this habit of presenting what amounts to a loose metaphor as if it were a tight scientific abstraction. He points out that

'Students of animal behavior feel free to use the behaviors of nonhuman species when making points about human behavior. For example, in a recent book, the chapter on human behavior cites the behaviors of many nonprimates to make important points. The possibility of atavistic behaviors in human beings is illustrated by a picture of a musk-ox in a defensive position. To show how peculiar this habit of proof really is, consider what the reaction would be if I sent to a zoological journal a paper on the musk-ox with defensive positions illustrated by the British squares at the Battle of Waterloo!'

Other studies in the field pivot their argument on a curious kind of inversion whereby a psychological concept and technique which commonly has human reference is used in such a way as to throw some light on animal behaviour. But the flow of information is strictly in that one direction and although such studies might be regarded as contributions to zoology they manifestly contribute nothing to psychology. An example from our chosen volume is the work of Kelly and Masterton (1977) who in their paper 'Auditory sensitivity of the albino rat' used a conditioning technique to test the hearing range of the animal. This does not tell us anything about 'conditioning' since it is simply using an already established technique of conditioning as a measuring device to discover something about the perceptual capacities of an animal. Another paper of an even more overtly restricted kind in the same journal is that of Daly (1977) on 'Some experimental tests of the functional significance of scent marking by gerbils'. Here the study achieves exactly what its title implies, in that it calls into question the hypothesis that scent marking by gerbils functions territorially and instead suggests that the primary targets are adult female gerbils. This is a useful enough thought if we are specifically interested in gerbil behaviour, but only by crediting the *psychologically* vast and vacuous statement 'man is an animal' can we believe that it will cast light on the psychology of humankind.

THE SEDUCTIVE ASPECT OF ANIMAL EXPERIMENTATION

If the very notion of 'animal' psychology is questionable and if its practice has been fraught with contradictions, then we are faced with the question of wherein lies its powerful attraction—why has it formed such a major part of modern psychology. One possible answer is that it is a way of avoiding the central issue of reflexivity (Kelly, 1955; Oliver and Landfield, 1963; Bannister 1970b). Psychology, unlike the natural sciences, needs to be reflexive—it must reflect upon itself. This demand can take many forms. Since psychological theorizing and experimentation is itself a form of human behaviour then a psychological theory needs to account for its own construction and the behaviour of the psychologist is not only a legitimate but a necessary part of his or her own subject matter. If we fail to meet the demand for reflexivity in psychology then we will fall into the paradox of making propositions which are negated by the very fact of their being made. The most common manifestation of such a paradox in psychology is the assertion that human behaviour is entirely 'caused' while presenting 'reasons' for this assertion. If we attempt to explain the behaviour of subjects on the grounds that they are being conditioned and if we are offering this as a universal explanation, we can be reasonably called to account for the conditioning which leads us to condition our subjects. Psychologists have not entirely managed to avoid the issue of reflexivity in much of their work. The studies of workers such as Orne (1962) and Rosenthal (1967) have sought to show that a psychological experiment is essentially a social situation in which people are interactively influencing people. Hence, there is a tendency for subjects to behave in the manner predicted by different experimenters even when the different experimenters are making different and contradictory predictions. 'Experimenters' implicitly communicate with 'subjects'. Interestingly enough this seems to apply even when the subjects are rats. However, such has been the eagerness of psychologists to avoid confronting the reflexivity issue, that they have treated this kind of finding as merely one more methodological problem to be somehow partialled out of the experimental situation. They have not pondered it as, perhaps, the heart of the matter.

Significantly, one of the most startling examples of a psychologist trapped in the paradox of reflexivity hinges on the issue of 'scientific' cruelty. Milgram (1965 *et seq.*) conducted a series of experiments in which his subjects were deceived into believing they were assisting a chief experimenter in studies of the effect of pain on behaviour. Thus they were placed 'in control' of an electric shock apparatus and stooge victims acted out the role of pain-battered subjects each time the 'assistants' 'administered' a shock. It was found that many ordinary people were prepared to inflict what they believed to be intense suffering given the sanction that they were assisting in a 'scientific experiment'. Clearly, whatever is said at debriefing, the 'assistants' were made conscious of their own capacity to inflict suffering and Milgram seems only confusedly aware that in thus painfully disabusing people of their faith in their own innate kindness he was himself being cruel in the name of science. He *himself* was unwittingly providing an additional proof that 'science' can loosen ethical restraints—he had failed from the beginning to see the reflexive implications of his own experimental undertaking.

THE POLITICS OF ANIMAL PSYCHOLOGY

Animal psychology offers its practitioners a domain within which they can personally and publicly avoid the issue of reflexivity in all its forms. It has served as an undercover way of introducing and maintaining mechanical models in psychology because it is easy to be mechanistic about animals but more difficult to be mechanistic about one's fellow human beings. Not surprisingly, the current wave of humanistic psychology which insists that we see human beings as purposive and creative and that psychology acknowledge human concerns, received much of its initial impetus from a rejection of 'rat' psychology.

Working within the socially remote world of animal experimentation has sheltered many psychologists from the kind of political questioning to which psychology is now rightly subject. There is a debate within modern psychology about the degree to which values are inevitably involved in its practices. For example, the debate, at its sharpest, has led to attacks on the industry of intelligence testing, the kinds of credence given to alleged racial differences in intelligence, and the way in which intelligence testing is used to categorize and control people within our educational system. The psychologist who stays rigidly within the animal laboratory is less open to attacks of this kind, though not necessarily less deserving of them.

If we look forward to developing debates within psychology then, already over the horizon, are issues arising from the feminist charge that historically we have created a masculine science of psychology—a psychology not only statistically and administratively dominated almost entirely by men but a psychology which has helped to enforce masculine stereotypes of women. Faced with such an issue, what will we gain from an animal-based psychology that defines sexual differences (and thereby implies sexual values) in pseudo-biological terms?

Part of the urge to evade the reflexivity issue stems from a desire to mimic, in a concretistic way, the natural sciences: to earn for ourselves the title of 'scientist' and be rewarded with the prestige attached to that title. Thus psychologists have sought the kind of 'precision' which they see as being the hallmark of science. We have sought the appearance of precision by hiding tenuous argument behind elaborate statistical design (quoting figures, whose conceptual referents are doubtful, to the third decimal place) and by using impressive instrumentation. However, so long as our experiments involve people, our peers, with their capacity to outwit and seduce us and to see beyond the experiment, then our precision is often set at nought. If we use speechless animals—our infinitely manipulable *property*—then we can achieve a kind of spurious precision: we can be precise within the confines of our experiment even though the wider implications of the experiment are enormously imprecise.

ETHICS AND LOGIC

Perhaps the greatest gift offered by animal experimentation to the hard-pressed psychologist has been that it enables him or her to maintain a sharp, trade union demarcation line between 'scientist' and 'subject'. So long as our subjects are human beings then we may be forced to face the fact that we are simply persons trying to understand persons. However systematic we attempt to be, we are all confined within the inevitable mysteries that lie between people. If we work on rats or cats or pigeons then they will not confuse us by talking our language nor will they be insulted by us nor tempt us into a human relationship with them. We can maintain a closed shop for scientists and sit safely on our side of what we believe to be the enormous gulf that lies between us and animals.

There is a growing recognition in psychology generally that we may have to abandon our simple mimicry of the natural scientist and recognize that we cannot usefully experiment *on* our subjects, we may have to experiment co-operatively *with* them. Experimenting *on* animals offers us a way of delaying the day of that recognition.

We can now begin to see that in psychology the ethical issues involved in experimenting on animals are not separate from the scientific issues. It is not simply being argued here that it is an unkind practice to experiment on animals and that at the same time *it happens to be* a not very clever practice. The unkindness and the foolishness stem from the same source, that is, from a particular notion of what 'being a scientist' is about in psychology. If psychologists continue to believe that a 'scientific' psychology must be 'objective' in the manipulative and non-reflexive sense of that term, then they will use those strategies which favour that kind of 'science'. Animal experimentation is such a strategy. It allows the psychologist to ride on the back of existing cultural and ethical permissions about, and gulfs between, species.

Picture the kind of psychologist who sits, notebook in hand, watching rats drown in a water-filled glass maze as they desperately strive to find the exit, thereby increasing his or her knowledge of 'learning under stress'. He or she is not simply personally indifferent to suffering but is trapped, as surely as the rat, within a total view of the nature of science and of his or her own nature as a scientist. Psychology has failed in that it has given such a person no psychological view of his or her own character or of the nature of his or her predicament.

If the arguments proposed here are even broadly tenable then the sad reflection must be that the countless animals who have died in psychological experiments have died not only cruelly, but in vain.

REFERENCES

Albert, D. J. and Richmond, S. E. (1977). 'Reactivity and aggression in the rat', *Journal of Comparative and Physiological Psychology*, **91**, 886–896.

Bannister, D. (1968). 'The myth of physiological psychology', *Bulletin of the British Psychological Society*, **21**, 229–231.

Bannister, D. (1970a). 'Comment on explanation and the concept of personality', in *Explanation in the behavioural sciences*, (R. Borger and F. Cioffi, eds), Cambridge University Press, Cambridge, pp. 411–418.

Bannister, D. (1970b). 'Science through the looking glass', in *Perspectives in personal construct theory* (D. Bannister, ed.), Academic Press, London, pp. 47–62.

Coover, G. D., Sutton, B. R., and Heybach, J. P. (1977). 'Conditioning decrease in plasma corticosterone level in rats by pairing stimuli with daily feedings', *Journal of Comparative and Physiological Psychology*, **91**, 716–726.

Daly, M. (1977). 'Some experimental tests of the functional significance of scent marking by gerbils', *Journal of Comparative and Physiological Psychology*, **91**, 1082–1094.

Floody, O. R. and Pfaff, T. W. (1977). 'Communication among hamsters by high frequency acoustic signals', *Journal of Comparative and Physiological Psychology*, **91**, 794–806.

Haskins, R. (1977). 'Effect of kitten vocalisation on maternal behaviour', *Journal of Comparative and Physiological Psychology*, **91**, 830–838.

Kelly, G. A. (1955). *The psychology of personal constructs*, Vols I and II, Norton, New York.

Kelly, J. P. and Masterton, B. (1977). 'Auditory sensitivity of the albino rat', *Journal of Comparative and Physiological Psychology*, **91**, 930–936.

Milgram, S. (1965). 'Some conditions of obedience and disobedience to authority', *Human Relations*, **18**, 57–76.

Oliver, W. D. and Landfield, A. W. (1962). 'Reflexivity: an unfaced issue in psychology', *Journal of Individual Psychology*, **18**, 114–124.

Orne, M. T. (1962). 'On the social psychology of the psychological experiment: with particular reference to demand characteristics and their implications', *American Psychologist*, **17**, 776–783.

Phoenix, C. H. (1977). 'Factors influencing sexual performance in male rhesus monkeys', *Journal of Comparative and Physiological Psychology*, **91**, 697–710.

Rosenthal, R. (1967). 'Covert communication in the psychological experiment', *Psychological Bulletin*, **67**, 356–367.

Washburn, S. L. (1978). 'Human behavior and the behavior of other animals', *American Psychologist*, **33**, 405–418.

Waters, R. H., Rethlingshafer, D. A. and Caldwell, W. E. (1960). *Principles of comparative psychology*, McGraw-Hill, New York.

Animals in Research
Edited by David Sperlinger
© 1981 John Wiley & Sons Ltd.

Chapter 15

Why Knowledge Matters

MARY MIDGLEY

I shall say nothing here about how we should deal with situations in which animals are killed for vital human interests, for life and limb, as in essential medical research. I shall concentrate instead on asking how we should value all that range of research which does *not* affect those vital interests. What sacrifices should be made for it? More generally, what sort of justification does knowledge itself, pursued for its own sake, provide for sacrificing anything, including animals? This topic may look like a soft option, but I think it has to be handled before the tougher and rarer direct conflicts can be approached. Extreme moral oppositions tend to produce a sense of unreality, and their difficulty often springs simply from their extremeness. A scientist who is asked suddenly to abandon his whole work may reasonably complain of culture shock, even if he does not write his critics off as ignorant and unrealistic. The notion that animals can have any serious claim against people's real needs is not at present clearly integrated into our morality. But *unnecessary* research is a problem which everybody already recognizes. Whatever we may guess to be its extent, some of it exists. In opposing it, zoophiles find themselves side by side with a wide spectrum of people who, for a variety of reasons, just don't like waste. And an increasing number of these people are now willing to consider wasted suffering as a form of waste which is rather specially objectionable. Since reform calls for wide agreement, this is the sort of area where it can most easily start. There are, too, very interesting questions, not often enough considered, about the various kinds of value of which knowledge itself has, and what it must be like if it is to justify research which doesn't claim to be useful. I shall not therefore question here the common principle that, where the serious interests of its own species are at stake, any creature may properly sacrifice the interests of other species. I shall simply ask what makes knowledge such a serious interest.

The question is, then, where does knowledge stand in the hierarchy of human values? Are there any limits to the price we ought to pay for it? Someone from another planet, glancing over our civilization, would see at once that we do prize it highly. Still, most of us would say that there are limits, that it must take its place among other values. What is that place?

Since we certainly do want to place it high, let us start by looking at the extreme position which George Steiner took in his Bronowski lecture called 'Has truth a future?' (Steiner, 1978). Steiner there celebrated the intense disinterested search for theoretical truth which is one characteristic of our culture. He distinguished this search from the mere prudent collecting of useful knowledge for practical convenience. Knowledge, he said, may not be useful at all, it may even be dangerous, but if we are really disinterested, that danger ought not to stop us pursuing it. He told the story of Thales, who, it is said, tumbled down a well because he did not mind his feet while he was star-gazing, and urged us to follow that example, to pursue the enquiries before us even if they lead, as he concedes they quite well may, to the destruction of the human race, because, as he puts it, 'the truth matters more than man'. Examples of what he takes to be such enquiries are genetic engineering and the politically explosive comparison of the IQs of different races.

Now part of this is true and important. Knowledge *is* an end to be pursued for its own sake. But it is not the only end; there are others. What could commit us to making *unlimited* sacrifices of all those other ends for knowledge? For example, we may ask, ought such sacrifices to be vicarious? Thales, after all, fell down the well himself. He did not push other people down in order to measure their falling-time or to see how long they continued to yell from the bottom.

Again, in Norse mythology, Odin gave his right eye for wisdom, but it was his own right eye. It is not heroic to sacrifice other people, even if we leave animals out of the picture. In the second place, even scientists who experiment on themselves have to avoid suicide if they are serious in their search, since there can be no knowledge if there is nobody left to own it. Thales did not *deliberately* fall down wells; if he had, his astronomical knowledge would have died with him. And if he had continued not to watch his feet after the first fall, he would have been a bad and careless astronomer. Similarly, in Steiner's more melodramatic case, if the human race goes—or even if civilization goes—the truth evidently goes too, so the suggestion that truth 'matters more' is an obscure one. There is no knowledge without a knower, and the notion of knowledge makes no sense in abstraction from the *kind* of knower to whom it is valuable. The knowledge which (say) Shakespeare or Tolstoy had of human life simply could not be bequeathed to beings which did not share human interests. They could not understand it, so for them it would not be knowledge at all. These truths would simply be lost. Unless we bring in God (who is not likely to back such suggestions) no human value can 'matter more than man'. They all matter because of their place in human life.

Now this relativity of knowledge to knowers has a profound effect on the notion of *disinterested* knowledge. Certainly the search for knowledge should be free from irrelevant inducements like ambition or cash. But it cannot be free from interest in the sense in which 'interest' is opposed to the boring and the pointless. Someone who incessantly counts the sand on the beach, and collects and weighs pebbles, and calculates the relative frequency of different shapes among them just for the hell of it is certainly 'pursuing knowledge for its own sake'. He passes Steiner's grandiloquent test for the scholar; 'his addiction is with the abstract, the inapplicable, the sovereignly useless'. But this addiction will not make him a scholar. Uselessness alone is not enough. The sort of knowledge which is worth pursuing is not just miscellaneous units of information. It is understanding. Real enquiry is highly selective. Obsession is often its servant, but never its master. It does not aim at collecting indiscriminately all the facts there are. (There are an infinite number, so if it did, the number still uncollected would never grow less.) It aims at making life more *intelligible* by finding explanatory structures which underlie and shape its apparent confusion. In a clear sense therefore enquiry cannot be, and should not try to be, totally disinterested. It has to be directed to some questions rather than others, and the ones it ought to choose are those centrally important to the human race, those required for understanding the things which we most need to understand. We have to find the central questions, and distinguish them from the trivial ones.

I have paid attention first to discussing the kind of value knowledge has, because that seems necessary before we can ask what other valuable things ought to be given up for it. Few people, perhaps, will want to sign up for Steiner's extreme and romantic vision of the final victory of knowledge, of the whole human race well lost for the solution to a few problems in genetic engineering or comparative intelligence-testing. But during both these debates I have heard people give defences quite as extreme of the right of scientists to pursue, quite unhindered and regardless of consequences, any enquiry which they happen to have taken up. They find this position plausible because they unthinkingly take for granted the view which leads Steiner to his crazy conclusion—namely, that all truths are of equal and incomparable value. When we are considering the cost of research, whether in money, in animal suffering, or in any other kind of resource, we tend to speak in the abstract about our aims. What justifies this sacrifice, we say, is Science, Discovery, Research, the Advancement of Knowledge. We oppose the particular price that must be paid directly to these large abstract values. How can the interest of a few rats—or even a few human deaths in epidemics—possibly matter when weighed against such sublime ends? But we need also to ask about the importance of the actual limited enquiry involved, about the centrality of *this* particular issue, and about whether this experiment is the best or only way of illuminating it. To an alarming extent, judgments about this are determined by habit, by the methods that have become familiar in recent research and by the tradition of the journals.

What then should our view be when a particular piece of research is in fact trivial?

Non-scientists may be surprised at this question; whatever may be said of the arts, they may say, surely scientific research is never trivial? Scientists will not be surprised at it. Every serious scientist knows that there is a great deal of trivial research going on, not his own, but other people's. This is not surprising. Experiments are trivial if they are designed to test hypotheses which are themselves trivial, or hypotheses which are important but whose truth or falsity is already sufficiently established, or if they make unwarrantable background assumptions which vitiate their method. They are also trivial if they are badly designed, if they will not prove what they are meant to prove, or if what they are meant to prove is itself something obscure and incoherent, an idea not properly worked out by its begetters. Avoiding all these disasters is very hard, and the skills needed for it are not prominent in the education of scientists.

There is, unfortunately, no single unifying entity called Science, which inevitably gains by all scientific work and whose gain is always transmitted to the human race. We have a real dilemma here. We hesitate to prune. We find it natural to think that, as Mill urged in his *Essay on Liberty*, every important and life-enhancing activity should be allowed to proliferate as widely and luxuriantly as possible. But there must be some limits. Moreover, activities do not always thrive on this treatment. The example of the US cancer research programme, which has had virtually unlimited funds, is not encouraging. There has been a great deal of waste and corruption, and the cost, naturally, is not only in terms of money:

'The General Accounting Office also found that staff, equipment and animals paid for by the National Cancer Institute were used on private contracts. Finally, Eppley bred far too many animals; of 84,300 bred during 1976, 50,015 were killed without any research use. Yet, until GAO stepped in, NCI was about to fund a substantial increase in Eppley breeding facilities. On top of all this, the GAO found that NCI hardly looked at Eppley's results.' (Report in *New Scientist*, 16 March 1978)

One could mention also the duplication of drugs by competing firms and the statutory tests for poisons. Examples in his own field will probably leap to the mind of any working scientist. All research produces knowledge, but a good deal of it does not seem worth producing.

Steiner's article is interesting and unusual because he has the courage of his convictions about this difficulty. He openly rejects the challenge to relate, to view science responsibly. He puts with childish clarity the childish position to which people who claim unlimited freedom for science are committed, but which as a rule they discreetly conceal. *No* justification is needed, he says, for any enquiry. People are right, he insists, to 'pursue the truth for its own sake . . . in a passionate autism, for no reason but its own, its beauty or that sharp edge of beauty which we call difficulty . . .'. Okay so far, if rather technicolour, but the same could be said of art, of exploration, of religion, of love, even of sport. Resources still have to be divided. What limit must each of these claims accept to make room for the rest? No limit, says Steiner; that is just his point. there are doors immediately in front of current research which are marked "too dangerous to open", which would, if we were to force them, open on chaos and inhumanity.' But we are still bound, he says, to open them, and, after this act of romantic suicide, to drown with a good conscience in the resulting chaos and inhumanity, cheered to the last by the sound of our own unhesitating applause. 'The conspicuous consumption of economic resources and personal existence on behalf of abstruse, useless truth could be our singular dignity . . . the best excuse there is for man.'

This luscious Bluebeard scenario, however, is a mere distraction, a fantasy which cannot possibly result from the idea of truth pursued disinterestedly, for its own sake. For its own sake, *all* truth is equally attractive—the number of pebbles or sand-grains on the beach, the history of the Greek alphabet, the archaeology of Ur, textual criticism of *Gammer Gurton's Needle*. In this perspective, any enquiry is as good as any other; there can be no urgent 'doors immediately in front of current research'. If we talk of such doors, we must already have made a selection in accordance with our own interests. Those interests supply the urgency. Steiner has of course done this, as anyone must, in accordance with his personal principle of selection. This leads him to concentrate on questions which are trendy, melodramatic, dangerous, and expensive. (He prefers physics to biology for reasons which are far from clear, but which seem to centre on its higher cost and its greater likelihood of blowing us all up.) This is a journalist's interest, not a scholar's; the principle of selection is a simple preference for disaster. But for all interests the same point holds; there must be some statable principle of selection. If the aim of enquiry really were just the impartial collection of facts, any facts would do, and we should stand helpless, like Buridan's Ass, before their multiplicity. We should be quite at a loss, not merely about how to divide resources among different activities, or different lines of enquiry, but about how, within any one enquiry, to decide where to go next, which issue to put foremost. Science would indeed be reduced to sand-counting and pebble-collecting.

The relation between my starting-point and the general theme of this book should by now be growing plain. It has two stages. (1) If science *were* just indiscriminate pebble-collecting—if it really required all facts equally, then the huge body of facts which can only be gathered by inflicting suffering would be as necessary to it as all the others—though of course no more so. And (2) if the demands of science really were absolute—if we had to meet its every demand—if we had no way of arbitrating between its claims and those of other elements in life, then it would be our duty to gather *all* those facts. But both these suppositions are simply confusions.

The first one—the indiscriminacy of science—makes no sense because, as I have pointed out, it would leave us with no principle of selection. It would turn science into a mere vast disordered memory-store, a mass of miscellaneous information, whereas the very notion of science centres on finding an intelligible system. The second—the supremacy of science—is entirely mysterious because the notion of an absolute claim needs arguing, and we are given no argument to show why science should have a walkover. Of course individuals *can* become single-mindedly obsessed with science to the point of sacrificing their own and other people's whole lives for it. But then, they can equally well become single-mindedly obsessed in this way by love or ambition or art, by religion, cello-playing, gambling, avarice, theft, constipation, spite, stamp-collecting, or football. Obsession itself is nothing noble, and certainly has no claim on public support. Nobody recommends mandatory grants for obsessives. We judge such involvements, not according to their intensity, but according to the importance of the activity concerned. This importance depends on their catering for a real serious human interest—not just an isolated impulse, but a need central to our nature.

Now curiosity certainly is such a need. But there are two aspects to curiosity. There is detailed curiosity, the wish to acquire particular bits of information, and universal curiosity, the wish to *understand* experience, to draw it together, trace its system, and see it as a whole. Both are essential to us, and they must work together. People capable only of the first are confused people and will make bad scientists. There is no necessary conflict between the two, nor between either of them and the rest of our nature. Local conflicts of course arise, but in principle the right way to deal with them is by more thinking, not by romantically emphasizing them and revelling in conflict for its own sake. Human resources must somehow be so divided as to give some sort of appropriate satisfaction to all people's various deep needs, and we do need—apart from survival—love, friendship, art, laughter, work, creativity, security, and many other things besides information and understanding. We have to arbitrate somehow among such needs, and our main tool for this arbitration is provided by the universal branch of our curiosity—our wish to understand life as a whole, to make sense of the world and of our own nature. The right order for serious enquiry is therefore, generally speaking, that the interests of the understanding should come first. Questions arise as we attempt to make sense of experience and are answered by detailed observation, rather than being picked quite at random in response to chance impulses of curiosity. Certainly this rule is not infallible; chance impulses may be inspired, hunches may be heaven-sent. But the rule must in general be right, simply because of the problem I have mentioned about pebble-collecting. The mass of experimentable questions surrounding us is literally infinite; uncharted masses of results trawled from it are unusable and will simply be wasted. The more clearly theories are worked out, the plainer the lines of investigation become, and the better the chance becomes of furthering both sides of our intellectual interests.

I am suggesting that the interests of science itself demand more emphasis on thought and less on action, that it would actually *pay* scientists to experiment less—to argue better, think harder, and express themselves more carefully, with a view to cutting down the need for experiment. A bonus resulting from this move would be a sharp decrease in the number of experiments which inflict suffering on animals, since more careful thought would show that these are specially likely to be unnecessary and misleading. There is, I shall suggest, a special factor likely to lead people who do not know quite what they are about to choose this sort of experiment, particularly in psychology.

Bonuses always arouse suspicion, but there is a real reason for this link between bad science and cruel experimentation, a reason connected with the nature of understanding. *If you pay the wrong sort of price, you tend to get the wrong sort of knowledge.* This will, I hope, emerge more clearly in the long run. My immediate point, however, rests on the distinction I have been making between important issues and trivial ones. Some issues, we all agree, are more important than others. And most of us would also agree that it is wrong to cause suffering for an entirely trivial issue. (For instance, in the case of the US cancer programme, it seems pretty uncontroversial that it is wrong to commission the performing of experiments on animals when you care so little about the results that you scarcely bother to look at them.) Justification must therefore rest on importance. But *an important issue is by definition a pervasive one.* It is not an isolated matter, it is something far-reaching which crops up in many contexts and has many widely varied effects. If this is so, it can be tested in many ways. So it is impossible that tests which involve inflicting suffering on animals are the only ones, and unlikely that they are the best ones, by which an important issue can be settled.

An obvious example of this is the series of isolation experiments on infant monkeys, carried out by Harry S. Harlow and his colleagues from 1961 onwards, which established the presence of strong and specific social tendencies in these babies, and showed how the frustration of those tendencies in solitude could permanently warp the creatures' nature and destroy their sanity. These experiments were originally of great interest because they played a large part in breaking the hold of unrealistic behaviourist theories, widely held throughout the social sciences, which attributed social development both in men and the higher animals entirely to conditioning. That was an important issue. The error that was exploded was a serious one, damaging both in theory and in practice. Did this automatically justify all that was done to the monkeys? To do so, it is not enough to show that the research proved its point. We need also to show that it was the best or only available way to prove it. But because of the very generality and importance of the point proved, it could not be the only way. Crude behaviourism was so bad a theory, so thoroughly at odds with experience, that there were countless other ways of refuting it. What seems to have been needed in the first place was an advance in understanding, a clear, logical argument to show the incoherence of the theory, a critique of its basic concepts. Chomsky provided this when he pointed out that the capacity for speech must have an innate basis. Beyond this, there was also a need to show how behaviourism conflicted with the ordinary observed facts of life. Anyone with experience of children or of other young animals could have done this. But social scientists did not readily listen to such people. Common observation had to be strengthened by thoroughly systematic studies of spontaneous behaviour, supplemented by non-brutal experiments where these were actually necessary. This was in fact done by many observers of human children, such as Eibl-Eibesfeldt, Bowlby, and Blurton-Jones. The skills needed to observe spontaneous behaviour methodically had already been worked out by Konrad Lorenz and his followers; the notion that such observation must be merely 'anecdotal' was already exploded, and ethologists already knew enough about the behaviour of young animals to make the points which Harlow and his collaborators made. This information could easily have been supplemented, where necessary, both from further studies in the wild and from less drastic studies of caged animals.

It may seem natural to object that this offered a less direct way to the truth than the deprivation experiments. But that depends on *which* truth you are after—the general truth about how social instincts work, both in animal and man, or the detailed truth-pebble, the truth about how rhesus monkeys react to certain selected forms of psychological torture—the large truth about the world, or the small truth about events in the laboratory. If the isolation experiments had provided a sharp, final, theoretical breakthrough, and had then at once given place to different and more subtle methods, suitable for studying more detailed questions about the social instincts whose existence they established, they might have been good strategy. But in fact, like all successful experiments, this series produced imitators rather than successors and developers. Isolation and deprivation became standard favoured techniques for the study of instinct, and have remained so. Instead of being abandoned, they spread and prosper, because they are simple to apply, and because in this direction as much as in any other there is always an infinite number of further truth-pebbles to be gathered. Each experiment always seems to need another to supplement it. Directness has *not* effected any saving in the cumulative mass of research. The only factor that at present seems likely to slow this self-perpetuating industry is the rising cost of monkeys. Even the prolific macaques are now becoming harder to get, while all the higher apes are already in danger of extinction. What view future researchers are expected to take when this material has ceased to be available for them is not too clear.

This is the kind of thing I mean by saying that the wrong sort of price gets you the wrong sort of knowledge. The social instincts, both in animals and man, are very complex, and need to be studied over the full range of their normal positive manifestations—that is, in normal development and under normal circumstances—rather than merely by concentration on the one negative issue of discovering how hard it is to stunt and destroy them. The point seems obvious. People studying pottery do not start by reaching for a hammer and seeing how hard it is to smash it. But there are still many people who cannot accept the alternative. They regard observation, however precise and methodical, as inevitably less 'scientific' than experiment. They also tend to think that experiments, to be rigorous, ought to be drastically simple, ought to exclude irrelevant factors, not just in the sense of allowing for them, but of actually taking place in their absence. Thus they can fail to see that the simple but profound field experiments of that most careful and scientific researcher, Tinbergen, are *more* rigorous than drastic laboratory experiments, expensively performed with 'clean' animals—that is, ones uncontaminated by normal life, and therefore already socially crippled.

What rigour demands is not simplicity but clear thinking. In physics and chemistry, the ideal of completely isolating subjects is suitable and easy to apply. With any living thing it gets harder, and by the time we reach social creatures it is virtually unworkable. At this point, the separation between various factors in the objects studied can only be partial; distinguishing them properly must be the work of the mind. Certainly this is troublesome, but then more interesting subject matter does tend to be more troublesome to study. The simpler method distorts and destroys elements in the very behaviour you are investigating. Scientific method in these areas, therefore, demands that we avoid the simple and drastic.

It is hard to see this because of a strong, but partly accidental, twist towards drastic methods in our scientific tradition. At the Renaissance, it was necessary to point out the value of directed experiment as against mere casual observation. The third possibility—methodical observation—was not then a factor to be considered; it had not been invented. And physiology, in particular, when pursued with crude instruments and before the development of anaesthetics, did seem to require drastic and ruthless methods. At this point, Descartes put forward the view that animals were actually unconscious automata. It is not clear how far he or the scientists who followed him literally believed this and how far they merely thought it necessary for the development of science to behave as if it were true, but either way, this began to be adopted as the proper attitude for their enquiries; its correctness was treated almost as an established fact. Along with it, however, there grew up also a peculiar dramatization of science as something hostile and predatory, wresting truth violently from a world which should be regarded as a conquered enemy. This drama was current in the seventeenth century and is still popular today. Thus Bacon, in asserting the need for experiment rather than casual observation, wrote that it is our business to 'put Nature to the question'. But by this he did not mean just that we need organized questioning to guide our experiments. The phrase 'put to the question' means 'put to the torture'. As an experienced Jacobean statesman, Bacon understood that phrase very well and used it advisedly. Now no doubt this dramatization, like other dramatizations, has its uses. It *is* sometimes necessary to nerve a scientist for persistent and dispassionate enquiry into something he is accustomed to leave unquestioned because he regards it as sacred. The metaphor of conquest and predation may achieve this. But it does so at the expense of forming a disastrous habit in him of despising his subject-matter and supposing that he has already fully sized it up, of assuming that there is nothing in it that he does not understand, nothing that he would do better to approach with awe, wonder, interest, and respect. He reckons he knows all about it. Now this is a barrier to enquiry. The reason why people were slow to discover Evolution is that they already knew all about the Creation and the reason they were slow to notice the complexity of their sexual instincts before Freud was that they already knew all about those. Contempt breeds apathy.

This seems very largely to have happened about animals. Although the naturalists (including Darwin) who observed them in natural conditions did treat them with respect, laboratory experimenters typically did not. And in spite of the central importance of Darwin, the identikit notion of 'a scientist' became increasingly that of an experimenter in a laboratory. Naturalists were patronized as amateurish and anecdotal. Only in the last few decades has this situation been rectified, and the enormous importance of information about spontaneous behaviour begun to be fully appreciated. A neat but by no means trivial example of this strange history emerges from a casual remark in Peter Crowcroft's fascinating book *Mice all over* (Crowcroft, 1966). Crowcroft there describes his careful pioneering factual study of the spontaneous behaviour of mice, a study made with a view to understanding how best to keep them from infesting granaries. When starting his work he looked, as a good zoologist should, for earlier material, and was pleased to find that an immense mass of work already existed on the behaviour of mice. When he began to read it, however, he found it was virtually useless to him, since nearly all of it dealt (as he flatly says) with the behaviour of mice under torture. This is not sentimental or emotive language; the description is factual and is necessary to make his point. Nobody had bothered to study the normal behaviour of mice before setting about distorting it by electric shocks and similar methods familiar to those ingenious fellows who, in various countries, work in the cellars of the secret police. The effect of the distortion can therefore never have been properly understood. Much of the immense mass of experimentation on the behaviour of mice, rats, and other rodents undertaken a few decades back on the assumption that they had no nature of their own, and were therefore adequate stand-ins for human beings, was simply work down the drain. Ethologists have since forced some attention to these issues. Yet the Baconian prejudice remains. There are many researchers who still do not feel that what goes on outside a laboratory can be science, or that what goes on inside a laboratory is quite conclusive unless it involves serious stress and injury. There is something dreadfully conclusive about a ruined monkey. Drama *is* convincing. Yet to be convinced only by drama is not at all the proper response of scientist; undramatic truths can be just as important or more so. On the issue of drama, science really should be neutral.

Neutrality about drama is indeed an essential part of the notion of science. This emerges well from a comment Darwin made in a letter about animal experiments. Vivisection, wrote Darwin in much distress, is justified if it is done for the sake of science, but never for 'damnable and detestable curiosity'. What is the difference? Curiosity of some kind, after all, *is* the proper motive for science; what marks off the damnable and detestable kind? The simple and natural answer seems to be, its limited object, and its bias towards drama. Curiosity about pain and destruction *for their own sake*—rather than as aspects in some larger topic—is identical with cruelty. Children pulling flies to pieces are genuinely curious; they really do want to know what will happen next, and they are in a way quite disinterested. It is the topic that makes their curiosity illicit, and there are plenty of other examples of this. I may be genuinely curious about your private life. I may really want quite badly to read your letters, listen to your conversations, test your pain threshold, and find out how you react to simulated bad news, and this simply for the sake of it, without expecting any advantage. But it is my business to control this feeling. Curiosity in itself gives no sort of general licence for action, and the expectation of excitement makes it worse, not better.

This point will probably look surprising today, not only to scientists but to academics generally, because the thrust of most public debate on this question is to distinguish pure from applied research and to exalt the pure kind. Against powerful commercial and political pressures, intellectuals have quite rightly and repeatedly insisted that knowledge has direct value as an end, not just as a means. They may well be inclined to think it follows that we ought to pursue every kind of knowledge. The examples I have just given should make it plain that it does not. Parallel cases will show that this is not at all surprising. Pleasure, peace, and fulfilment too have, equally with knowledge, their value as ends in themselves. But the man who finds pleasure, peace, and fulfilment either in interminably counting pebbles or in working as a torturer has chosen badly. Quite apart from enquiries which are politically dangerous, like those mentioned earlier, there are plenty which are genuinely trivial and valueless, and—still more remarkably—others which are intrinsically iniquitous *simply from the topic*. A professional torturer, for instance, may (though his employers usually are not) be motivated simply by disinterested curiosity. He may actually acquire a great deal of physiological and psychological knowledge about strains and endurances, and an intellectual interest in these may really be the main source of his job satisfaction. Since, however, curiosity here runs counter to every other value we recognize, we condemn his way of life completely. And this condemnation is not based merely on his being practically dangerous; it protests against his curiosity as such, against the direction of his attention. This becomes clear if he grows arthritic and has to retire, but is still allowed to frequent his old workshop as a spectator on condition of remaining quiet. (He might even have been in this position in the first place.) If we think of him as passing on his hard-won knowledge to others, we may be inclined to hesitate, and to wonder whether that vindicates his investigations. But actually this is only one of many cases where good accidentally comes out of evil, and it can give no justification. The part of his knowledge which it is proper for him to pass on could have been acquired in other ways and would have been if that had been his intention. The other part, which is simply knowledge about the best ways to torture, constitutes that rather surprising thing, knowledge which nobody ought ever to have had or wanted in the first place. People are often inclined to think that we need this knowledge in order to *oppose* torturers. This is a mistake, and one which has been ruinous to many such opponents. (The CIA (see Marks, 1979) were not alone in making it.) Very few details of the procedures involved are needed, and these can all be collected from the victims. To acquire more, it is necessary to put oneself at the operators' point of view, and thereby to adopt their moral position.

I mention this extreme, but by no means isolated, case of 'bad knowledge' simply to complete the argument which, in its more familiar stretches, deals mostly with triviality and worthlessness. Researchers usually meet this charge by some variation of the 'spin-off' argument, pointing out—justly—that discoveries of real practical and theoretical value have often resulted by chance from enquiries which did not in themselves look at all important. There is much in this, but the trouble is that it proves too much. If what we are talking of is not a hunch about some real, specific, possible application, but pure, unadulterated, blind luck, then it might hit us in the course of any enquiry whatever, and it seems to follow that no research project should ever be rejected or abandoned. Everything must be investigated. We might simply draw lots for laboratory space, or concentrate on pebble-counting, or (alternatively) do as I have been suggesting and favour particularly projects whose conceptual relevance is fully and carefully argued. If we are really gambling, our expectations are no less in one case than in the other. But of course we are not just gambling. In the hot competition which reigns between projects, there have to be priority systems and standards of choice. And it is the principles of these which we are now discussing.

The spin-off argument cannot excuse research which does not even pretend to have a point. Scientists sometimes like to boast that their work is useless, and if they only mean that it has no practical application, this can be quite in order. If it means lack of theoretical application, it cannot. Long shots are legitimate; shots quite at random are not. The opening and closing sections of scientific papers, in which the importance of the work is discussed, ought to be extremely carefully thought out and extremely rigorously criticized. This sounds uncontroversial. But in practice the standard is often amazingly low. A remarkable, but by no means exceptional example is Suomi and Harlow's article on 'Depressive behaviour in young monkeys subjected to vertical chamber confinement' (Suomi and Harlow, 1972). This describes the isolation of infant monkeys in what the authors call 'the well of despair'—that is, a vertical stainless steel chamber, in which the monkey is left entirely alone for 45 days. Just what theoretical problem the experimenters were trying to solve when they designed this particular apparatus never clearly emerges. They speak initially of 'the implications of these findings for the production *and study* of depressive behaviour in monkeys', but throughout they write as if production rather than study were their central business, as though they were primarily technicians designing apparatus to produce something already agreed to be obviously desirable:

'These (earlier) findings indicated that the vertical chamber apparatus had potential for the production of depressive-like behaviours, and the following study was performed to further investigate the chamber's effectiveness in production of psychopathology . . .'

Not surprisingly, this treatment reduces the monkeys to a state of incurable social paralysis much deeper even than that found in controls who had merely been isolated in wire cages. When released, they show symptoms such as 'increases in self-clasp and huddle, decreases in locomotion and exploration, and a paucity of activity directed towards peers' which are also found in human infants who have lost a parent or parent-substitute. This for some reason surprises the experimenters, who comment, 'it is intriguing that a non-social manipulation can apparently produce behavioural components paralleling those resulting from a manipulation clearly social in nature'. Since these are simply typical, general responses to misery in solitary young primates, and ordinary symptoms of regression, one would like to know what intrigues them. They are not, however, sufficiently intrigued to explain the presuppositions which make this behaviour seem surprising, or to try to devise others which might make it understandable. They do not discuss the relation between an animal's social instincts and the rest of its nature at all. If the point of the experiment is to distinguish between social and other kinds of deprivation, it should surely proceed by putting members of one group in a full natural environment—in woodland—but singly, without company, and those of the other in monotonous confinement, but together. Now that they are intrigued, if not before, you might expect the experimenters to do this, but their conclusion is far simpler—just more of the same. I quote their last paragraph:

'Clearly, chamber confinement early in life rapidly and effectively produces profound and persistent deficits of a depressive nature in young monkeys. Whether this capability can be traced specifically to variables such as chamber size, duration of confinement, age at time of confinement, prior and/or subsequent social environment . . . remains the subject of further research.'

Anyone who is inclined to think unguided, spontaneous curiosity the best guide to the choice of a research topic might like to ponder its tendency to roll instantly, like this, down the groove provided for it.

I should make plain that I am not at the moment criticizing the brutality, but the futility of this experiment. It is hard to make clear one's *bona fides* in doing this without quoting far more of the work than I have room for. An account which looks like the one I have just given can of course be misleading; one can make serious research sound trivial to outsiders by omitting the context. The unfortunate thing is that a converse phenomenon sets in; people unfamiliar with a field tend to assume that there must be a serious context giving sense to apparently pointless work, or it would not be going on. I can only strongly recommend that readers interested in this question look at the articles. (References to the others can be found in the one I have cited.) I must add that when I did so myself, I was dumbstruck not to find better arguments. I had always assumed that a deeper structure of thought must be present, that no scientist could show so little interest in what he was destroying, or in the background of possible conceptual schemes on which he appeared to be commenting. The emphasis is always placed—just as it used to be with mice and rats—on the changes which can be produced in the laboratory. But what happens in a laboratory can only matter because of the specific light it throws on what happens outside. If one asks where that light is to be thrown, the answer clearly is never 'on the general nature and behaviour of monkeys'. This background is not considered. No question thrown up and left unanswered by the many, careful observers of full-scale, natural monkey behaviour is discussed, reformulated, and used in designing the experiments. The very interesting observations which people like Jane Goodall have made of damaged and defective young primates in the wild are not used. Harlow and his colleagues do not seem to think either that they need the thorough grounding, the long observation of normal behaviour, which any ethologist would think necessary for the interpretation of the abnormal. Instead, they show extraordinary confidence in the shortest possible observation periods. They observe their animals for only '*two five-minute periods, five days a week*' and their controls for only one such five-minute period. As far as interest in monkeys is concerned, they seem dominated by just one single question—what is the quickest way to drive them mad? Though some theoretical questions are certainly involved in solving this puzzle, it is hard to see obsession with it as an example of disinterested theoretical curiosity at any level.

This attitude can of course to some extent be explained by the fact that their real interest is not in monkeys at all, but in people. As they say in the article under discussion,

'The separation paradigm has been utilized to develop a monkey model of human anaclitic depression, since both the precipitating situation and the resulting behaviours are seemingly identical for monkey and human infants alike.'

The metal chamber is described simply as 'an alternative approach' to this problem, without explanation of what is alternative about it, or of why an alternative is needed. To 'develop a model' should mean, one would think, to examine rigorously what the points of likeness and unlikeness are and what they mean; further experiments ought to be designed to clear up doubts arising in this process. Instead, they say merely,

'Depression in humans has been characterized as embodying a state of "helplessness and hopelessness, sunken in a well of despair" (Schmale 1970), and the device was designed on an intuitive basis to reproduce such a well both physically and psychologically for monkey subjects.'

But why is *this* the best way of studying human depression? Psychologists, after all, do not usually rush about constructing physical models to reproduce in concrete form every interesting metaphor that is used in describing their subject. The word 'intuition' in the slack, casual sense which it evidently has here means just 'vague, unmethodical thought'. But what we need here is something extremely methodical. It is all right to introduce one's *solution* as guesswork, but not one's problem. There should be a sharp, positive account of the alternative ways of thinking about depression which emerge from studying the human scene, and of how these have been pursued so far by normal, obvious direct methods to the point where something needs settling which cannot be settled in this way; then, finally, of why the metal chamber device is the appropriate way to settle it. Some such specific question is needed because the experiment itself is so specific. It is not possible to treat it as just a general, exploratory procedure which will reveal a range of possibilities to be investigated—and again, if it were such a procedure, the need to do just that ought to be explained. On the other hand, it is not (in spite of a certain superficial resemblance) really at all like the medical experiments in which a particular physical disease is deliberately induced, either so as to verify a hypothesis about what causes it or to test a cure. There is no suggestion that human depression is caused by sitting in metal wells, and no cure is attempted.

More generally—neither depression nor psychopathology is a clearly identifiable disease, like measles, which is essentially the same in any species which it attacks. Both are wide ranges of conditions. Both are identified in human life in various ways, among which figure many references to subjective factors and to numerous other aspects of human sociality. To cast light on them in anything but the vaguest way by comparisons with another species, one would need the fullest possible background of comparative psychology in all the relevant respects. Without this, only the broadest and grossest analogies can be drawn. Since there actually is a considerable likeness between the nervous and emotional systems of all primates, even these analogies do throw up striking parallels—striking, that is, to people who have not before considered the probability that they would occur. Rhesus monkeys, like ourselves, are fully equipped, nervously and emotionally, for the arduous and complex business of responding to parental care and affection—for the long, helpless infancy which must be fully used to develop the intelligence of so advanced and sensitive a creature. Crashing into this subtle and delicate structure with a sledge-hammer, Harlow and his colleagues do indeed reveal, in passing, aspects of it which are genuinely important, but none which could not have been better understood by other methods. Destroying a creature emotionally can never be the best way to understand its nature.

At this point the question of triviality leads us back to the moral question, and does so quite properly because, as I suggested at the start, if the charge of triviality cannot be met, our quite ordinary, unpretentious morality finds the infliction of suffering objectionable. Anybody who treated monkeys like this outside a laboratory would go to prison for it. If it is different when psychologists do it, that difference must rest on the real, serious contribution which they make to knowledge. But the argument that there is such a serious contribution rests, in cases like this, on the experimental animal's being extremely close to man in the quality of his suffering. Thus, the existence of the species barrier confronts experimenters like Harlow with only two clear alternatives:

(1) Human beings and rhesus monkeys are indeed very closely comparable emotionally. In this case his results, though slight, may have some validity for human beings, and he is guilty of cruelty so enormous that hardly any theoretical advance could justify it.

(2) Human beings and rhesus monkeys are not closely comparable emotionally. In this case he may be guilty only of callousness (their peculiar capacity for suffering would remain to be investigated) but is convicted of enormous and wasteful intellectual confusion, and his results are void.

That the truth may lie somewhere between, combining both evils, does not mend the matter. The fact that some useful conclusions have been drawn from these experiments in such fields as autism counts, as far as it goes, in favour of alternative 1. But, as my earlier discussion of spin-off showed, this kind of luck is no vindication of badly designed experiments.

To sum up—I have not attempted to find a completely right course in the difficult area of animal experimentation. I have no confidence in attempts of this sort. Instead, I have tried to do what I think we must do about all real moral problems—to point out certain things which are plainly *wrong*, and to suggest principles on which we might avoid them. I am looking for ways of eliminating what is actually disgraceful, and so moving from the very bad, past the merely bad, to the morally tolerable.

In this process, it seems to me that we do start with something plainly wrong, namely, experiments which cause suffering and do no good to anybody. People have defended even these, however, on the principle that all knowledge is valuable, and sometimes on the even stronger principle that it is infinitely valuable, valuable beyond comparison. This strong principle is, I have tried to show, absurd and indefensible. The milder principle—that all knowledge has some value—I accept. But I add that this kind of value, like any other kind, has to work for its living, to show reason why it should prevail over the other elements in human life. In this competition, mere casual curiosity about isolated facts should, I suggest, rate quite low. It has nothing sacred about it. By contrast, the desire for *understanding*—for a better general map of the universe, a fuller and deeper grasp of its workings, does seem to me really valuable, and in fact one of the most valuable things in life. Philosophers like Plato, Aristotle, and Spinoza were not being folish when they spoke of contemplation as the central goal of life. They may sometimes have undervalued other elements, but the point in itself is sound. The contemporary exaltation of science contains, along with many meaner elements, a real recognition of this fact, and we cannot simply dismiss it as an error.

It follows that (apart from useful goals like medicine) research becomes more important the greater its relevance to our general understanding and the wider the consequences that can be drawn from it. Of course this relevance is not always easy to estimate, but the job can be attempted, and this would eliminate disgracefully trivial experiments. It seems a reasonable requirement that the use of animals should be treated as scarce and expensive resources already are, as something available only for research of proved importance. It ought not to be, what it is at present, the normal fodder of every project that can get itself accepted in a laboratory at all, the universal stuff of doctoral theses. To prove this importance ought not to be hard because, as I have suggested, important issues are pervasive ones, and must inevitably have other manifestations.

If this were done—if only important theoretical issues were being investigated—would human beings thereby be entirely justified in sacrificing other creatures, who don't stand to profit by knowledge, to their purely human aim? I understand the position of those who deny this, but I don't think I agree with them.

I do feel quite differently about experiments like Tinbergen's from about something like the Harlow series. Tinbergen's enquiry is always an important one, and he experiments only where it is actually necessary, when observation and thought have been fully used. He then devises his experiment so as to get the maximum of useful information with the minimum disturbance to the animals. Moreover, since he understands their normal way of life, he can keep this disturbance well in the range that falls within it. Many herring gull chicks and eggs get lost and eaten in any case, and overpopulation leads to cannibalism. By contrast, what happens to the rhesus monkeys in the deprivation experiments is a fate of quite peculiar horror, which is entirely a human artefact.

I am inclined to suggest provisionally that a core of really well-thought out experimenting of Tinbergen's kind may fall into the same category as Crowcroft's anti-mouse devices—that is, it may be regarded as part of the unavoidable clash which continually arises between the essential interests of different species. This involves counting contemplation or understanding of the physical world as an essential interest of the human species. I think that this is in fact the sort of consideration on which serious scientists rely. It may be wrong, but it deserves to be clearly stated and defended.

This attitude, however, would be bound to narrow the existing area of experiment a great deal. Understanding, unlike casual curiosity, involves reverence. The last thing which observers like Tinbergen and Lorenz want is to destroy by drastic interference the complex wholes which it is their business to study. Observation, if it is to be fruitful, needs to be restrained. We owe the understanding which we are now at last getting of the life of primates to the endless patience and courage of observers like Jane Goodall and George

Schaller, who have overcome the powerful European urge to begin study of these creatures by shooting them or putting them in solitary confinement. Seen on their own ground, rather than in zoos or laboratories, the great apes have, as all these observers report, a formidable dignity. This is true of other animals as well. Observers can take them seriously as experimenters cannot. The sharp division of the world into *us*—the experimenters—and *them*—the experimental material—then seems less compelling. It becomes possible, for the first time in human history, to take seriously our continuity with other species without finding it degrading, and thereby to understand the dignity of our own animal nature.

REFERENCES

Crowcroft, P. (1966). *Mice all over*, Foulis, London.

Marks, J. (1979). *The search for the Manchurian candidate*, Allen Lane, London.

Schmale, A. (1970). 'The role of depression in health and disease', paper presented at meeting of the American Association for the Advancement of Science, Chicago, December 1970.

Steiner, G. (1978). 'Has truth a future?', *The Listener*, **99**, 42–46.

Suomi, S. and Harlow, H. S. (1972). 'Depressive behavior in young monkeys subjected to vertical chamber confinement', *Journal of Comparative and Physiological Psychology*, **80**, 11–18.

Tinbergen, N. (1953). *The herring gull's world*, Collins, London.

Animals in Research
Edited by David Sperlinger
© 1981 John Wiley & Sons Ltd.

Chapter 16

Experimenting on Animals:
A Problem in Ethics

CORA DIAMOND

INTRODUCTION

My aim in this chapter is to clarify our present disputes about experimentation on animals. Someone from Mars trying to make sense of it all—the televised debates, the speeches at public meetings and addresses to learned societies, the demonstrations, the lobbying, the books and articles—might at first take it that on one side there were defenders of the freedom of investigators, and on the other, the defenders of animals. That would be one way of dividing up the combatants, but it is oversimple, and I want to look at the dispute differently. I want to represent it as having two 'sides', but each of the two sides is meant to include a range of views.

THE TWO SIDES OF THE DISPUTE

I can represent one side by using this diagram:

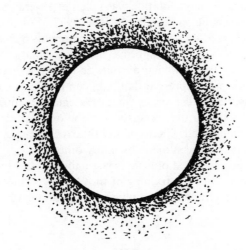

The circle represents the limits of the justifiable use of animals in experiments. Outside the circle are to be found various sorts of unjustifiable modes of treatment of animals. One is positively sadistic treatment, and another would be wanton indifference to the animals' welfare. Also outside the realm of the justifiable, on this view, would be the handling of animals by those without real technical competence, and, in general, neglect due to ignorance or inadequate training.

The side of the dispute which I want to characterize using the diagram can be put this way: So long as a scientist's use of animals remains *within the circle*, there is no real basis for criticism from the point of view of ethics. On this view, then, there are certain minimal standards which animal experimentation has to meet to be justifiable, but once those standards are met, there is no further need for the scientist to try, for example, to minimize his use of animals. His own professional judgement is to be relied on here as it is in his decision to use any piece of complicated and expensive equipment. Sir John Eccles (1971) has stated such a view very clearly. He contrasts the conditions which must be met if we are to carry out experiments on human beings with those which must be met if an experiment on animals is to be justified. In the case of experiments on human beings, we must consider the relation between risks to the subject and the possible benefits to the subject and others; we must be certain the importance of the objective is in proportion to the risk to the subject. But such considerations have no place at all, he claims, in connection with experiments on animals. We must indeed take precautions about pain (presumably with the qualification: so far as is consistent with the purposes of the investigation), but that is all we have to do: beyond that we can do what we wish on the animal. Thus Eccles's example: the experimental implantation of electrodes in the brains of human beings is unethical because it is destructive and not at all therapeutic, but exactly the same sort of investigation carried out on primates is highly desirable. It involves no pain, the animals can be experimented on, if well looked after, for weeks or months, and there should be far more of such experimentation. A similar view of animal experimentation was taken by several people in the 1977 BBC television programme on the subject (BBC 1977). For example, the Home Office spokesman claimed that callousness in experiments on animals is unlikely because one's colleagues would disapprove. The idea is that pressure from one's peers would keep one from moving off outside the circle. Again, in the BBC programme, the view was expressed that pain can muck up your experimental results, so experimenters will naturally want to safeguard against it as far as they can. Here we have the idea that departures in the outward direction from the circle will result in diminished effectiveness of the experiment as well as increased costs; the same point has been made by many others.

The conception just sketched of what is justifiable in experiments on animals goes with a certain view of the training of those engaged in such experiments. They ought to know the fundamental physiological and psychological needs of the kinds of animals they will work with, and the ways in which various kinds of stress, discomfort, pain, and so on can arise and how they can be prevented, just as they ought to know the proper use and care of any kind of equipment they will use. But there is no idea here that they ought to be trained in such things as how to avoid using animals. An animal on this view is fundamentally a delicate and expensive piece of equipment, and apart from any economies there might be in replacing it with something else, there is no particular reason to think one ought to use, or try so far as possible to use, something else instead. *Which* instrument to use is fundamentally a matter for scientific and economic considerations to decide, assuming one stays within the circle—and in any case there are scientific and economic pressures to stay within it.

People who accept this view of animal experimentation (which I shall call the First View) may nevertheless be divided on whether there is any need for change in the laws covering research (or for change in social policy more generally—including here the organization of medical, veterinary, or scientific education). Some may be reformers. For example, they may think that the current legislation is outdated and inadequate to ensure against abuses, or that something should be done to ensure that the training of those who will deal with animals is not wholly inadequate: 'a very delicate piece of machinery needs *specialists* to look after it' (BBC 1977). Others, though, who accept the First View may feel that any attempt to strengthen regulations dealing with animal experimentation is the thin end of the anti-vivisectionist wedge. John Dewey was *still*—in 1975—being quoted with approval on this:

'The point at issue on the subjection of animal experimentation to special supervision and legislation is thus deeper than at first sight appears. In principle it involves the revival of the animosity to discovery and to the application to life of the fruits of discovery which, upon the whole, has been the chief foe of human progress. It behooves every thoughtful individual to be constantly on the alert against every revival of this spirit, in whatever guise it presents itself.' (Visscher, 1975, pp 78–79)

Someone who accepts Dewey's view might then recognize that there are some—or even many—things done to animals in the name of science which ought not to be done, but he would nevertheless be opposed to any kind of regulation to prevent such abuses. Any tidying up in the house of science that needs to be done should not be imposed from without. So there is room for quite a lot of *practical* disagreement on the part of those who share the same fundamental view of animal experimentation.

I shall represent the opposed view of animal experimentation by the diagram on page 340. Here the sphere of justifiable experimentation on animals is under a sort of pressure or tension, tending to narrow the sphere of what is allowable so far as possible. And an animal is perceived here, not as a delicate instrument, but as a creature with a life of its own, so that the systematic interference by us with an animal's life for our ends, and equally the creation of such life for our ends, is seen as quite unlike the manufacture and use for our ends of a microscope, say, or

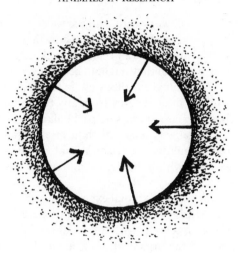

other piece of technical equipment. There are on this Second View questions which can appropriately be raised about how we are justified in interfering with an animal's life or creating such life for our ends, questions which cannot be raised about the use and manufacture of machinery. Even if an experiment would lie within the allowable sphere of animal experiments on the First View, the Second View would require us to ask further questions before we could conclude that the procedure was justified: could we at perhaps some increased cost substitute a procedure which did not use animals or which did not subject them to such-and-such treatment? Is the value of the result in terms of human knowledge (or whatever other benefit may be in view) really sufficient to justify the kind of interference that is proposed?—which is not a question which science itself can answer. Such questions imply a mode of thinking about animals quite different from that implicit in the First View. On the First View, once certain standards of attention to the welfare of the animal have been met, the only further questions which arise are ones for the professional judgement of the scientist; on the Second View, questions do arise about the animal's treatment which are not for the scientist as a scientist, and which take the animal as a centre of moral claims in something *like* the way a human being is. Again, Sir John Eccles provides a way of characterizing the distinction between the two sides: experimentation on human beings involving risk of any sort should be diminished so far as possible; on the First View (his), nothing corresponding can be said about animal experimentation. On the Second View, something of the same sort can and indeed should be said about animal experimentation.

Just as there can be wide practical disagreement among those who share the First View, so can there be among those who hold the Second. Just as holders of the John Dewey version of the First View are in principle opposed to any governmental supervision of animal experiments as anti-scientific and the thin end of the anti-vivisectionist wedge, so holders of an equally extreme version of the Second View (far fewer in number than they once were) are in principle opposed to any governmental supervision of animal experiments short of abolition—as an implicit recognition of the principle that animals *may* justifiably be used merely as instruments in the pursuit of scientific knowledge. So the extremes meet—but only because each takes *any* governmental regulation as granting some recognition to the other side.

In summary, then, we have these two ways of looking at the use of animals in experiments:

Within certain limits, experimental animals may be regarded as delicate instruments, or as analogous to them, and are to be used efficiently and cared for properly, but no more than that is demanded.	Within certain limits, animals may be regarded as sources of moral claims. These claims arise from their capacity for an independent life, or perhaps from their sentience, but in either case the moral position of animals is seen as having analogies with that of human beings.

THE DISPUTE NOT *OVER* A MORAL ISSUE BUT OVER WHETHER THERE *IS* A MORAL ISSUE

I have mentioned that on the First View of animal experimentation, there is no moral need to justify the use of animals in experimentation, provided that certain kinds of abuse (arising from wilful sadism or ignorance) are avoided. This means that from the First point of view, holders of the Second View, who do regard at least some of what goes on 'within the circle' as morally unjustifiable, are not so much mistaken in the *particular* moral view they take, but in making or trying to make a moral issue where there is none. In this part of the paper I want to show the significance in the First View of the idea that there *is no moral issue* about the use of animals 'within the circle' and its connection with other characteristic features of the First View.

The dispute between holders of the First View and holders of the Second should *not* be conceived as a dispute between those who attach greater and those who attach lesser weight to the interests of animals in the clash between their interests and ours. Let me explain by considering a contrasting case, in which animals are used in scientific investigation, but in a setting in which their interests are taken into account—given weight, but less weight than ours. Sled dogs have been used in the course of scientific work in the Antarctic, during which they have undergone serious risks, considerable pain, and prolonged discomfort. Most people would of course think that if, during such an expedition, a situation arose in which a dog's interests and a person's clashed (say, only one could be rescued from a crevasse) the dog's would properly be sacrificed. But many people would *also* think that at the end of the expedition the dogs should not be killed (however painlessly) merely to save the money it would cost to ship them back. They might think in terms of something 'owed' to the dogs in such a case, and indeed in the actual case I have in mind all the dogs involved were adopted. What then is the difference between the treatment of such animals and the treatment of dogs in experiments? It is not that the latter are necessarily treated *worse* or are subjected to more *pain*. This need not be so at all. The point is that in the experimental setting, the dog may come to be thought of merely as a useful and disposable object; we may come automatically to take it that there simply is *no room* for thinking of it as a being with a life of its own. We do not then see it (for example) as a being to which something may be owed, in at least that minimal sense in which we may feel something—analogous, at least, to gratitude—*is* owed to the sled dog after what *it* has gone through. The animal seen in the laboratory setting becomes something we may respond to in accordance with quite a different set of ideas from those which are natural and quite common with the sled dog. The person who thinks this way need not deny that it is wrong to be cruel to dogs or indeed to other animals. But he believes that there *is* cruelty only when the laboratory animal is subject to some treatment *outside* the circle, like wanton abuse. On this view, when the animal is treated in the laboratory as a delicate instrument, to be used properly and disposed of painlessly, no question of cruelty, no moral question at all, arises.

This way of thinking of the laboratory context, and the animal viewed as an instrument in it, is in some ways analogous to the mode of thought of such slave-holding societies as the *ante bellum* American South. Slave-holders may recognize some modes of treatment of slaves as morally wrong, but these will be *outside* what they take to be the circle of appropriate economic use of the slave, and would include wanton cruelty. As in the case of laboratory animals, there is economic pressure to stay within the circle. From the slave-holder's point of view, modes of treatment which are within the circle, and belong to the proper use of the slave as economic tool, can raise no moral issue, and do not properly speaking involve *cruelty*. In both cases, people who may be perfectly kindly set apart an area of practical activity in which beings of a sort which they themselves regard in *some* contexts as proper objects of moral concern, can be treated as practicality dictates, and no moral concern, it is held, is appropriate, provided the treatment involves no *gratuitous* suffering.

The First View then does not differ from the Second in giving *less* weight to the costs borne by laboratory animals: provided the minimal conditions for justifiable animal experimentation are met, the costs borne by the animals are seen as having no moral significance. The idea is not of a clash between their interests and ours, in which ours must take precedence, but of an area in which we can simply get on with the job of asking and answering scientific questions, treating animals solely as our instruments in doing so.

This feature of the First View is reflected not only in direct assertions that animal experimentation does not really raise any moral questions but also in other ways; I shall look at five of them.

(a) Later on, I shall discuss the accusation made by those who hold the First View that the other side is guilty of *sentimentality*. What is relevant now is that one basis on which that charge is made is that holders of the Second View do precisely regard experimental animals as properly objects of moral concern, even within the setting of well-conducted experiments. They are thought of as sentimental because they bring a type of concern for animals which might be appropriate in the sled dog context into what is, on the First View, a wholly inappropriate context for *that* sort of concern. The basis of the accusation is not that they care *as much* about animals as about people—they mostly do not—but that they make a moral issue where, on the First View, there is none.

(b) If you conceive of a certain sort of situation as one in which two groups of beings come into conflict, and if you think that such situations are likely to occur frequently, then a test whether you really do think that there is at least some weight on *both* sides is: are you concerned to find some way of avoiding such conflicts? Thus, for example, in the case of abortion: If you think of the situation as one in which the mother's interests outweigh those of the fetus, but you think the interests of the fetus do count for *something*, you might show that that was what you believed by, say, an interest in making adoption an easily available and attractive option. On the other hand, in so far as you are totally uninterested in preventing the occurrence of situations in which there is a clash between the interests of the mother and those of the fetus, situations, that is, in which one's interests cannot be respected unless the other's are sacrificed, you tend to show that you do not take abortion as raising any real moral questions, with real costs to be borne on either side.—This argument is based on a discussion of moral dilemmas by Ruth Barcan Marcus, and I have drawn on her use of the abortion example. She has argued that there is a fundamental moral principle: 'as rational agents with some control over our lives and institutions, we ought to conduct our lives and arrange our institutions so as to minimize predicaments of moral conflict' (Marcus, 1980). If that is indeed a fundamental principle, then there is a strong implication that a total *lack* of concern with institutional arrangements intended to prevent conflict of some sort must tend to show that such conflicts are not being seen as ones in which there is serious moral weight on both sides. Thus, someone might hold that in a conflict between the interests of mice and those of people, the interests of people must win, but if he *also* regards it as quite unnecessary, or even wrong-headed, that training for medical or scientific research be such as to incline people to use alternatives if they possibly can, to design experiments to limit the number of animals used or the kind of interference involved—this would tend to show that there is on his view no real moral conflict. The First View goes with the idea that it is not worth human ingenuity or resources or concern to work out ways of diminishing animal experimentation, and that it is not worth trying to alter social institutions like scientific or medical education so that people are inclined to look for alternatives and to use them when they can. Anyone who accepts that does *not* think of animals as having merely *less* moral significance than people, but thinks of the context of scientific experimentation as one in which the animals' lives have no moral significance. This is an important point because proponents of the First View often attempt to enlist support by appealing to the generally accepted idea that a rat's life or suffering, or a dog's, matters less than a person's. But someone who holds that generally accepted view should not on that account alone be willing to support the First View. The idea that an animal's life or suffering matters less than a person's is entirely compatible with the Second View.

(c) If, as I have said, the holder of the First View does not regard animal experimentation as a moral issue, this will be reflected in the way he looks at statements of the Second View. He will tend *not* to regard them as *moral claims* with which he disagrees (for that would go with the idea that animal experimentation within the circle raises moral issues on which there are different positions) but will put them in some other category. Usually they will be seen as expressions of irrational or non-rational likes and dislikes, but occasionally they may be seen as expressions of an eccentric theological or metaphysical position. Again the comparison with the case of abortion may be useful. Those who believe the interests of the fetus to have *some* moral weight will see the expression of their own position as one within what we might call the arena of moral discussion. But those for whom the fetus is no more than a collection of tissues which may or may not be wanted by its bearer will see that the fetus may well be the centre of great *emotions* of different sorts for different people, but cannot (cannot usually, that is) take seriously that the view opposed to their own is a view in the *moral* arena at all. It appears as *merely* an emotional response (or a religious reflex), and any attempt to protect the interests of the fetus is seen as a fundamentally illegitimate attempt to interfere, on merely emotional or religious grounds, in what is a personal decision for the pregnant woman. To recognize someone else's position as *moral* disapproval of a course of action one favours is to accept at least this much, that the kind of conduct falls into the arena of moral discussion. And this the holders of the First View often show they do not do, in their ways of describing their opponents' ideas. Thus for example in the 1977 BBC programme, the advantages and limits of tissue culture as an alternative to the use of animals in medical research were being discussed; the advantages were said to be that it was cheap, and pleasing for those who find animal experimentation distasteful. The use of the words 'pleasing' and 'distasteful' here takes for granted that the issue is one in which *taste* is all that is involved. (Compare a discussion of war in which pacifists were described as finding killing *distasteful* and other methods of settling disputes more *pleasing*.)

(d) Holders of the First View insist that the choice whether or not to use animals in some experiment is a choice which can be judged appropriate or inappropriate only by an *expert*. This idea is closely related to the last: describing one's opponents as merely expressing their emotions is one way of emphasizing that the issue is not in the moral arena at all, as far as you can see, and the idea that it is a field *for experts* is simply the other side of the same notion. Once an area is recognized as one in which there are genuine moral problems, it is *eo ipso* recognized as one not to be left to experts in the particular field. If (for example) there *are* moral problems concerning euthanasia, it follows that the position and training of the doctor does not *as such* qualify someone to decide such matters; if it is urged that the expertise of scientific investigators *is* sufficient and necessary for judging what use of animals in science is appropriate, it is being assumed that there is no significant moral issue. It is interesting that the people doing research for the BBC Horizon programme on animal experiments had assumed, beforehand, that the licensing procedure for such experiments involved some kind of judgement whether the likely benefits of the result of some proposed experiment balanced the suffering for the animal subjects. But they found to their surprise that the licensing procedure involves no such judgement at any point, and 'the scientist is not licensed to conduct a given experiment at all, but is licensed for a period of time to *carry out certain procedures on animals*', and indeed may alter the programme of investigation without needing any change of license (Harris, 1977). We can see in the present licensing system a reflection of the idea that once certain minimal standards are met in animal experiments, no further issues arise in which society might properly be interested. On the other hand, the suggestion made by the Littlewood Committee in 1965, and strongly supported by many people in animal welfare groups, that there be a standing body *involving four laymen* to advise the Home Office on matters to do with the use of animals in scientific work, and concerned not only with the ways in which animals are treated but also with the *purposes* for which they are used—this recommendation reflects a position opposed to the First View.

(e) One further reflection of the First View can be seen by anyone who looks at the literature in the field now called bioethics. It is certainly true that articles concerned with animal experimentation turn up *occasionally* in the professional journals concerned with ethics and the life sciences. But for many working in the field, the phrases 'ethical problems posed by research' and 'ethical problems posed by research *on human subjects*' are treated as simply interchangeable. Again, a large (79-page) bibliography on society, ethics, and the life sciences, described by its publishers as containing the most pertinent references on precisely such subjects as experimentation, contains nine pages on ethical and legal problems of experimentation, including, besides general material, sections specifically on experimentation on fetuses, prisoners, mental patients, and children—but does not include references to the ethical problems of animal experimentation (Sollitte and Veatch, 1978). The explanation I was offered of this omission is that references to a particular problem would be put either in a section devoted to the subject—but this would be done only if the bibliographers found 10–15 relevant pieces, and apparently they had not done so—or under some more general heading—but this had not been done either: Peter Singer's *Animal Liberation*, for example, 'fell between the cracks in the structure of the bibliography' (R. M. Veatch, personal communication). But major discussions *can* fall between the cracks, the articles (which certainly do exist, on both sides) *can* fail to be noticed by bibliographers, because the subject itself is perceived as merely a peripheral one in bioethics—at best. For many scientists with strong interests in bioethics, it simply is not an ethical problem at all, and the professional 'ethicist' often shows by ignoring the issue that he tacitly shares the view that it is not one of the things we need to be thinking about. This is striking, because in a sheer numerical sense, animal experimentation is an enormously bigger thing than human experimentation (far bigger in the number of experimental subjects involved, in the number of experimenters and technicians and of people indirectly involved, and in the amount of money spent) or indeed than most of the other problems bioethicists discuss. It is not merely big in numerical terms, but a feature of so many parts of our social life: it is hard to think of anything produced in our society except jokes that is not tried out on animals! It is partly the very *normality* of animal experimentation that helps make it invisible to the bioethicist, but it is also the acceptance by many of some version of the First View, and the prestige the view gains by its connection with the research establishment.

NEITHER SIDE IS MORE RATIONAL THAN THE OTHER

People who take the First View do sometimes tend to regard it as more rational. But how could this be shown? What we can see right off is that there are the two different analogies which play a role in thinking about animal experiments—the delicate-instrument analogy, and the person-with-moral-claims analogy. A laboratory rat *is neither* a machine *nor* a person; if it really were one or the other there would be no problem how to draw the boundaries of morality. Just simply looking at the two views as two differet views, it cannot be said that one is rational and the other merely emotional, or that one is a matter of a taste and the other appropriately scientific. Until something more is said, there is a considerable symmetry: we have the two views.

One thing that is often said by holders of the First View is that the Second cannot be held to consistently, or that if its implications were clearly seen it would lose all popular appeal. The idea is this: In many parts of our lives, we *do* treat animals as mere instruments. Most significantly, we manipulate their lives and kill them off for food—food which is not essential to our health. If we think that to please our palates, we may justifiably treat animals pretty much as mere objects serving our needs and wishes, how can we think that the free use of them in scientific experimentation should be questioned? As one well-known experimeter on primates has put it: 'there is no more need to seek alternatives to the use of animals for medical research than to search for non-flesh substitutions for meat in our diets' (White, 1971). So the idea is that only a vegetarian could *consistently* object to the use of animals in research, and, if this is once clear, ordinary folk should see that, given their own basic views (including their views about the justifiability of eating meat), they have no reason to treat the use of animals in science as a moral issue.

To clarify the issues which this line of argument raises, recall the summary description of the two sides. As I stated the First View, it was that *within limits* it is appropriate to treat animals as any other instrument that might be used in scientific investigation, and the Second View was that *within limits* it is appropriate to treat animals as sources of moral claims and in that respect like human beings. As a matter of fact, there are people who take an extreme version of the First View and do not recognize *any* limits on the circumstances in which animals may be regarded as mere things giving rise to no moral claims on their own account, and there are people who take an extreme version of the Second View and do not recognize *any* limits on the circumstances in which an animal is to be treated as a source of moral claims analogous to those made on us by another human being. *Most* people taking the First View, though, do not hold the extreme version of that view—nor do most taking the Second View hold the extreme version of *it*. It is perhaps worth noting how very unattractive the extreme version of each is. The extreme version of the Second View involves holding that there is no duty to frighten off a wolf (say) stalking a human baby, or to try to get the baby out of harm's way, *unless* there is a comparable duty to frighten off a wolf stalking some other sort of animal, or to try to get *it* out of harm's way. Anyone who thinks that one might have a duty to a human being in such circumstances but that one does not have a comparable duty with respect to other animals does not hold the extreme version of the Second View. The extreme version of the First View holds that we may have a duty to an animal's owner not to molest it, or again we may have a duty to other human beings not to treat an animal in a way which would increase the likelihood that we would treat *human beings* badly, but that just considering the animal itself and leaving aside all the ways injury to an animal indirectly injures human beings, we have *no* reason at all to think there are moral constraints (of kindness or justice or of any other sort) on how the animal should be treated. On this extreme view, if you and a dog were, let us say, the only surviving living things on the planet—so nothing you could do to it could injure anyone else—and if it amuses you—helps you pass the time—to torment the dog, there is no possible moral objection to this. Indeed if your amusement takes the form of tying tin cans and so on to the dog while it is asleep, so that it does not take in at all that you are the source of its daily miseries and it never gives up its pathetic affection for you—this refinement is equally morally all right.—I do not think one needs to try to decide *which* extreme view is more unacceptable. The point is that most people who hold the First View have something important in common with most people who hold the Second: they regard animals as sources of moral claims analogous to those arising from human beings in *some* circumstances but not others. Holding such a view is not *itself* a sign of inconsistency on either side. We can no more assume that a holder of the Second View who eats meat is inconsistent than we can assume that there must be inconsistency if a holder of the First View says to a child tormenting an animal, 'How would you like it if someone did that to you?' in the same spirit in which he might say it if the victim had been a smaller child. There is, you could say, a built-in tension in our modes of treatment of animals. From childhood on, we are familiar with animals treated sometimes as within the sphere of morality, and sometimes as mere things, sometimes as companions, sometimes as lamb chops on the hoof—and so on. (See Susan Isaacs (1930, pp. 160–170) and David Sperlinger's discussion of these issues in his chapter of this book.) One may, on reflection, alter some of one's views about animals; for example, one might go from thinking that using animals for food raises no real moral issues to thinking it does—or the other way. That is, I am not suggesting that whatever we are brought up to accept we should go on thinking is acceptable. I *am* suggesting that the complex of beliefs which the typical holder of the First View might accept is not in any obvious way a more rational collection than that which might be accepted by his typical opponent; there is no reason at all to think that considerations of consistency ought to drive one to give up the Second View. Further, if we distinguish, as we should, between consistency in moral response and mere singlemindedness, we should also see that *neither* extreme view is forced on us by the demands of consistency.

The accusation of inconsistency brought against the non-vegetarian holder of the Second View may be based on the different idea that the cases—of the use of animals for food, and their use as laboratory instruments—cannot be morally distinguished on any reasonable grounds. A vegetarian myself, I should nevertheless argue that the accusation is ill-founded, and that it can be shown to be so by appeal to examples. One such example would be the views of G. K. Chesterton, which I shall not discuss in detail. His moral position, a coherent and in many ways attractive one, includes opposition to animal experiments *and* to vegetarianism—his views on both subjects being connected with his hatred of snobbery (see, for example, Chesterton, 1910, pp. 79–86.)

THE SPECIALNESS OF HUMAN BEINGS

Proponents of the First View often claim that it is based on our recognition of the immense gulf that separates human beings from all other animals, while many of those who hold the Second View argue that what makes a being an appropriate object of moral concern is its possession of sentience or of the capacity to live its own life—things, that is, shared by human beings and animals. Their argument continues: What makes it morally questionable to carry out risky experiments on people is not their rationality but the fact that they have interests, interests to which we may not be giving adequate consideration, and animals equally having interests, there must be a moral question whether *theirs* may be overridden. A common and closely related argument is that there is *no* feature of human beings, such as rationality of such-and-such a level, which is actually shared by *all* human beings and *no* other animals, and which could be used as the basis for giving human beings a specially privileged position in morality. The basic idea in this argument is that if you put the level of rationality low enough to include the senile or hopelessly retarded among us, many animals such as primates (at the very least) will come out *higher*. How then can we treat experimentation on primates as perfectly all right because of the great gap between them and us in rationality, and experiments on seriously retarded people as a great big moral problem? Again, we *ignore* differences in rationality when we consider whether a human being is a proper object of moral concern; the Nazis are criticized for (among other things) taking the senile and insane as things that were a mere drain on the nation's resources and that should be painlessly disposed of. If we regard such people as entitled to the same moral concern as other people, despite their absence of rationality, how can we say that animals may be treated in scientific experimentation as mere instruments for the satisfaction of our wants and needs, because they have not got *our* rationality? It has been further argued on the same lines that only if we invoke some sort of *theological* doctrine can we give any justification for taking the boundary between human beings and other animals as of any moral significance. If you believe in an immortal soul that people and not animals have, that idea might back up the claim that people, being very special, can use other animals, which are not at all special, for experimentation, but cannot use the retarded, the senile or the insane. But then the criticism is that the idea of the specialness of *human* life cannot be maintained without its theological backing.

I believe that neither sort of argument will do: The First View cannot be supported by arguments from the gulf between animals and ourselves, and the Second View cannot be supported by the arguments that without theological support, there is no basis on which to defend a morally significant distinction between human beings and other animals, and that men and many animals share those empirical characteristics (like sentience) in virtue of which moral concern about what is done to them is justified. We do not need to discuss *either* how far Mozart is from the most musical whales, *or* how far Washoe the chimpanzee is ahead of some severely retarded children. There are two claims I should wish to make against these bad arguments offered by both sides.

(a) It is possible to combine a recognition of human life as terribly special with the Second View—they are perfectly compatible. The bad arguments offered by both sides share the false assumption that the First View is supported by the idea of human life as special—special in some morally significant way.

(b) The bad arguments offered on both sides involve the same confused picture of morality. It is a confusion about what would be involved in any attempt to justify treating human beings and animals differently. The bad arguments suppose that any attempted justification would have two elements. You would have to be able to point to some empirical difference between human beings and animals—say, that human beings have x and animals do not—, and you would further have to accept some general moral principle to the effect that *any* beings with x can properly be treated differently in such-and-such circumstances from *any* beings which lack it. Take a simple parallel to illustrate the general point: the idea is that we might offer a justification of 'Kicking stones is okay and kicking people is not' by (i) pointing out that stones do not feel pain when kicked and people do, and then (ii) appealing to the principle that things that feel pain ought not to be treated in ways which cause them pain. The bad argument for the First View is an attempt to give such a justification for distinguishing between how people and how animals should be treated, and the bad argument for the Second is that any such attempted justification would have to appeal to a principle justifying experiments on lower grade people, and that there is *no* principle which really would be acceptable, if its consequences were thought through, to proponents of the First View and which would justify treating animals as they think we should.

The heart of the shared idea of morality held by those putting forward such arguments on both sides is that what is involved in moral thought is knowledge of empirical similarities and differences, and the testing and application of general principles of evaluation. The maker of a moral evaluation is 'the moral agent', and the essence of a moral agent is that it can find out facts and understand and accept general principles and apply them to actual cases. Thus the idea is that if a Moral Agent believes it is wrong to torment a cat, this must depend on his having some such general principle as that The Suffering of a Sentient Being is a Bad Thing and that one ought therefore to prevent such bad states of affairs from existing.

Against this, I should claim that we are not Moral Agents in that sense at all—and thank heavens for it. We have resources in our moral thinking which the story I have just told forgets all about. We can, for example, ask what human beings have *made of* the difference between human beings and animals. A difference like that may indeed start out as a biological difference, but it becomes something for human thought through being taken up and made something *of*—by generations of human beings, in their practices, their art, their literature, their religion, their ethics. This is true of any difference. If you want to know whether it is a good thing to treat men and women differently, or to treat the old with respect simply because they are old, or to treat dogs differently from other animals, or cows differently from other animals—it is absurd to think these are questions you should try to answer in some sort of totally general terms, quite independently of seeing what particular human sense people have *actually* made out of the differences or similarities you are concerned with. And this is not predictable. If the Nuer, for example, had not actually made something humanly remarkable out of giving cows a treatment quite different from that accorded other animals, one could not know that 'singling cows out for special treatment' could come to that. It would be foolish to raise the question 'Should cows be treated differently from other animals?' as a general abstract question, because one would not know what it could come to to do so. As the Nuer have made something quite remarkable out of their special relationship with their cattle, so our culture has made something (which we might recognize as remarkable if it were not so familiar) of treating dogs, and to a lesser extent cats and horses, differently from other animals. Much more important, of course, we have made something of human significance and great depth out of the difference between men and women—and we have never made anything humanly valuable of the differences between races, the mythology of the American South notwithstanding.

The view I am putting forward is that we are never confronted merely with the existence of 'beings' with discoverable empirical similarities and differences, towards which we must act, with the aid of general principles about beings with such-and-such properties deserving so-and-so. The modes of life and thought of our ancestors, including their moral thinking, have *made* the differences and similarities which are now available for us to use in our thinking and our emotions and decisions. We have available to us, not for example the mere biological notion of difference-of-sex, but a human notion of that difference, made by literature, art, and common thought and life over centuries—and we can make something more or less of it in our turn. Far from it being the case that we have moral principles which can be given in purely general terms, and which are then applied in a world of empirically given similarities and differences, what differences there are *for us in our thought* is a matter created in part by past moral thought, marking and making something with human sense of such things as male/female, human/non-human. What we have made of these differences may be valueless or valuable or a mixture of good and bad.

That was intended as a general point about morality. The specific point is that we have made something of great human significance out of the very notion of *human life*. People think the invention of the wheel was a great thing. But so was the invention of the idea of human life as something special. (I do not mean 'invention' to imply that someone sat down and came up with it as a good idea! I mean that it is something made by us.) Chesterton expresses the idea I have in mind this way:

'Ordinary things are more valuable than extraordinary things; nay, they are more extraordinary. Man is something more awful than men; something more strange. The sense of the miracle of humanity itself should be always more vivid to us than any marvels of power, intellect, art, or civilization. The mere man on two legs, as such, should be felt as something more heartbreaking than any music and more startling than any caricature. Death is more tragic even than death by starvation. Having a nose is more comic even than having a Norman nose.' (Chesterton, 1959, pp. 46–47)

The bad arguments for both the First and Second Views suggest that you could justify making a moral distinction between how people are to be treated and how animals are to be if there were some specially significant characteristic of human beings that animals lack. Against this I am claiming that making such distinctions in our mode of treatment is not grounded on a prior recognition of differences but is itself one of the complex of human activities through which we come to have the concept of the heartbreaking specialness of human life, of which Chesterton spoke.

Two further brief points:

(i) We do not need *theology* to back up the idea of the specialness of human life, any more than we need it to back up the idea of the human nose as comic. The idea of human life as poignantly special did not rest, for Chesterton, on his religious views. The relation between that idea of human life and Christian doctrine may be seen very clearly in Hopkins's 'That Nature is a Heraclitean Fire'. The poet exclaims with pity and indignation at the disappearance in the fire of nature of man and all the marks of his mind that shine out, but then is comforted by the resurrection. The significance of the resurrection here depends on a prior sense of the tragedy of death as such.

(ii) The capacity to feel the heartbreakingness of the mere man on two legs as such—this is a main source of that moral sensibility which we may *then* be able to extend to animals. We can come to think of killing an animal as in some circumstances at least similar to homicide, but the significance of doing so depends on our already having an idea of what it is to kill a man; and for us (as opposed to abstract Moral Agents) the idea of what it is to kill a man does depend on the sense of human life as special, as something set apart from what else happens on the planet. There is then a possibility of extending that sense of what life is to animals, without this involving any going back on the idea of human life as special. I should want to claim (though I am not going to argue for this here) that the idea of the specialness of human life is not merely compatible with the Second View, but essential to all except the most unattractively simple-minded versions of that view.

AN ACCUSATION AGAINST THE FIRST VIEW

I have argued in the two preceding sections that neither view is more rational than the other nor more securely grounded in the actual empirical similarities or differences between human beings and animals. Is there any more fruitful approach to the dispute? Yes; and it requires us to notice one difference between the two views that I have not yet touched on at all. Those who hold the Second Views usually regard experimentation on animals as likely to harden those who engage in it, and to do the same to others as well, for whom it will serve as a model—leading, it is suggested, to a general increase in callousness. Proponents of the First View usually deny that this is so. It is worth attending to this issue; it will lead us to the heart of the dispute about animal experimentation. In the rest of this section, I shall explain the charge of callousness as conceived from the Second View, and in the following section I shall show its significance for us.

The first thing to note is that the question whether animal experimentation makes one more callous is not a purely empirical one. As I pointed out on page 342, there are ways of treating animals which will be regarded as cruel on the Second View but not on the First (in which there is cruelty only when pain or distress is imposed on an animal through sadism or neglect or ignorance). Thus, for example, the removal of a monkey from its natural habitat to a cage in a laboratory might be regarded as cruel on the Second View but not on the First. In the absence of agreement on what counts as cruelty, there can clearly be no agreement on what counts as increasing people's callousness (their indifference to cruelty, or willingness to treat it as acceptable), and so it is no wonder the two sides disagree whether animal experimentation does so.

It may be useful to consider an actual example in which the importance of what one takes to be callousness comes out. In 1912, Stephen Paget, the founder and secretary of the Research Defence Society, published a collection of excerpts from the testimony given to the Royal Commission on Vivisection, which had just completed its work (Paget, 1912). Among his main aims was to show that experimenters were not inhumane and that virtually all experimentation met the tests of humanity towards animals. One passage which he singled out for inclusion was from the testimony of the eminent pharmacologist Sir Lauder Brunton. Brunton himself was concerned to bring out that much experimentation involves no pain to speak of, and to illustrate his point he described a dog on whom he had made a gastric fistula, and 'which never showed the slightest sign of pain'. Brunton went on that whenever he went to examine the dog,

'it showed great delight—just like a dog that has been sitting about the house, and wants to run out for a walk. When it saw that I was going to look into its stomach, it frisked about in the same way as if I was going to take it out for a walk.' (Paget, 1912, p. 90)

These daily examinations were not only painless but, Brunton explained, eagerly anticipated by the dog, because it liked to be made much of and shown round.—Now one possible reaction to this story is that it is a miserable life for a dog to have nothing to look forward to but the daily examination of the interior of its stomach. Of course such things happen to people too, in hospital—but it is pathetic in either case. It is clear that neither Brunton nor Paget (who selected Brunton's testimony for the collection) saw the pathos of the case, the possibility of its eliciting not, 'Oh good, no pain', but, 'What a miserable life for an animal'. Paget, note, was concerned especially to show that animal experimentation was carried out by humane men. Humane men in his society, including (let us grant) himself and Brunton, would no doubt have regarded it as deplorable, *outside* the context of scientific experimentation, to deprive a dog of a normal sort of dog's life, and would not think that 'no pain' made it *un*deplorable. The charge, that animal experimentation makes one callous, is (among other things) the charge that one can come to think like Brunton and Paget, that is, can cease to take in that what one is doing to animals is something which one would oneself in other circumstances regard as deplorable, and which one therefore should, at the very least, notice, be concerned about, regret, and regard as something one should try to avoid, and so on. It is possible for humane men to set apart the area of experimentation as one in which one simply cannot bring in the sorts of consideration that play a role in judging how animals are treated outside that area. The judgement, 'What a pathetic life!', which they themselves might make in connection with a confined animal outside the sphere of experimentation they do not make, and do not see as appropriate, when the confinement and deprivation of normal life is for the sake of scientific knowledge. But the animal's life is not made any *less* pathetic by the fact that its deprivations are for the sake of science. We may think, given that we *are going* to treat animals this way, it is impractical to respond as the humane man would in other cases of confinement and deprivation of normal life. This is to say that the humane person who is involved in experimentation on animals had better become inured to it, had better set up these compartments, as it were, in his mind—so that he comes quite naturally to apply quite different standards of what counts as humane treatment in the context of scientific experimentation from those he uses in other contexts. He no longer thinks, 'How pathetic!' of treatment like that of Brunton's dog. It is part of the normal life of animals in laboratories; and one simply does not see this normal laboratory life as itself raising any questions.

What I have tried to bring out with this example is how the same case can be seen (as by Brunton and Paget) as meeting all the requirements of humanity in the treatment of animals, and (from a different point of view) as illustrative of the way animal experimentation can make experimenters callous by encouraging a compartmentalization of mind in which the experimenter can simply get on with the job. Once you have accepted this sort of compartmentalization, you simply do not *look at* the treatment of animals in science as you otherwise might: here is an animal, here is how it is treated, and my gracious, is it really justifiable to do *this?* The 'this' in the particular case is something we no longer bring to realization. Our powers of imagination and judgement are not brought to bear on the case; and this (it is suggested) is a form of callousness. We do without fully thinking what we do.

Supporters of the First View sometimes mention that 'even' Albert Schweitzer allowed that animal experimentation could not be given up. But it is important to see that his position is opposed to the First View at what really is its central point: that we can give a *general* justification for animal experimentation and then *get on with the job*, provided only we meet the minimal conditions. The heart of

Schweitzer's view is that no such *general* justification can be given, but only justifications which take fully into account what we are doing to particular animals in the particular case, justifications that recognize them as individual animals with their own lives. In other words, to experiment in the spirit of Schweitzer, one would have fully to recognize the morally problematic character of even routine experiments—not to speak of heroic ones; one would have to recognize also the need to fight against a tendency to separate one's way of judging the treatment of animals in science from the way one would judge it in other contexts.

I have tried to show that the accusation made by holders of the Second View, that animal experimentation makes for callousness, should be connected with the idea that it leads to a harmful compartmentalization of mental life, in which one does not bring to full imaginative realization what one is doing. I want now to turn to the other part of the accusation: the increase in callousness is supposed to affect not just the experimenter but others. I have in front of me, as I write, a popular scientific journal put out by the Museum of Natural History in New York. In an article on why food rots, there is a description, quoted from a pharmacological journal, of an experiment in which mouldy maize was fed to farm animals. Several refused it, and post-mortem examination showed that they were in fact starving rather than eating; others were force fed and got very ill (showing depression, weakness, and bloating, and undergoing seizures) before dying (Janzen, 1979). What I am interested in is the way the experiment is described, both in the original report and by the writer of the popular article incorporating the quotation. The language is the language of the normal, of the not-in-need-of-any-comment even. The reader is not supposed to think, as he would, reading in the newspaper of a farmer who treated his animals so, 'How horrible!' He is not supposed to think that treating animals like *that* must need some very weighty justification. The misery caused the animals was unavoidable, given the nature of the experiment, and so the investigation falls 'within the circle'. The reader is expected, or, one might say, invited, by the language of the description, to share the compartmentalization of mind in which there is 'no point' in allowing oneself to realize fully in imagination what was being done.

This compartmentalization of mind is also imparted in schools, to an extent which is perhaps difficult to appreciate in Britain, where there has never been the American combination of absence of legislative control and affluence capable of providing large numbers of living animals for children to learn science on. In my school, for example, each pair of 13-year-olds in the biology classes had a living but pithed frog to cut up. Two years after doing this in school, I wanted to give younger children at a summer camp a biology lesson. Incapable of pithing anything, I battered a snake's head—it seemed about the same—and then showed the children all the pieces inside the snake. I could do this because I thought of it as science. Heaven knows it was not, but what a child can pick up all too easily is the idea that in 'science' one can irresponsibly treat the world and what is in it as *interesting things*. I want to emphasize that my claim is not that the scientific spirit involves treating animals as *things* to be investigated in a state of moral anaesthesia, but that the practice of animal experimentation is in our society accompanied by the compartmentalization I have spoken of, and that biology teaching using living animals imparts, besides impressive— unforgettable—knowledge of what a beating heart looks like, the idea, impressive in a different way, that if you are doing science, you are in a situation in which you do not take seriously the lives of the animals used—do not see them as individual animals with lives of their own at all.

Consider a different example, again an American one. 'Send a mouse to college' was the official slogan for a recent cancer research collection in my community. 'Grow a tumour on a mouse' would not have quite done. If holders of the First View can claim that experimentation is not connected with callousness, how can the apparent callousness of such a slogan be taken? The answer would be that a humane person has in the end to be more concerned about the sufferings of human beings from cancer than about the sufferings of mice. But that is no answer; a humane person should indeed be more concerned about men than about mice, but does not therefore have to take mice as joke animals, with no more reality to them than the happy pigs dancing on a sausage van.

It has been claimed that the result of contact with laboratory animals in one's work is heightened respect for them, 'likely to be followed by considerate treatment and a general enrichment of life for both man and animals'. This from Dr Lane-Petter (1976), one of the very few scientists working with animals who has written or spoken with real sympathy for the Second View. There is no doubt that for many scientists such contact teaches respect for the marvellous complexity of animals: of their physiology and their behaviour; and such respect may breed contempt for those who treat such marvellous things sloppily or carelessly. But considerate and feeling treatment is different, and has a different source—imagination, not characteristically scientific imagination, but the imagination exemplified here:

'now and again I caught and brought home in a glass bottle some miserable minnow or stickleback. A wooden bucket—if not a large wooden tub—would then be filled with the best of cool well water fresh pumped; a stone or two, a crock or two, would be dropped in to make a comfortably uneven floor and there—Why would the wretched stickleback never live? Ah—why rather did it linger so long? The laceration of its mouth by the hook was probably the least of the horrors. Cold and motionless well water replaced the limpid freedom of the running water; and the company of other fishes, and all the homeliness of a well known environment, had been lost. What knowledge the fish had was no longer of any use to it; it might as well have been blinded. In vain I dropped worms into the water. No longer served by its own senses, the lonely captive swam round and round, turned whitish; and at last, after hours, died. And I never dreamt that the whole enterprise on my part was sheer cruelty from beginning to end . . . . But intentional cruelty horrified me if I had eyes to see it.' (Sturt, 1977)

THE DISAGREEMENT EXPLORED FURTHER

In the preceding section I explained the charge brought by holders of the Second View that animal experimentation makes for an increase in callousness, in experimenters and others. The charge should not be taken to mean that people become less humane *in general* but that they set apart an area insulated from the kinds of thought which they themselves (granting now that they are indeed humane people in general) would bring to bear on at any rate some other situations involving the treatment of animals.

Holders of the First View do not just deny that animal experimentation increases callousness. There is a counter-accusation they bring: that the Second View is essentially sentimental. One version of this charge is comparatively unimportant: it may be claimed that holders of the Second View are sentimental in that their attacks on animal experimentation reflect a fantasy of what animals are. This might be called the Bambi version of the sentimentality charge. It applies, no doubt, to some anti-vivisectionists, especially Victorian ones, but is no more applicable to most than is the charge of 'wanton sadism' to most experimenters. Far more important, and worth taking seriously, as the Bambi accusation is not, is a charge of sentimentality on other and indeed hardly compatible grounds: the accusation is that holders of the Second View think it appropriate to go in for imaginative realization of what the individual animals experimented on go through or are deprived of, when it is sentimentally self-indulgent to bring this to mind at all. Here the 'How pathetic!' said of Brunton's dog because it had nothing in its life but a daily stomach examination would be seen as the bringing to bear on that dog of considerations appropriate to *pet* animals but totally inappropriate to it, and even more glaringly out of place, let us say, in connection with such laboratory creations as the nude mouse.

It is confused to reply to such charges in the traditional style of defenders of the Second View. The traditional reply is that sentimentality consists in *not* looking at reality, but in self-indulgently inventing what you prefer. The holder of the Second View then adds: I am not sentimental; for I look at the reality of what is done to animals: I bring it to attention, if I can, that these are ways of treating animals that no humane person can view with equanimity.

The confusion of such a reply lies in not seeing the heart of the accusation, which is not that one has been putting fantasies of what animals are for facts, but that one has been treating certain facts as worth attention and concern when they should be treated as things we should learn to take in our stride, as the way things have to be done. It is sentimental, the accusation is, to take these animals' lives seriously, when the serious business of our lives requires that we treat the area of experimentation as one in which such considerations are simply irrelevant. The accusation is indeed that the holder of the Second View substitutes a fantasy for reality, not a fantasy about animals but a fantasy about human life. He fails to treat scientific investigation in an appropriate and realistic way, taking into account its enormous significance to us.

I said this is the counter-accusation to the charge that experimentation on animals leads to callousness. I can now state the reason. From the First View, the imaginative realization of what is done in experimentation to the individual animals is self-indulgent sentimentality; from the Second View, it is a form of callousness to set this area apart as one in which imaginative attention to what is done is out of place. One view, that is, takes to be callous exactly what the other takes to be required by the practicalities of scientific investigation. One side takes to be sentimental exactly what the other takes to be required by humanity.

I should emphasize that by 'imaginative realization' I do not mean 'thinking about all that pain'. Pain is not what the problem is about, though it comes in. A neurophysiologist once mentioned in a lecture to philosophers about his work that he was distressed by all the death of animals that was part of that work—important as he recognized his work to be. When I mentioned this to a colleague of his, *he* was bothered that any fellow-worker should feel that or say so. The idea was that a scientist should not take the deaths so, should not think 'All these deaths!' and be distressed by their inseparability from the kind of work he does.

What I hope is clear is the very deep disagreement that is reflected in the two accusations; a disagreement about how people should live, about the place of science in life and the place of imagination in it, and the role of the spirit of practicality, of getting on with the job. It is this disagreement which lies at the root of the question discussed on pages 341–346: whether the use of animals 'within the circle' should be thought of as a moral issue.

The matter need not be left with a mere statement of the depth and extent of the disagreement. For we can also ask: *is* it sane and practical, or is it a kind of callousness, to be untroubled by the enormous amount of animal experimentation that is 'within the circle'? The idea that it is sane and practical has two sources, and I shall, very briefly, criticize each.

The first source is the conception of scientific investigation itself as a special sort of activity, whose special character is marked by (among other things) its immunity from some sorts of ordinary moral criticism. This immunity is not seen simply in connection with the use of animals. Social scientists, for example, routinely use deception in experiments on people, with the idea that normal standards of honesty in dealing with other people do not apply if one is conducting a scientific experiment meeting certain minimal conditions. That is, there is the idea here too of a circle (experiments 'within the circle' here are those not involving serious risk of harm to the subject)—and lies to the subject when the experiment lies 'within the circle' are not thought of by the experimenter as involving *dishonesty*. They occur in what has become for him a moral enclave, analogous to the enclave of animal experimentation 'within the circle'. And indeed—no accident—critics of such experimentation claim it increases the callousness of those who engage in it and of others influenced by them. The source of the callousness is the same in both cases: the conception of scientific research as conferring a special moral status on what one does. The point was made in Victorian language by George Rolleston, Professor of Anatomy and Physiology at Oxford, and a member of the group of distinguished scientists from whom nineteenth century experimenters expected support on the vivisection issue:

'. . . I must say that with regard to all absorbing studies, that it is the besetting sin of them, and of original research, that they lift a man so entirely above the ordinary sphere of daily duty that they betray him into selfishness and unscrupulous neglect of duty' (Royal Commission, 1876, qn. 1287; cf. also French, 1975, Chapter 4).

The removal from the 'ordinary sphere' which Rolleston speaks of is removal *to* a sphere in which the things the scientist is investigating exist in their wonderful complexity, their endlessly complicated and fascinating relationships. In this sphere, one can attend with the highest and truest scientific respect—or even awe—to the reality of the things one is investigating, and nevertheless fail to take the things one is studying seriously *except* as things it is fascinating and rewarding, or frustrating, to study. That is, in this sphere, attention to their reality is scientific attention, in the fullest sense of that term, which indeed has an ethical element: but it is quite distinct from ordinary ethics, which ceases fully to apply—or so it seems. For the scientist using laboratory animals, they provide no moral counterweight to his own wish to pursue his investigations; the only considerations which are seen as applying are scientific ones. Rolleston, speaking with the splendid certainty of Victorian ethics, implies that it is a bad thing for any man to regard any of his activities as taking place in a special sphere, free from the sorts of consideration which otherwise apply to human activities. The criticism, though, is not grounded in some special feature of Victorian thought, but in common human sense—which rejects the pretension of any activity to special moral status. The most forceful arguments for the Second View are those which rest on this bit of common sense.

What is being criticized, note, is not the desire for knowledge, but the significance which that motive may be given, and the view that actions from that motive have a special and superior status: the view that a man is not *dishonest* who lies to others only in the course of experiments, or *hard* who would only deafen birds because we do not know how they learn to sing. Some defenders of the freedom of scientists have construed any such criticism as Rolleston's as an attack on *science*—but this is balderdash, of the same sort as that which identifies patriotism with the doctrine that ordinary moral considerations fail to apply to what is done for the sake of one's country. Again, it may be argued that treating scientific investigation as having some morally special status is justified by its results. But what are *its* results? It is a complete *non sequitur* to argue from the desirable results of scientific investigations which have used animals, or which have involved deception, to the desirability of continuing to view scientific investigation as an activity insulated from certain kinds of moral criticism. Is the idea that we should probably have far fewer valuable scientific discoveries if scientists were encouraged (by legislation, education, and the structure of institutions) to resist the temptation to view research as a special sort of sphere, morally insulated in the way I have been discussing? If that is the idea, I do not see that it has ever been seriously argued for.

I said that the view that it is not a kind of callousness but good sense to be untroubled by what is done to animals in experiments 'within the circle' had two sources. One is the idea that scientific investigation is properly marked off as a special sort of activity, within which one's subject matter may be viewed purely in a scientific spirit. The other source is quite different. Here the idea is not that science, as a human activity, is a special one and properly treated differently, but that certain kinds of moral consideration are inappropriate in the practical conduct of life, of which scientific activities are part. This second line of thought is especially important when it is physiological investigations using animals that are being defended; the first is more important in the defence of animal experimentation where practical applications are not seriously in view. But the two lines of thought are often combined.

What is this idea that it is a virtue to be practical and hard-headed? Practicality *is* a virtue, if what is meant is that one does not *just* sit about and consider moral difficulties, but does what one can, what needs doing, as well. But what is sometimes called 'practicality' and boasted of is not a virtue at all: when it means getting on with the job, ignoring questions of any moral complexity, putting any difficulties out of view, treating them as a sort of luxury to be left to those who like that sort of thing and have the time for it. Thus Albert Schweitzer, who would count as a deeply practical man if the term is used in the first sense, has been described as *impractical*—which indeed he was, in the second sense. Scientists do sometimes boast of practicality in this second sense, in which it is nothing to boast of. Thus if a biologist, working on *in vitro* fertilization, is aware that it is regarded as a morally problematic area by many, he may say or think with some complacency, 'These are not problems for me; I cannot try to resolve them. I am basically a biologist, I have no moral qualms about this' (Colen, 1979).—This conception of practicality does not involve the rejection of moral considerations altogether, but drastic oversimplification: the side of humanity is the side of the battle against disease, deformity, and death, so let us get on with the job. With blinkers on. One example of what I mean is the nineteenth and twentieth century history of medical support for social policies intended to fight venereal disease, policies distinguished by insensitivity to any other considerations than public health, notably to liberty and respect for persons (see, for example, Amos, 1875; Yanovsky, 1972).

For those who use animals in experiments and who see practicality in this sense as central in their own work, consciousness of their own humanitarian motives may be combined with impatience at those who insist on bringing up problems—problems that may appear to the practical humanitarian spirit as so much gas, when there is *all this good* to be done by moving full speed ahead. Thus for example brisk impatience at restrictions on animal experimentation is sometimes combined with similar impatience at those restrictions on human experimentation (requiring consent, say, or barring experiments on children or prisoners) which may seem to be merely hindrances to the achievement of great good, and unethical on that account.

In a somewhat different and extreme form, the ideal of practicality involves the relegating of moral concern and imagination entirely to the peripheries of life: on this view it would be sentimental and unrealistic to criticize the administration of a colony, say, because it failed to adhere to standards of justice in its treatment of colonial subjects, or the conduct of a war, because it failed to respect the humanity of the enemy. This conception of practicality as an ideal is implicitly cruel: it involves being prepared to overlook cruelty, to call it necessity, and to regard its recognition as sentimental or hypocritical; Chesterton aptly called such a view brutalitarian. As an ideal of character it is contemptible, but the same cannot be said of the ideal of simplistic practical humanitarianism mentioned earlier. But *it* takes one-eyedness for a virtue and treats deadness of imagination as admirable.

CONCLUSION

In the second section, 'The two sides of the dispute', I characterized two ways of looking at the use of animals in experiments. The heart of the First View was the idea that within certain limits, laboratory animals may properly be regarded as delicate instruments; the heart of the Second View was that even in laboratories, animals are to be regarded as sources of moral claims, analogous to those of which human beings are the source. Each view goes with an accusation: the First with the accusation that opponents of animal experimentation are sentimental, precisely in insisting that laboratory animals not be thought of merely as instruments, and the Second with the accusation that animal experimentation makes for callousness. In the last two sections I tried to make clear the lines of thought involved in each accusation. In particular, I tried to show that the accusation of sentimentality has two sources: ideas about scientific activity which indeed underlie the First View. One is a view of scientific activity as special and set apart from other activities in the inapplicability to it of certain kinds of moral criticism. The scientist is seen as engaged in an objective investigation of reality; and by a *non sequitur* this suggests that he is not called upon to bring moral imagination to bear on what he does—to animals or, indeed, to other people. The other source was a simplistic ideal of practicality: *not* the idea that doing what needs doing is important, but the idea that a practical person need not bother overmuch with the problematic character of the area in which he acts, indeed need not try fully to see *what* he is doing, if that might slow him down. Worrying about it is for theologians, philosophers, and other such 'idealists', while biological science is 'immutably practical' (cf. for example White, 1971,

p. 507). These views of science and of the practical underlie the First View and the idea of an area, scientific work 'within the circle', in which the imaginative realization of what one is doing is unnecessary and inappropriate. They further underlie the idea that discussion of animal experimentation as a moral issue is itself unnecessary: that there is no serious issue to be discussed. To reject these views—indeed to reject the idea of there being any area to which we should not bring our thought and imagination as best we can—is not to deny in any way the value of scientific work, either as a rewarding human activity or as the source of great practical benefits. It is simply to say that there is a great self-indulgent cop-out in treating what is done to animals in science as nothing worth thinking about.

REFERENCES

Amos, S. (1875). *A comparative survey of laws in force for the prohibition, regulation and licensing of vice in England and other countries,* Stevens and Son, London.

BBC (1977). 'The guinea-pig and the law'. 'Horizon' programme, broadcast 2 February 1977. (All references to this programme are based on notes taken by the author during the programme.)

Chesterton, G. K. (1910). *George Bernard Shaw,* John Lane, New York.

Chesterton, G. K. (1959). *Orthodoxy,* Doubleday, New York.

Colen, B. L. (1979). 'Norfolk doctors ready to implant embryo', *Washington Post,* 28 September.

Eccles, J. C. (1971). 'Animal experimentation versus human experimentation', in *Defining the laboratory animal* (International Committee on Laboratory Animals and the Institute of Laboratory Animal Resources, National Research Council), National Academy of Sciences, Washington, DC, pp. 285–293.

French, R. D. (1975). *Antivivisection and medical science in Victorian society,* Princeton University Press, Princeton.

Harris, S. (1977). 'The guinea-pig and the law', *The Listener,* **97,** 226–228.

Isaacs, Susan (1930). *Intellectual growth in young children,* Routledge, London.

Janzen, D. H. (1979). 'Why food rots', *Natural History,* **88,** 60–64.

Lane-Petter, W. (1976). 'The ethics of animal experimentation', *Journal of Medical Ethics,* **2,** 118–126.

Marcus, Ruth Barcan (1980). 'Moral dilemmas and consistency', *Journal of Philosophy,* **77,** 121–136.

Paget, S. (1912). *For and against experiments on animals: evidence before the Royal Commission on Vivisection,* Paul B. Hoeber, New York.

Royal Commission (1876). 'Report of the Royal Commission on the practice of subjecting live animals to experiments for scientific purposes', *Parliamentary Papers* C. 1397, **xli,** 277.

Sollitte, S. and Veatch, R. M. (1978). *Bibliography of society, ethics, and the life sciences, 1979–80,* The Hastings Center, Hastings-on-Hudson.

Sturt, G. (1977). *A small boy in the Sixties,* Harvester, Hassocks, Sussex.

Visscher, M. B. (1975). *Ethical constraints and imperatives in medical research,* Charles C. Thomas, Springfield, Ill.

White, R. J. (1971). 'Antivivisection: the reluctant hydra', *The American Scholar,* **40,** 503–507.

Yanovsky, B. (1972). *The dark fields of Venus: from a doctor's logbook.* Harcourt Brace Jovanovich, New York.

Author Index

363

Subject Index